Numerical syntaxonomy

Advances in vegetation science 10

Edited by
EDDY VAN DER MAAREL

Numerical syntaxonomy

Edited by

L. MUCINA and M. B. DALE

Reprinted from Vegetatio, volume 81

Kluwer Academic Publishers
DORDRECHT/BOSTON/LONDON

ISBN 0-7923-0388-1

Published by Kluwer Academic Publishers,
P.O. Box 17, 3300 AA Dordrecht, The Netherlands.

Kluwer Academic Publishers incorporates
the publishing programmes of
D. Reidel, Martinus Nijhoff, Dr W. Junk and MTP Press.

Sold and distributed in the U.S.A. and Canada
by Kluwer Academic Publishers,
101 Philip Drive, Norwell, MA 02061, U.S.A.

In all other countries, sold and distributed
by Kluwer Academic Publishers Group,
P.O. Box 322, 3300 AH Dordrecht, The Netherlands.

Printed on acid free paper

Printed in Belgium

Contents

Vegetatio **81**: 1–15, 1989.

Twenty years of numerical syntaxonomy

Ladislav Mucina[1] & Eddy van der Maarel[2]
[1]*Department of Vegetation Ecology and Nature Conservation, Institute of Plant Physiology, University of Vienna, Althanstrasse 14, A-1091 Wien, Austria;* [2]*Institute of Ecological Botany, Uppsala University, Box 559, 751 22 Uppsala, Sweden*

Accepted 1.2.1989

Keywords: Cluster analysis, Databank, Ordination, Resemblance, Table sorting, Transformation

Abstract

The development of numerical syntaxonomy during its first 20 yr is reviewed. The use of methods of numerical classification and ordination is the dominating feature of the development. National and local phytosociological data banks were established, large data sets handled and many important vegetation monographs were methodically based on multivariate data analysis. Particularly the development in Italy, the Netherlands, Czechoslovakia, and Sweden contributed to new theoretical elements of numerical syntaxonomy. Ordination became a common tool of searching for reticulate synsystematic relations between community types. The most popular ordination techniques have been Principal Components Analysis and Detrended Correspondence Analysis. Hierarchical agglomerative techniques of clustering still prevail in classification, although the divisive strategy of TWINSPAN has also become an effective tool for phytosociological clustering and table sorting. Extensive program packages, also for personal computers have now become standard equipment for many vegetation scientists.

Introduction

It is now 20 years ago that European phytosociologists started to use computers for their syntaxonomical work, both for the effective storage of relevés and environmental information, and for the treatment of their data. Numerical phytosociology germinated in 1964, when Ellenberg & Cristofolini (1964) introduced the use of punched cards in phytosociology and when, during the International Botanical Congress in Edinburgh, Benninghoff & Southworth (1964) presented the first computer-ordered phytosociological table we know of; van der Maarel *et al.* (1964) presented some outlines of numerical phytosociology, and Lambert & Dale (1964) reviewed the use of statistics in phytosociology. The notion of a quantification of phytosociological concepts goes back to German plant sociology in the 1930s (Schwickerath 1931; Tüxen & Ellenberg 1937; see Westhoff & van der Maarel 1973 for a review).

At the initiative of Italian colleagues (Pignatti *et al.* 1968) a Working Group for Data-Processing was founded during the 1969 symposium of the International Association for Vegetation Science in Rinteln. The first, and for a long time the main aims of the group were clearly phytosociological: to establish a system of coding and storing phytosociological material, to adapt already existing methods, and to develop new methods for classification and ordination of relevés and composite units, and finally to establish regular cooperation between centres and individuals willing to give their phytosociological work a numerical basis (see van der Maarel 1974 for a short history of the Working Group's first years).

This development must be seen in the light of the aspirations to establish a Prodrome of European plant communities. The idea for this Prodrome came from and was for some years heavily promoted by the great phytosociologist Reinhold Tüxen (1971), who at that time was at the top of his creativity, organising power and influence. Clearly, this would have been an ideal combination of efforts and as shown in at least one early case, that of *Spartina* salt marshes, the numerical approach (Kortekaas *et al.* 1976) proved to be helpful for the classical syntaxonomists to the extent that at least one new syntaxon was suggested by the numerical results and accepted (Beeftink & Géhu 1973). Unfortunately neither did the Prodrome produce many volumes in the planned series of Prodrome monographs, nor did numerical phytosociologists and ecologists succeed in convincing many classically operating colleagues in adopting a 'numerical syntaxonomy'.

In the meantime, many vegetation scientists, both within the traditional European approaches and on other continents, adopted numerical techniques and a wealth of papers, monographs and even handbooks has been published. However, relatively few of them deal with a formal description and ecological characterization of plant community types and most of them concern local studies of vegetational variation in relation to (usually gradient-type) variation in the environment. Most of the new ideas we have received through this development concern gradient theory and models of plant species performance (e.g., Austin 1980, 1985, 1987). Another clear tendency was the theoretical development of classification and ordination as numerical techniques and the comparison and further development of resemblance measures and multivariate techniques. The vast literature from the last 25 years can be summarised here only briefly. In addition, we may refer to a number of textbooks (Whittaker 1973; Gauch 1982; Greig-Smith 1983; Legendre & Legendre 1983; Feoli *et al.* 1984; Orlóci & Kenkel 1985; Digby & Kempton 1987; Jongman *et al.* 1987; Ludwig & Reynolds 1988) and surveys and compendia which treat the development of numerical vegetation science and ecology and comment also on relevant aspects of numerical syntaxonomy (Goodall 1973a,b, 1986; Dale & Webb 1975; van der Maarel *et al.* 1976, 1980; Noy-Meir & Whittaker 1977; van der Maarel 1979a, 1980a, 1981; Greig-Smith 1980; Legendre 1983; Feoli 1984; Wildi 1986).

After the Working Group had published the main results in a number of Vegetatio papers collected in van der Maarel *et al.* (1980), numerical syntaxonomy continued to be practised, but mainly outside the sphere of influence of the Braun-Blanquet approach to vegetation systematics. The Galanta symposium was in fact the first meeting on this subject after a period of relative silence. The present volume reports on a number of case studies and on theoretical developments, especially in the field of classification, which were put forward in Galanta. We also included some numerical vegetation studies which only partly concern syntaxonomy. (Vegetation scientists involved in numerical syntaxonomy have a broader interest nowadays and they also, or even preferably, study theoretical plant ecological problems).

We hope that this volume will put the perspectives of numerical syntaxonomy in the focus of interest again.

Elements of numerical syntaxonomy

Numerical syntaxonomy was defined by van der Maarel (1981) as 'the phytosociological discipline employed to establish a syntaxonomical system of plant communities with the help of numerical methods'. He recognised four perspectives which imply the basic elements of the numeric-syntaxonomical methodology:

(1) the more objective way of arriving at a hierarchy of types;

(2) the possibility of calculating homotoneity

parameters for both lower and higher units and of much more objectively evaluating their information content;
(3) the possibility of relating the levels of a syntaxonomical hierarchy to the numerical levels of the corresponding dendrogram;
(4) the representation of reticulate relations between types and the improvement of the ecological interpretation of the variation in plant communities.

Numerical syntaxonomy is an integrative approach within vegetation systematics (synsystematics) aimed at the derivation and characterization of vegetation types using numerical techniques. In order to be able to review the developments in this field and to set up new perspectives we wish to address primarily two aspects of numerical synsystematics: (1) the numerical derivation of vegetation types and (2) numerically-aided interpretation of the units.

Several important prerequisites of numerical vegetation analysis will have to be treated as well, such as data storage and transformation, selection of variables, and problems of homogeneity.

Prerequisites of numerical syntaxonomy

Data banks and expert systems

One of the outcomes of the work of the Working Group for Data-Processing was the creation of a storage system for relevés and a coding system for plant species. Relevés are stored table-wise because most relevés are obtained from publications of internal reports where they are already grouped in tables (Cristofolini *et al.* 1969). Species get a four-digit genus and a three-digit species code (Pignatti 1976). The Working Group has in this way built up an archive of over 7000 salt marsh relevés and a code for ca. 6000 plant species (including the checking of synonyms). This system has not been followed in other approaches; it has not even been discussed.

Various other, mostly national, data banks have been developed, usually together with an own software. A very big data bank is that of the National Vegetation Classification of Great Britain (Huntley *et al.* 1981). Other examples are data banks in Trieste (Lagonegro *et al.* 1982), Birmensdorf (Sommerhalder *et al.* 1986) and Strasbourg (de Ruffay *et al.* 1985).

Expert systems were introduced in vegetation science by Noble (1987) and discussed in a syntaxonomical context during the Galanta symposium by Sabo & Mucina (unpubl. lect.).

Transformation of data

Phytosociologists usually collect data on the of occurrence of species in relevés using some ordinal cover-abundance scale (Braun-Blanquet, Hult-Sernander-DuRietz, Domin). Van der Maarel (1979b, also in Westhoff & van der Maarel 1973) suggested an ordinal transformation of the Braun-Blanquet scale into a quasi-linear scale ranging between 0 and 9. This scale has become popular, but many other scales are in use as well. Van der Maarel (1979b) reviewed other attempts in this field and provided a scheme for transformation of many scales into each other. Van der Maarel (1979b) and Clymo (1980) published general transformation formulas enabling the weighting of the original input (cover or cover-abundance) data along a continuous scale. Podani (1984) presented a graphical representation of this process.

As was shown by Noy-Meir (1973) and Noy-Meir *et al.* (1975) and later confirmed in numeric-syntaxonomic studies (Jensén 1978; Grabherr 1985; Mucina & Zaliberová 1984) ecologically and syntaxonomically interpretable structures of different character can be obtained by various transformations. In was often suggested (e.g., Smartt *et al.* 1976) that an emphasis on quantitative differences would become more and more useful for detecting classification structures on the lower levels of dendrograms. However, Jensén (1978) found that quantitative differences were equally important on higher classification levels.

Similar transformation problems occur while treating so-called synoptic (composite) tables comprising various communities. Each synoptic unit is usually represented by constancy-class values (Westhoff & van der Maarel 1973), which as such can be subjected to numerical analyses. Constancy-class values are comparable to presence/absence values in relevés. This is not free from problems as constancy-class values are dependent on the number of relevés in a table (Flintrop 1984). Gunnlaugsdóttir (1985) and van der Maarel *et al.* (1987) combined frequency and average cover of species in a synoptic table and provided a realistic synoptic value suitable for 'quantitative' numerical treatment of synoptic tables.

Redundancy of vegetation and selection of variables

Vegetation can be seen as a 'redundant system' (Orlóci 1978; Feoli *et al.* 1984; Grabherr 1985). This means that it does not make much difference in results when certain species are removed or a selection of relevés is made from the same data set (e.g. Goff & Mitchell 1975; Feoli & Feoli Chiapella 1979; Jancey 1979). Bouxin (1975) found, however, that ordination based on qualitative data was phytosociologically less powerful than that based on quantitative data. The reduction of species was originally motivated by limiting computer capacity or budget. Nowadays these procedures, also known as ranking (see below), serve as powerful tools in trend seeking by removing possible sources of noise. A review of methods for species selection is provided by Dale *et al.* (1986).

Avena *et al.* (1981), applying Analysis of Concentration (Feoli & Orlóci 1979) to species-rich oak woods of Central Italy found that binary data were more predictive than quantitative data in terms of derived syntaxonomic schemes. Selected sets of species (Feoli 1975), higher-taxonomic categories (van der Maarel 1972; Elsol & Clifford in press), growth- or life-forms (Jensén & van der Maarel 1980; Bemmerlein 1986), chorological spectra (Feoli *et al.* 1984; Nimis 1984; Feoli & Ganis 1985); various combinations of these (Feoli 1984) or hierarchically-nested character sets (Orlóci & Orlóci 1985) have also been proved useful in numeric-syntaxonomical studies.

Homotoneity of data and outliers

According to a qualitative definition a phytosociological table is homotoneous when all its relevés belong to the same vegetation type (Guinochet 1973). It is one of the ambitions of numerical syntaxonomy to quantify homotoneity, to determine sources of heterotoneity (deviating relevés) and to evaluate the influence of heterotoneity on data treatment and interpretation of results.

Distribution of species in constancy classes as defined by Braun-Blanquet (1964), also expressed in Raunkiaer's (1934) curve can be used as an empirical assessment of the table homotoneity (see also Westhoff & van der Maarel 1973). A J-shaped curve is considered as an indication of a homotoneous (one-community) table. Godron (1970), Guinochet (1973), Schmitt (1978) and de Foucault (1979) attempted to approximate the Raunkiaer curve by mathematical functions, particularly those based on information theory. De Foucault (1979) further showed that homotoneity is a function of the distribution area of samples combined in the table.

Homotoneity tests, primarily serving the detection of aberrant samples, consider relations between average and extreme numbers of species per relevé, or ratios between characteristic and total species (e.g., Tüxen 1970; Moravec 1971; for a review see Westhoff & van der Maarel 1973).

Homotoneity of a table can also be based on pooled floristic similarity of its relevés (Češka 1966; Neuhäusl 1977). This approach is incorporated in the table structuring program TABORD (van der Maarel *et al.* 1978) and was used in comparing homotoneity level with syntaxonomic rank (Tyler 1979).

Another approach is based on measurements of diversity (Dahl 1980; Tyler 1979), which are especially sensitive to presence of so-called species tails, sets of low-frequent species occurring only in one particular relevé or another. Identification of aberrant relevés containing such tails is particularly useful when comparing relevé sets of different origin. Removal of rare species (optional in programs such as TWINSPAN and

DECORANA, Hill 1979a,b) can be suggested when the analysis should concentrate on more frequent species, or when a reduction of local influences in large-scale studies is preferred (Kissling *et al.* 1988). Otherwise, one of the ranking procedures mentioned above is recommended for reducing noise in vegetation data.

Aberrant samples can also be detected by a probabilistic approach (Goodall 1966, 1969; Goodall *et al.* 1987). Goodall & Feoli (1988) suggested that the homotoneity of a table should be tested before further analyses are done.

Derivation of vegetation types

Automatic table sorting

Ever since Benninghoff & Southworth (1964) introduced computerised table sorting similar programs have been developed (see van der Maarel *et al.* 1978 for an early survey). Two more elaborate programs became available in the late 1970s. TABORD is a program for agglomerative clustering with relocation of relevés and clusters and subsequent one-dimensional arrangement of clusters in a structured table (van der Maarel *et al.* 1978). TWINSPAN, two-way indicator species anlysis, is a program for divisive clustering of relevé sets on the basis of indicator species differentiating relevé groups towards each other (Hill *et al.* 1975), followed by an arrangement of both relevé groups and species groups in a structured table (Hill 1979b).

A review on computer programs and approaches to the sorting of phytosociological tables was presented by van der Maarel (1982). Most of the critical comments made there are still actual, notably those on the effects of species weighting (which are not always obvious from the beginning) and on the relatively poor structure of many phytosociological tables produced by objective methods. These many pragmatical contributions have no doubt promoted the extended use of computers, both main-frame and personal, in everyday phytosociological work. But most of them have not led to major developments in the theory of phytosociological classification or any large-scale phytosociological revision.

The 1980's are characterised by a boom of microcomputer and personal-computer technology which enabled the implementation of new shuffling program packages in a more readily accessible form (e.g., Kuhn 1983).

Both TABORD and TWINSPAN became available for personal computers, and especially the latter program (partly because of its versatility and possibility to handle large data sets in a relatively short time) became widely spread. Other automatised programs, usually embedded in a data-analysis package, include Wildi & Orlóci (1983), Lagonegro & Feoli (1984) and van Tongeren (1986).

Resemblance measures

In all agglomerative clustering methods the choice of a resemblance measure is important because the resulting cluster structure may vary much with the coefficient chosen. The merits and pitfalls of the many resemblance functions (see e.g., Orlóci 1972; Goodall 1973b) have been discussed in the contexts of both classification and ordination. The Similarity Ratio was found effective in the TABORD program and in numerical phytosociology in general (Westhoff & van der Maarel 1973; Janssen 1975; van der Maarel *et al.* 1978; van der Maarel 1979a). However, Campbell (1978), who tested the properties of some similarity measures in connection with Braun-Blanquet table sorting, found Similarity Ratio and Euclidean Distance moderately successful and Czekanowski's Percentage Similarity superior. Hajdu (1981) compared the sensitivity of some measures regarding the occurrence of common versus differentiating species and concluded that Similarity Ratio and Percentage Similarity were both robust.

The similarity between species and measures for ecological association were reviewed by Janson & Vegelius (1981), who preferred the indices of Ochai, Dice and Jaccard (the latter being related to the Similarity Ratio as its presence/absence version). Presence/absence coefficients were compared in a biogeographical context by Kenkel & Booth (1987), who found again Ochai and Jaccard suitable, especially in ordinations.

Another aspect of similarity measures was studied by Faith *et al.* (1987), viz. their robustness as a measure of ecological distance. From their comparison it appears that the robust measures are those based on the difference between species scores in relation to score sums, or on the minimum scores of comparison pairs in relation to score sums, and Percentage Similarity is one of them. Noest & van der Maarel (in prep.) found such a measure also more effective in classification and preferable to Similarity Ratio.

Numerical syntaxonomists may be confused by all these comparisons and would certainly appreciate a 'key' for choosing a measure (or its family) depending on the type of data analysed and the type of treatment (e.g., Legendre & Legendre 1983: Ch. 6).

Classification methods

Despite the wealth of numerical classification methods (for a review see e.g., Legendre & Legendre 1987) vegetation scientists used only a limited number of well-established clustering methods. Their choice was probably guided by the availability of computer programs rather than by the suitability of algorithms. TWINSPAN (Hill 1979b) is a good example here. TWINSPAN as a classification procedure has been appealing to phytosociologists for its versatility (many options to manipulate the result) and may be also its black-box character (some of the options are preprogrammed as 'default' to aid the choice).

On the other hand, Gauch & Whittaker (1981), using simulated data to test the resolution power of several agglomerative and divisive strategies, found that in many cases TWINSPAN really was the most successful approach.

Theoretical analyses of the pro's and con's of particular classification algorithms were summarised by Goodall (1973a), Orlóci (1978), Wildi (1986), Jongman *et al.* (1987), Digby & Kempton (1987) and many others. Nevertheless, empirical (comparisons) and pragmatical (ability to reproduce intuitive classifications) criteria have remained more important in making choices.

Ordination methods

Over the last 20 years a rapid development of ordination methods can be noticed, both in ecology and phytosociology, with the thesis of van Groenewoud (1965) as a precursor. Developments during the 1970's have been reviewed by Noy-Meir & Whittaker (1977), Orlóci (1978), van der Maarel (1979a) and several authors in van der Maarel (1980a). The introduction of Detrended Correspondence Analysis (Hill & Gauch 1980) and the following discussions of the merits of Correspondence Analysis versus the main alternatives: Principal Component Analysis and Multidimensional Scaling, prompted a new series of tests and reviews (e.g. Gauch *et al.* 1981; Wilson 1981; Beals 1984; Kenkel & Orlóci 1986; Ezcurra 1987; Minchin 1987; ter Braak & Prentice 1988). Important issues are:

(1) the development of canonical forms of PCA, RA and DCA in vegetation science (ter Braak 1987) in order to adjust compositional gradients to effective environmental gradients;

(2) the critical evaluation of the detrending procedure (Oksanen 1987, 1988; Wartenberg *et al.* 1988; Peet *et al.* 1988);

(3) the promotion and distribution of non-metric ordination strategies (Kenkel & Orlóci 1986; Minchin 1987).

Achievements of numerical syntaxonomy

The early work of the Working Group for Data-Processing, especially the numerical treatment of the extensive European data set on salt marsh vegetation may be mentioned as a first major achievement. It was clearly demonstrated that new vegetation types could be derived through numerical analysis which had hitherto remained undiscovered in the relevé material (Kortekaas *et al.* 1976). Also, the coherence of higher units in the classification system could be elucidated by numerical approaches (Lausi & Feoli 1979). See van der Maarel *et al.* (1980) for a summary of this early phase.

Because Trieste, Nijmegen and Prague were active centres in this early development more numerical vegetation studies were initiated in or around these centres. Large-scale Italian numerical studies include those of alpine grasslands by Pignatti & Pignatti (1975), of higher units of aquatic vegetation by Feoli & Gerdol (1982), of beech woods by Feoli & Lagonegro (1982), of epiphytic lichen communities by Nimis (1982), of wetland communities by Camiz *et al.* (1984) and of wall vegetation by Hruška (1987).

Dutch examples are the doctoral theses of van Gils (1978) on the spatially complex fringe communities of the *Trifolio-Geranietea* and of Sýkora (1983) on the temporally complex (i.e. fluctuating) communities of the *Lolio-Potentillion*, and of Gremmen (1982) on subantarctic vegetation. Many of these studies were wholly or partly published in Vegetatio, besides the above-mentioned Italian studies also van Gils & Kovács (1977), and the *Calthion palustris* study by van Schaik & Hogeweg (1977), but also other journals did accept such studies, e.g. Sýkora (1982) and the Utrecht study of European heathlands (Barendrecht 1982).

Czechoslovakian (and related East-European) studies have concentrated on ruderal vegetation (Mucina 1982; Mucina & Zaliberová 1984, 1986; Mucina *et al.* 1987) and woodlands (Neuhäusl 1977; Dzwonko 1986). In the Soviet Union the numerical traditions of the Leningrad and Tartu centres (see Aleksandrova 1973 for a review) were further developed and linked to the Braun-Blanquet approach in Ufa, Astrachan' and Kiev (see Mirkin & Naumova 1983; Mirkin 1986).

A new country where numerical vegetation studies became a standard approach is Sweden. Both vegetation types of the country itself were numerically described, e.g. rich fens (Tyler 1979, 1981) and limestone grassland (Krahulec *et al.* 1986), and studies in other parts of the world were performed with Uppsala as a centre, ranging from Icelandic heath vegetation (Gunnlaugsdóttir 1985) to tropical montane grasslands (Zerihun 1985), Mexican coastal sand dune vegetation (Moreno-Casasola & Espejel 1986; van der Maarel *et al.* 1987) and savanna communities (Skarpe 1986).

Numerical phytosociology became also usual in North America, especially of alpine and arctic vegetation, with all interesting opportunities to compare American with European communities (first of all Komárková 1979, 1980; but also Orlóci & Stanek 1979; Cooper 1986; Odasz 1988). The National Vegetation Classification of Great Britain is an example of a (large-scale) numerical phytosociology in a country where the Braun-Blanquet approach did not have prominent representatives (see, however, Adam 1977; Adam *et al.* 1975). Judging from the outline of the data-processing (Huntley *et al.* 1981) and a first example of community description (Malloch 1985) this classification system will be of immediate significance for European syntaxonomy.

Evaluation of results

The Braun-Blanquet approach and numerical classification

The introduction of numerical classification in vegetation science led to a discussion of the merits of the so-called 'classical syntaxonomy' or 'floristic syntaxonomy' (Feoli & Lausi 1981), notably how the Braun-Blanquet approach relates to numerical agglomerative and divisive strategies. The comparisons were attempted mainly from the point of view of 'ecological interpretability' of the results, i.e. plausibility of the classification solutions. The early development of this field was reviewed by van der Maarel (1979a). The major points of these studies can be summarised in the following issues:

(a) Polythetic methods reproduce Braun-Blanquet tabular sorting better than monothetic strategies (Werger 1973; Coetzee & Werger 1973,

1975; Stanek 1973; Coetzee 1974). Only Feoli *et al.* (1981) seem to favour monothetic classification.

(b) Among the agglomerative methods, Sum-of-Squares Clustering and Average and Complete Linkage produce results most similar to tabular sorting (Coetzee & Werger 1973; Stanek 1973; Mucina 1982; Kovář & Lepš 1986). Feoli & Gerdol (1982) tested many clustering techniques to reproduce 2 syntaxonomic systems of aquatic vegetation (based on qualitative and quantitative relations of variables, respectively) and found Median Sorting a successful method. Still, none of these methods can be considered a formalised form of the Braun-Blanquet table sorting. Also polythetic divisive strategies, notably TWINSPAN, may closely resemble the Braun-Blanquet approach (Hill *et al.* 1975; Grabherr 1985).

(c) Regarding similarity coefficient and their effectiveness in reproducing the results of Braun-Blanquet table sorting, various measurements have been found satisfactory, including Czekanowski's Percentage Similarity (Campbell 1978) and the Similarity Ratio (van der Maarel 1979a). In view of recent considerations (Faith *et al.* 1987; Noest & van der Maarel, in prep.) Czekanowski-type measures may be preferred. Qualitative coefficients are supposed to be more powerful to reveal syntaxonomic structures on higher levels of hierarchies (Campbell 1978). This is in line with the observation that towards higher syntaxonomic levels the importance of floristic criteria (and especially quantitative relations) retreat in favour of non-floristic criteria (Mucina 1982).

(d) Numerical similarity levels may be linked to particular syntaxonomic hierarchy levels (e.g., Kortekaas *et al.* 1976; Neuhäusl 1977; see van der Maarel 1979a for a review). The results are varied, however. Kortekaas *et al.* (1976) found high similarity levels for lower units of salt marsh vegetation. Komárková (1980) concluded that for alpine vegetation the 30% and 37% similarity levels could be satisfactorily linked to the levels of order and alliance, but it was not possible to specify a general similarity for the class level. Neuhäusl (1977) found that the variability in similarity among forest associations of the *Carpinion betuli* is high and unpredictable.

We may conclude that there is no general way of quantifying numerical-syntaxonomical similarity since similarity levels vary with:

1. species richness (which is specific for the vegetation complex under study)
2. biogeography (i.e. there is often a high similarity between different syntaxa but from the same phytogeographic region); and
3. classification criteria (see the above discussion on their importance for the erection of hierarchies).

Dendrogram evaluation

Dendrograms serve as geometric images of hierarchical vegetation classifications. The interpretation of these structures can be based on predictivity analysis (Feoli & Lagonegro 1982) who defined predictivity as the capacity of a classification to predict states of external variables. The dendrograms can be inspected for predictivity regarding a tabular sorting of the same data using the Braun-Blanquet approach (e.g., Mucina 1982; Dzwonko 1986). More generally we need to adopt methods of comparing dendrograms as developed in idiotaxonomy and morphometrics, but neglected in vegetation science (see Podani & Dickinson 1984 for a review).

Phytosociologists paid more attention to the use of stopping rules and splitting levels in dendrograms in order to transform the hierarchical classifications into non-hierarchical ones (van der Maarel 1969; van der Maarel *et al.* 1978) and to define and interprete pools (Pielou 1984) or partitions (Podani 1986).

Stopping rules in vegetation classification are most frequently based on the relation between within-cluster and between-cluster similarity, in order to define an 'optimal' combination of homogeneous (tight) clusters discontinuities between them (e.g., Hogeweg 1976; van Schaik & Hogeweg 1976; Lausi & Feoli 1979; Feoli & Lausi 1980; Ratliff & Pieper 1981; Popma *et al.* 1983; Ratkowsky 1984). An extensive review and evaluation of the methods for definition of clusters as well as stopping rules in numerical classification was published recently by Dale (1988).

Evaluation of table structure and fidelity

Two-way structured phytosociological tables can be checked with an Analysis of Concentration (Feoli & Orlóci 1979) in order to judge upon the sharpness of species-relevé blocks, using so-called internal variables (compositional variation within a block) and a χ^2 test. Predictivity of external variables (e.g. environmental or historical) towards an existing table classification was suggested by Feoli (1976).

Having performed different classifications on the same data set, the results may be compared for predictivity using several methods for the comparison of partitions (e.g., Mucina 1982; Kovář & Lepš 1986; see Podani 1986 for a review). Feoli & Ganis (1986) used multi-way contingency tables (based on combined non-floristic variables) to derive vegetation types. An ordination of these types was found to be more predictive than a floristic ordination.

The most frequent step in the evaluation of structured tables is the definition of phytosociologically important species (species groups). This implies formalised definitions of character, differential, constant, diagnostic, faithful (Westhoff & van der Maarel 1973), effective (Williams & Bunt 1980), specific (Orlóci & Stanek 1979), or discriminatory species (Dale *et al.* 1986). Cristofolini *et al.* (1970) used standard deviations from the frequency distribution type of companion species as a measure of the characteristic value of species. Feoli (1976) argued that differential species are those related to a classification at a χ^2 probability level $\leqslant 0.05$ and used this criterion later in a numeric-syntaxonomical study of Italian beechwoods (Feoli & Lagonegro 1982). Jancey (1979) devised a ranking method based on existing classification and ranked species in fact according to discriminative power towards a cluster level in hierarchical classification. Kovář & Lepš (1986) based their definition of character-species on both species frequency in the table and on a criterion based on contingency-table relations; character-dominants were derived on the basis of a ratio of mean importance values within and between relevé groups. Orlóci & Stanek (1979) defined character and specific species using species frequencies within and outside the considered group. Odasz (1988) defined character species as having an average frequency and a high uniformity values (according to Dahl *et al.* 1981).

Identification

Assigning new relevé material to existing classifications using a numerical approach has received little attention. Batista (1988) used discriminant analysis for this purpose, a method in use since long in taxonomy (Sneath & Sokal 1973). Perspectives in this field may be expected particularly from expert systems (Noble 1987).

Syntaxonomic interpretation with the help of ordination plots

Procedures of interpretation of ordination plots in syntaxonomic terms were summarised by van der Maarel 1979a, 1980b). The issue here is how well classification is reflected by an ordination pattern. Many ordination methods have been found suitable for this purpose: Principal Components Analysis (e.g. van der Maarel 1969, 1972; Werger *et al.* 1978; Feoli-Chiapella & Feoli 1977; Gunnlaugsdóttir 1985); Principal Coordinate Analysis (Dzwonko & Kozłowski 1980; Podani 1985); Reciprocal Averaging (e.g., Mucina 1982; Mucina & Zaliberová 1986) and Detrended Correspondence Analysis (e.g., Moreno-Casasola & Espejel 1986). Wildi (1979) devised an algorithm for searching natural clusters in ordination spaces (see also Zumbühl 1983).

An alternative approach is the plotting of representation values of diagnostic species groups on ('isocenes') ordination planes (van der Maarel 1969; Feoli-Chiapella & Feoli 1977; Mucina

1982). The superimposed values of the shared representation of diagnostic species can be approximated by continuous functions. Thus they can be viewed as fuzzy sets (Roberts 1986) and illustrated for instance by ellipses of equal concentration (Jonasson 1981; Lagonegro & Feoli 1985). Fitting the overlays using trend surfaces (Dargie 1984), kriging, connecting noda of vegetation types or samples by a Minimum Spanning Tree (e.g., Bergeron & Bouchard 1983), or testing for randomness of dispersion of the superimposed variables (Feoli & Ganis 1986) are also promising in seeking trends or coenoclines in ordinated vegetation data.

As an example, by comparing the rank of ordination (RA and PCA) axes and syntaxonomic structures emerging on the respective ordination plots Mucina & Zaliberová (1986) found associations to have been separated along axis 1 while low-ranked syntaxa (geographic races) were separated along axis 2.

Conclusions

Much attention has been devoted to the development of table-sorting methods, but only few of them included novel numerical ideas (intersection analysis, seriation). Agglomerative clustering methods have been preferred for small data sets, while automatic table sorting based on divisive strategies (TWINSPAN) became popular for the treatment of large data sets. PCA, RA and later also DCA were the most frequently used ordination methods; any preference for an ordination seems to be related to the availability of computer programs. Despite the wealth of methods for the numerical evaluation of classification and ordination results, the interpretation of results was mainly based on ecological reality.

New methodical tools for data analysis should be adopted, particularly for the formal comparison of results, automatic interpretation of dendrograms and ordination plots, non-metric ordination techniques and fuzzy set theory. More consideration should be given to the choice of variables for numerical analysis (species, species groups on higher taxa, versus structural data).

Phytosociologists should be encouraged to include numerical syntaxonomy in their large-scale descriptions of vegetation. Numerical syntaxonomy can be seen as an improvement (in terms of formalisation, accuracy and speed) of classical (largely floristic-sociological) approaches to vegetation classification.

Various of the perspectives mentioned here will be treated by one or more of the following papers!

Acknowledgements

The authors thank Dagmar Mucina for editing and typing the references.

References

Adam, P. On the phytosociological status of *Juncus maritimus* on British salt marshes. Vegetatio 35: 81–94.

Adam, P., Birks, H.J.B., Huntley, B. & Prentice, I.C. 1975. Phytosociological studies at Malham Tarn moss and fen, Yorkshire, England. Vegetatio 30: 117–132.

Aleksandrova, V.D. 1973. Russian approaches to classification of vegetation. In: Whittaker, R.H. (ed.), Ordination and classification of communities: 493–527. Junk, The Hague.

Austin, M.P. 1980. Searching for a model for use in vegetation analysis. Vegetatio 42: 11–21.

Austin, M.P. 1985. Continuum concept, ordination methods, and niche theory. Annu. Rev. Ecol. Syst. 16: 39–61.

Austin, M.P. 1987. Models for the analysis of species' response to environmental gradients. Vegetatio 69: 35–45.

Barendregt, A. 1982. The coastal heathland vegetation of the Netherlands and notes on inland *Empetrum* heathlands. Phytocoenologia 10: 425–462.

Batista, W.B. 1988. Relating new information to a previous vegetation classification: a case of discriminant coordinates analysis. Vegetatio 75: 153–158.

Beals, E.W. 1984. Bray-Curtis ordination: an effective strategy for analysis of multivariate data. Adv. Ecol. Res. 14: 1–55.

Beeftink, W.G. & Géhu, J.M. 1973. *Spartinetea maritimae*. Prodrome des groupements végétaux d'Europe. Vol. 1. J. Cramer, Lehre.

Bemmerlein, F.A. 1986. Bearbeitung von Lebensformengruppen mit numerischen Methoden. Untersuchungen an der Vegetation von Mauern in NW-Spanien. Tuexenia 6: 391–403.

Benninghoff, W.S. & Southworth, W.C. 1964. Ordering of tabular arrays of phytosociological data by digital computers. Abstr. 10. Int. Bot. Congr. Edinburgh: 331–332.

Bergeron, Y. & Bouchard, A. 1983. Use of ecological species groups in analysis and classification of plant communities in a section of western Québec. Vegetatio 56: 45–63.

Bouxin, G. 1975. Ordination of quantitative and qualitative data in a savanna vegetation (Rwanda, Central Africa). Vegetatio 30: 197–200.

Braun-Blanquet, J. 1964. Pflanzensoziologie. Grundzüge der Vegetationskunde. 3. Aufl. Springer-Verlag, Wien, N.Y.

Camiz, S., Pignatti, S. & Ubrizsy, A. 1984. Numerical syntaxonomy of the class *Agrostietea stoloniferae* Oberdorfer. Ann. Bot., Roma, 42: 135–147.

Campbell, B. 1978. Similarity coefficients for classifying relevés. Vegetatio 37: 101–109.

Češka, A. 1966. Estimation of the mean floristic similarity between and within sets of vegetational relevés. Fol. Geobot. Phytotax. 1: 93–100.

Clymo, R.S. 1980. Preliminary survey of the peat-bog Hummell Knowe Moss using various numerical methods. Vegetatio 42: 129–148.

Coetzee, B.J. 1974. Improvement of association analysis by Braun-Blanquet techniques. Bothalia 11: 324–328.

Coetzee, B.J. & Werger, M.J.A. 1973. On hierarchical syndrome analysis and the Zürich-Montpellier table method. Bothalia 11: 159–164.

Coetzee, B.J. & Werger, M.J.A. 1975. On association-analysis and the classification of plant communities. Vegetatio 30: 201–206.

Cooper, D.J. 1986. Arctic-alpine tundra vegetation of the Arrigetch Creek Valley, Brooks Range, Alaska. Phytocoenologia 14: 467–555.

Cristofolini, C. Lausi, D. & Pignatti, S. 1970. Über statistische Eigenschaften der Charakterarten und deren Verwertung zur Aufstellung einer empirischen Systematik der Pflanzengesellschaften. In: Tüxen, R. (ed.), Gesellschaftsmorphologie (Strukturforschung): 8–25. Junk, Den Haag.

Dahl, E. 1960. Some measures of uniformity in vegetation analysis. Ecology 41: 805–808.

Dahl, E., Prestvik, O. & Toftaker, H. 1981. En kvantifisering av karakterartbegrepet. K. Norsk Vidensk. Selsk. Mus. Rapp. Bot. Ser. 1981-5.

Dale, M.B. 1988. Knowing when to stop: cluster concept – concept cluster. Coenoses. 3: 11–32.

Dale, M.B., Beatrice, M., Venanzoni, R. & Ferrari, C. 1986. A comparison of some methods of selecting species in vegetation analysis. Coenoses 1: 35–52.

Dale, M.B. & Webb, L.J. 1975. Numerical methods for the establishment of associations. Vegetatio 30: 77–87.

Dargie, T.C.D. 1984. On the integrated interpretation of indirect site ordinations: a case study using semi-arid vegetation in southeastern Spain. Vegetatio 55: 37–55.

de Foucault, B. 1979. Elements pour une théorie statistique du tableau phytosociologique homogène. Vegetatio 40: 163–174.

de Ruffay, P. *et al.* 1985. La banque de données phytosociologiques de l'Institut de botanique de Strasbourg: description, premiers bilans. Invent. Faun. & Flor. 27: 1–143.

Digby, P.G.N. & Kempton, R.A. 1987. Multivariate analysis of ecological communities. Chapman & Hall, London.

Dzwonko, Z. 1986. Numerical classification of the Polish Carpathian forest communities. Fragm. Flor. Geobot., Kraków, 30: 93–167. (in Polish)

Dzwonko, Z. & Kozłowski, W.M. 1980. Principal coordinates analysis and its application in synecology. Wiad. Ekol., Warszawa, 26: 265–277. (in Polish)

Ellenberg, H. & Cristofolini, G. 1964. Sichtlochkarten als Hilfsmittel zur Ordnung und Auswertung von Vegetationsaufnahmen. Ber. Geobot. Inst. EHT, Stiftung Rübel, Zürich, 35: 124–134.

Elsol, J.A. & Clifford, H.T. in press. Higher taxonomic descriptors for vegetation classification. Vegetatio.

Ezcurra, E. 1987. A comparison of reciprocal averaging and non-centred principal components analysis. Vegetatio 71: 41–47.

Feoli, E. 1975. Sull'uso della combinazione specifica caratteristica nei confronti tra i tipi di vegetazione. Gior. Bot. Ital. 109: 87–96.

Feoli, E. 1976. Correlation between single ecological variables and vegetation by means of cluster analysis. Not. Fitosoc. 12: 77–82.

Feoli, E. 1984. Some aspects of classification and ordination of vegetation data in perspective. Stud. Geobot. 4: 7–21.

Feoli, E. & Feoli Chiapella, L. 1979. Relevé ranking based on a sum of squares criterion. Vegetatio 39: 123–125.

Feoli, E. & Ganis, P. 1985. Comparison of floristic vegetation types by multiway contingency tables. An example of application of an integrated information system to some beech wood types of the Apennines. Abstr. Bot., Budapest, 9: 1–15.

Feoli, E. & Ganis, P. 1986. Autocorrelation for measuring predictivity in community ecology: an example with structural and chorological data from mixed forest types of NE Italy. Coenoses 1: 53–56.

Feoli, E. & Gerdol, R. 1982. Evaluation of syntaxonomic schemes of aquatic plant communities by cluster analysis. Vegetatio 49: 21–27.

Feoli, E. & Lagonegro, M. 1979. Intersection analysis in phytosociology: computer program and application. Vegetatio 40: 55–59.

Feoli, E. & Lagonegro, M. 1982. Syntaxonomical analysis of beech woods in the Apennines (Italy) using the program package IAHOPA. Vegetatio 50: 129–173.

Feoli, E., Lagonegro, M. & Biondani, F. 1981. Strategies in syntaxonomy: a discussion of two classifications of grassland of Friuli (Italy). In: Dierschke, H. (ed.), Syntaxonomie, pp. 95–107. J. Cramer, Vaduz.

Feoli, E., Lagonegro, M. & Orlóci, L. 1984. Information analysis of vegetation data. Junk, Dordrecht.

Feoli, E. & Lausi, D. 1980. Hierarchical levels in syntaxonomy based on information functions. Vegetatio 42: 113–115.

Feoli, E. & Lausi, D. 1981. The logical basis of syntaxonomy in vegetation science. In: Dierschke, H. (ed.), Syntaxonomie, pp. 35–42. J. Cramer, Vaduz.

Feoli, E. & Orlóci, L. 1979. Analysis of concentration and detection of underlying factors in structured tables. Vegetatio 40: 49–54.

Feoli-Chiapella, L. & Feoli, E. 1977. A numerical phytosociological study of the summits of the Majella massive (Italy). Vegetatio 34: 21–39.

Fewster, P.H. & Orlóci, L. 1983. On chosing a resemblance measure for non-linear predictive ordination. Vegetatio 54: 27–35.

Flintrop, T. 1984. Die Aussagekraft von Stetigkeitsangaben. Tuexenia 4: 293–295.

Gauch Jr, H.G. 1982. Multivariate analysis in community ecology. Cambridge Univ. Press., Cambridge.

Gauch Jr, H.G. & Whittaker, R.H. 1981. Hierarchical classification of community data. J. Ecol. 69: 537–557.

Gauch, H.G., Whittaker, R.H. & Singer, S. 1981. A comparative study of nonmetric ordinations, J. Ecol. 69: 135–152.

Godron, M. 1970. Application de la théorie de l'information à l'étude de l'homogeneité et de la structure de la végétation. In: Tüxen, R. (ed.), Gesellschaftsmorphologie (Strukturforschung), pp. 31–38. Junk, The Hague.

Goff, F.G. & Mitchell, R. 1975. A comparison of species ordination results from plot and stand data. Vegetatio 31: 15–22.

Goodall, D.W. 1966. Deviant index – a new tool for numerical taxonomy. Nature 210: 216.

Goodall, D.W. 1969. A procedure for recognition of uncommon species combinations in sets of vegetation samples. Vegetatio 18: 19–35.

Goodall, D.W. 1973a. Numerical classification. In: Whittaker, R.H. (ed.), Ordination and classification of communities, pp. 575–618. Junk, The Hague.

Goodall, D.W. 1973b. Sample similarity and species correlation. In: Whittaker, R.H. (ed.), Ordination and classification of communities, pp. 105–156. Junk, The Hague.

Goodall, D.W. 1986. Classification and ordination: their nature and role in taxonomy and community studies. Coenoses 1: 3–9.

Goodall, D.W. & Feoli, E. 1988. Application of probabilistic methods in the analysis of phytosociological data. Coenoses 3: 1–10.

Goodall, D.W., Ganis, P. & Feoli, E. 1987. Probabilistic methods in classification: a manual for seven computer programs. Grup. Elab. Automat. Dati Ecol. Quant., Trieste, 7: 1–51.

Grabherr, G. 1985. Numerische Klassifikation und Ordination in der alpinen Vegetationsökologie als Beitrag zur Verknüpfung moderner 'Computermethoden' mit der pflanzensoziologischen Tradition. Tuexenia 5: 181–190.

Greig-Smith, P. 1980. The development of numerical classification and ordination. Vegetatio 42: 1–9.

Greig-Smith, P. 1983. Quantitative plant ecology. 3rd ed. Blackwell, Oxford.

Gremmen, N.J.M. 1982. The vegetation of the subantarctic islands Marion and Prince Edward. Junk, The Hague.

Guinochet, M. 1973. Phytosociologie. Masson, Paris.

Gunnlaugsdóttir, E. 1985. Composition and dynamical status of heathland communities in Iceland in relation to recovery measures. Acta Phytogeogr. Suec. 75: 1–84.

Hajdu, L.J. 1981. Graphical comparison of resemblance measures in phytosociology. Vegetatio 48: 47–59.

Hermy, M. 1984. The creation and the analysis of data matrices in vegetation science. Bull. Soc. Roy. Bot. Belg. 117: 63–72.

Hill, M.O., Bunce, R.G.H. & Shaw, M.W. 1975. Indicator species analysis, a divisive polythetic method of classification, and its application to a survey of native pinewoods in Scotland. J. Ecol. 63: 597–613.

Hill, M.O. 1979a. DECORANA – a FORTRAN program for detrended correspondence analysis and reciprocal averaging. Ecology & Systematics, Cornell University, Ithaca.

Hill, M.O. 1979b. TWINSPAN – a FORTRAN program for arranging multivariate data in an ordered two way table by classification of individuals and attributes. Ecology & Systematics, Cornell University, Ithaca.

Hill, M.O. & Gauch Jr, H.G. 1980. Detrended correspondence analysis: an improved ordination technique. Vegetatio 42: 47–58.

Hogeweg, P. 1976. Iterative character weighting in numerical taxonomy. Comp. Biol. Med. 6: 199–211.

Hruška, K. 1987. Syntaxonomical study of Italian wall vegetation. Vegetatio 73: 13–20.

Huntley, B., Huntley, J.P. & Birks, H.J.B. 1981. PHYTOPAK. A suite of computer programs designed for the handling and analysis of phytosociological data. Vegetatio. 45: 85–95.

Jancey, R.C. 1979. Species ordering on a variance criterion. Vegetatio 39: 59–63.

Janson, S. & Vegelius, J. 1981. Measures of ecological association. Oecologia 49: 371–376.

Janssen, J.G.M. 1975. A simple clustering procedure for preliminary classification of very large sets of phytosociological relevés. Vegetatio 30: 67–71.

Jensén, S. 1978. Influences of transformation of cover values on classification and ordination of lake vegetation. Vegetatio 37: 19–31.

Jensén, S. & van der Maarel, E. 1980. Numerical approaches to lake classification with special reference to macrophyte communities. Vegetatio 42: 117–128.

Jonasson, S. 1981. Plant communities and species distribution of low alpine *Betula nana* heath in northernmost Sweden. Vegetatio 44: 51–64.

Jongman, R.H.G., ter Braak, C.J.F. & van Tongeren, O.F.R. (eds) 1987. Data analysis in community and landscape ecology. Pudoc, Wageningen.

Kenkel, N.C. & Booth, T. 1987. A comparison of presence-absence resemblance coefficients for use in biogeographical studies. Coenoses 2: 25–30.

Kenkel, N.C. & Orlóci, L. 1986. Applying metric and nonmetric multidimensional scaling to ecological studies: some new results. Ecology 6: 919–928.

Kissling, P., Kuhn, N. & Wildi, O. 1988. Le relevé mérocénotique et son application à l'étude du changement floristique en forêt. Bot. Helv. 98: 39–75.

Komárková, V. 1979. Alpine vegetation of the Indian Peaks area, Front Range, Colorado Rocky Mountains. J. Cramer, Vaduz.

Komárková, V. 1980. Classification and ordination in the Indian Peaks area, Colorado Rocky Mountains. Vegetatio 42: 149–163.

Kortekaas, W.M., Lausi, D., Beeftink, W.G. & van der Maarel, E. 1980. Survey of salt marsh relevés included in the data-bank of the Working-Group for Data-Processing. In: van der Maarel, E., Orlóci, L. & Pignatti, S. (eds), Data-processing in phytosociology, pp. 207–225. Junk, The Hague.

Kortekaas, W.M., van der Maarel, E. & Beeftink, W.G. 1976. A numerical classification of European *Spartina* communities. Vegetatio 33: 51–60.

Kovář, P. & Lepš, J. 1986. Ruderal communities of the railway station Česká Třebová (Eastern Bohemia, Czechoslovakia) – remarks on the application of classical and numerical methods of classification. Preslia, Praha 58: 141–163.

Krahulec, F., Rosén, E. & van der Maarel, E. 1986. Preliminary classification and ecology of dry grassland communities on Ölands Stora Alvar (Sweden). Nord. J. Bot. 6: 797–809.

Kuhn, N. 1983. VEGTAB, ein Computerprogramm als Hilfe zur tabellarischen Vegetationsgliederung. Tuexenia 3: 499–522.

Lagonegro, M. & Feoli, E. 1984. THREE-PAckages for classification and ordination of multivariate data. Libreria Goliardica, Trieste.

Lagonegro, M. & Feoli, E. 1985. The use of ellipses of equal concentration to analyse ordination vegetation patterns. Stud. Geobot. 5: 143–165.

Lagonegro, M. *et al.* 1982. Un software per banche dati di flore territoriali, estendibile alla vegetatione. CNR, AQ/5/38, Udine, 1–160.

Lambert, J.M. & Dale, M.B. 1964. The use of statistics in phytosociology. Adv. Ecol. Res. 2: 59–99.

Lausi, D. & Feoli, E. 1979. Hierarchical classification of European salt marsh vegetation based on numerical methods. Vegetatio 39: 171–184.

Legendre, P. 1983. Numerical ecology: developments and recent trends. In: Felsenstein, J. (ed.), Numerical taxonomy, pp. 505–523. Springer-Verlag, Berlin.

Legendre, L. & Legendre, P. 1983. Numerical ecology. Elsevier, Amsterdam.

Ludwig, J.A. & Reynolds, J.F. 1988. Statistical ecology. J. Wiley & Sons, New York.

Malloch, A. 1985. Plant communities on British sand dunes – the National Vegetation Classification. In Doody, P. (ed.), Sand dunes and their management, pp. 57–97. Nature Conservancy Council Deterborough.

Minchin P.R. 1987. An evaluation of the relative robustness of techniques for ecological ordination. Vegetatio 69: 89–107.

Mirkin, B.M. 1986. Paradigm change and vegetation classification in Soviet phytosociology. Vegetatio 68: 131–138.

Mirkin, B.M. & Naumova, L.G. 1983. Gradients analysis in vegetation. Usp. Sovrem. Biol., Moskva, 95: 304–318. (in Russian).

Moravec, J. 1971. A simple method for estimating homotoneity of sets of phytosociological relevés. Folia Geobot. Phytotax., Praha, 6: 147–170.

Moreno-Casasola, P. & Espejel, I. 1986. Classification and ordination of coastal sand dune vegetation along the Gulf and Caribbean Sea of Mexico. Vegetatio 66: 147–182.

Mucina, L. 1982. Numerical classification and ordination of ruderal plant communities (*Sisymbrietalia, Onopordetalia*) in the western part of Slovakia. Vegetatio 48: 267–275.

Mucina, L., Abramova, L.M. & Mirkin, B.M. 1987. An attempt towards a numeric-syntaxonomical revision vegetation of Bashkiria. In: Rozenberg, G.F. (ed.) Problems of theoretical and experimental phytosociology, pp. 58–64. Bashkirian Sci. Centre Ufa (in Russian).

Mucina, L. & Zaliberová, M. 1984. A numerical classification approach to vegetation zonation in a sludge pond. Acta Bot. Slov. Acad. Sci. Slov., Ser. A, Bratislava, Suppl. 1: 239–251.

Mucina, L. & Zaliberová, M. 1986. Communities of *Anthriscus caucalis* and *Asperugo procumbens* in Slovakia. Fol. Geobot. Phytotax., Praha, 21: 1–25.

Neuhäusl, R. 1977. Delimitation and ranking of floristic-sociological units on the basis of relevé similarity. Vegetatio 35: 115–122.

Nimis, P.L. 1982. The epiphytic lichen vegetation of the Trieste Province (north eastern Italy). Stud. Geobot. 169–191.

Nimis, P.L. 1984. Phytosociology, ecology and phytogeography of epiphytic lichen vegetation in the Calamone Lake area (N. Apennines, Italy). Stud. Geobot. 4: 109–127.

Noble, I.R. 1987. The role of expert systems in vegetation science. Vegetatio 69: 115–121.

Noy-Meir, I. 1971. The multivariate analysis of the semi-arid vegetation in South-eastern Australia: nodal ordination by component analysis. Proc. Ecol. Soc. Aust. 6: 159–193.

Noy-Meir, I. 1973. Data transformation in ecological ordination. I. Some advantages of non-centering. J. Ecol. 61: 329–341.

Noy-Meir, I., Walker, D.J. & Williams, W.T. 1975. Data transformation in ecological ordination. II. On the meaning of data standardization. J. Ecol. 63: 779–800.

Noy-Meir, I. & Whittaker, R.H. 1977. Continuous multivariate methods in community analysis: some problems and developments. Vegetatio 33: 79–98.

Odasz, A.M. 1988. Quantification of character-species in vegetation at the tree limit, Brooks Range, Northern Alaska. Vegetatio 75: 65–71.

Oksanen, J. 1987. Problems of joint display of species and site scores in correspondence analysis. Vegetatio 72: 51–57.

Oksanen, J. 1988. A note on the occasional instability of detrending in correspondence analysis. Vegetatio 74: 29–32.

Orlóci, L. 1972. On objective functions of phytosociological resemblance. Amer. Midl. Natur. 88: 28–55.

Orlóci, L. 1978. Multivariate analysis in vegetation research. 2nd ed. Junk, The Hague.

Orlóci, L. & Kenkel, N.C. 1985. Introduction to data analysis with examples from population and community ecology. Int. Cooperative House, Fairland, MD.

Orlóci, L. & Stanek, W. 1979. Vegetation survey of the Alaska Highway, Yukon territory: types and gradients. Vegetatio 41: 1–56.

Orlóci, L. & Orlóci, M. 1985. Comparison of communities without the use of species: model and example. Ann. Bot., Roma, 43: 275–285.

Peet, R.K. 1980. Ordination as a tool for analyzing complex data sets. Vegetatio 42: 171–174.

Peet, R.K., Knox, R.G., Case, J.S. & Allen, R.B. 1988. Putting things in order: the advantages of detrended correspondence analysis. Amer. Nat. 131: 924–934.

Pielou, E.C. 1984. The interpretation of ecological data: a primer on classification and ordination. J. Wiley & Sons, New York.

Pignatti, E. & Pignatti, S. 1975. Syntaxonomy of the *Sesleria varia*-grasslands of the calcareous Alps. Vegetatio 30: 5–14.

Pignatti, S. 1976. A system for coding plant species for data-processing in phytosociology. Vegetatio 33: 23–32.

Pignatti, S., Cristofolini, G. & Lausi, D. 1968. Verwendungsmöglichkeiten einer elektronischen Datenverarbeitungsanlage für die pflanzensoziologische Dokumentation. Discussion Paper 12th Symp. Int. Ass. Vegetation Sci. Rinteln.

Podani, J. 1984. Spatial processes in the analysis of vegetation: Theory and review. Acta Bot. Acad. Sci. Hung. 30: 75–118.

Podani, J. 1985. Syntaxonomic congruence in a small-scale vegetation survey. Abstr. Bot., Budapest, 9: 99–128.

Podani, J. 1986. Comparison of partitions in vegetation studies. Abstr. Bot., Budapest, 10: 235–290.

Podani, J. & Dickinson, T.A. 1984. Comparison of dendrograms: a multivariate approach. Can. J. Bot. 62: 2765–2778.

Popma, J., Mucina, L., van Tongeren, O.F.R. & van der Maarel, E. 1983. On the determination of optimal levels in phytosociological classification. Vegetatio 52: 65–75.

Ratkowsky, D.A. 1984. A stopping rule and clustering method of wide applicability. Bot. Gaz. 145: 518–523.

Ratliff, D. & Pieper, R.D. 1981. Deciding final clusters: An approach using intra- and intercluster distances. Vegetatio 48: 83–86.

Raunkiaer, C. 1934. The life forms of plants and statistical plant geography. Clarendon Press, Oxford.

Roberts, D.W. 1986. Ordination on the basis of fuzzy set theory. Vegetatio 66: 123–131.

Schmitt, A. 1978. Apport des methodes numériques à l'établissement de synthèses phytosociologiques régionales. Application aux forêts du Jura. Doc. Phytosoc., Lille, N.S. 2: 401–408.

Schwickerath, M. 1931. Die Gruppenabundanz (Gruppenmächtigkeit); ein Beitrag zur Begriffsbildung der Pflanzensoziologie. Bot. Jb. 64: 1–16.

Skarpe, C. 1986. Plant community structure in relation to grazing and environmental changes along a north-south transect in the western Kalahari. Vegetatio 68: 3–18.

Sneath, P.H.A. & Sokal, R.R. 1973. Numerical taxonomy. W.H. Freeman & Co., San Francisco.

Sommerhalder, R., Kuhn, N., Biland, H.-P., von Gunten, U. & Weidmann 1986. Eine vegetationskundliche Datenbank der Schweiz. Bot. Helv. 96: 77–93.

Stanek, W. 1973. A comparison of Braun-Blanquet's method with sum-of-squares agglomeration for vegetation classification. Vegetatio 27: 323–345.

Sýkora, K.V. 1982. Syntaxonomy and synecology of the *Lolio-Potentillion* in the Netherlands. Acta Bot. Neerl. 31: 65–95.

Sýkora, K.V. 1983. The *Lolio-Potentillion anserinae* R. Tüxen 1947 in the Northern part of the Atlantic domain. Thesis, Nijmegen University, Nijmegen.

ter Braak, C.J.F. 1987. The analysis of vegetation-environment relationships by canonical correspondence analysis. Vegetatio 69: 69–77.

ter Braak, C.J.F. & Prentice, I.C. 1988. A theory of gradient analysis. Adv. Ecol. Res. 18: 271–317.

Tüxen, R. 1970. Einige Bestandes- und Typenmerkmale in der Struktur der Pflanzengesellschaften. In: Tüxen, R. (ed.), Gesellschaftsmorphologie (Strukturforschung): 76–107. Junk, Den Haag.

Tüxen, R. 1971. Vorläufige Liste von Mitarbeitern am Prodromus der europäischen Pflanzengesellschaften. Vegetatio 22: 265–268.

Tüxen, R. & Ellenberg, H. 1937. Der systematische und ökologische Gruppenwert. Ein Beitrag zur Begriffsbildung und Methodik in der Pflanzensoziologie. Mitt. Flor.-Soz. Arbeitsgem. Niedersachsen, Hannover, 3: 171–184.

Tyler, C. 1979. Classification of *Schoenus* communities in South and Southeast Sweden. Vegetatio 41: 69–84.

Tyler, C. 1981. Geographical variation in Fennoscandian and Estonian *Schoenus* wetlands. Vegetatio 45: 165–182.

van der Maarel, E. 1969. On the use of ordination models in phytosociology. Vegetatio 19: 21–46.

van der Maarel, E. 1972. Ordination of plant communities on the basis of their genus, family and order relationships. In: van der Maarel, E. & Tüxen, R. (eds), Grundfragen und Methoden in der Pflanzensoziologie, pp. 183–192. Junk, The Hague.

van der Maarel, E. 1974. The Working Group for Data-processing of the International Society for Plant Geography and Ecology in 1972–1973. Vegetatio 29: 63–67.

van der Maarel, E. 1979a. Multivariate methods in phytosociology, with reference to the Netherlands. In: Werger,

M.J.A. (ed.), Study of vegetation, pp. 163–255. Junk, The Hague.

van der Maarel, E. 1979b. Transformation of cover-abundance values in phytosociology and its effects on community similarity. Vegetatio 39: 97–114.

van der Maarel, E. (ed.) 1980a. Classification and ordination. Junk, The Hague.

van der Maarel, E. 1980b. On the interpretability of ordination diagrams. Vegetatio 42: 43–45.

van der Maarel, E. 1981. Some perspectives of numerical methods in syntaxonomy. In: Dierschke, H. (ed.), Syntaxonomie, pp. 77–93. J. Cramer, Vaduz.

van der Maarel, E. 1982. On the manipulation and editing of phytosociological and ecological data. Vegetatio 50: 71–76.

van der Maarel, E., Boot, R., van Dorp, D. & Rijntjes, J. 1985. Vegetation succession on the dunes near Oostvoorne, The Netherlands; a comparison of the vegetation in 1959 and 1980. Vegetatio 58: 137–187.

van der Maarel, E., Espejel, I. & Moreno-Casasola, P. 1987. Two-step vegetation analysis based on very large data sets. Vegetatio 68: 139–143.

van der Maarel, E., Janssen, J.G.M. & Louppen, J.M.W. 1978. TABORD, a program for structuring phytosociological tables. Vegetatio 38: 143–156.

van der Maarel, E., Orlóci, L. & Pignatti, S. 1976. Data processing in phytosociology: retrospect and anticipation. Vegetatio 32: 65–72.

van der Maarel, E., Orlóci, L. & Pignatti, S. 1980. Data-processing in phytosociology. Junk, The Hague.

van der Maarel, E., Westhoff, V. & van Leeuwen, C.G. 1964. European approaches to the variation in vegetation. Paper 10. Int. Bot. Congr., Edinburgh.

van Gils, H.A.M.J. 1978. Spatial and temporal transitions between limestone grasslands and deciduous forests in Europe. Thesis Nijmegen.

van Gils, H.A.M.J. & Kovács, A.J. 1977. *Geranion sanguinei* communities in Transsylvania. Vegetatio 33: 175–186.

van Groenewoud, H. 1965. Ordination and classification of Swiss and Canadian coniferous forests by various biometric and other methods. Ber. Gobot. Inst. ETH, Stiftg. Rübel, Zürich, 36: 25–103.

van Schaik, C.P. & Hogeweg, P. 1977. A numerical-syntaxonomical study of the *Calthion palustris* Tx. 37 in the Netherlands. Vegetatio 35: 65–80.

van Tongeren, O.F.R. 1986. FLEXCLUS, an interactive program for classification and tabulation of ecological data. Acta Bot. Neerl. 35: 137–142.

Wartenberg, D., Ferson, S. & Rohlf, F.J. 1987. Putting things in order: a critique of detrended correspondence analysis. Amer. Nat. 129: 434–448.

Werger, M.J.A. 1973. On the use of association-analysis and principal component analysis in interpreting a Braun-Blanquet phytosociological table of a Dutch grassland. Vegetatio 28: 129–144.

Werger, M.J.A., Wild, H. & Drummond, B.R. 1976. Vegetation structure and substrate of the northern part of the Great Dyke, Rhodesia: Gradient analysis and dominance-diversity relationships. Vegetatio 37: 151–161.

Westhoff, V. & van der Maarel, E. 1973. The Braun-Blanquet approach. In: Whittaker, R.H. (ed.), Ordination and classification of communities, pp. 617–726. Junk, The Hague.

Whittaker, R.H. (ed.) 1973. Ordination and classification of communities. Handb. Veget. Sci. 5. Junk, The Hague.

Wildi, O. 1979. GRID – a space analysis for recognition of noda in vegetation samples. Vegetatio 41: 95–100.

Wildi, O. 1980. Management and multivariate analysis of large data sets in vegetation research. Vegetatio 42: 175–180.

Wildi, O. 1986. Analyse vegetationskundlicher Daten. Theorie und Einsatz statistischer Methoden. Veröff. Geobot. Inst. ETH, Stiftung Rübel, Zürich 90: 1–226.

Wildi, O. & Orlóci, L. 1983. Management and multivariate analysis of vegetation data (2nd ed.) Eidg. Aust. Forstl. Versuchswes. Ber. 215.

Williams, W.T. & Bunt, J.S. 1980. Studies in the analysis of data from Australian tidal forests ('Mangroves'). II. The use of an asymmetric monothetic divisive classificatory program. Aust. J. Ecol. 5: 391–396.

Wilson, M.W. 1981. A statistical test of the accuracy and consistency of ordinations. Ecology, 62: 8–12.

Zerihun Woldu 1985. Grassland vegetation on the central plateau of Shewa, Ethiopia, in relation to edaphic factors and grazing conditions. Diss. Bot. 84: 1–114.

Zumbühl, G. 1983. Pflanzensoziologisch-ökologische Untersuchungen von gemähten Magerrasen bei Davos. Veröff. Geobot. Inst. ETH, Stiftung Rübel, Zürich 81: 1–101.

Vegetatio **81**: 17–28, 1989.

An outline for data analysis in phytosociology: past and present

H.S. Fischer[1] & F.A. Bemmerlein[2]
[1]*Swiss Federal Institute of Technology Zürich, Department of Geography, Winterthurerstr. 190, CH-8057 Zürich, Switzerland;* [2]*Institut für angewandte ökologische Studien, IFANOS, Austr. 14, D-8500 Nürnberg, Federal Republic of Germany; Present address: Institut für Ökologie, TU-Berlin, Schmidt-Ott-Str. 1, 1000 Berlin (West) 41, Federal Republic of Germany*

Accepted 4.1.1989

Keywords: Community ecology, History of science, Numerical method, Vegetation science

Abstract

As vegetation science develops its formalized methodology increased in importance. At the beginning of the century traditional European schools of phytosociology applied an intuitive classification and ordering scheme to establish vegetation units. The introduction of computer-supported data analysis into vegetation science accelerated the formalization. The present paper presents a formal outline for phytosociology and community analysis. In this scheme we distinguish different 'data states' which are connected by active data handling. Data sampling and subsequent analysis can be understood as filtering processes, making theoretical assumptions in vegetation science very important. The model implied provides a basis for discussing potential problems and possibilities in vegetation science.

Abbreviations: PCA = Principal Component Analysis

Introduction

Phytosociology has been preoccupied with the description of vegetation. This focus on the descriptive phase is one of several reasons for the outlier position phytosociologists are thought to have in the scientific community. For example, results of phytosociological research can hardly be used for industrial exploitation. On the other hand, important new tasks emerged in nature conservation and landscape management. Methodological problems emerging on the community level seem to be quite different from those in experimental biology. Multivariate statistical techniques, which may contribute to a 'new look' in phytosociology, were neither developed for use in community ecology nor generally accepted by most Central European phytosociologists. Still, because of the general availability and use of computers, extensive use of multivariate techniques has become an option in phytosociology s.l.

The application of automatic data processing and multivariate methods have promoted the development of new approaches to understanding vegetation structure and dynamics. Notably ordination methods have now become a standard tool in Anglo-American ecology (McIntosh 1985). Similarly, the measurement and simulation of matter and energy fluxes have fostered the development of ecosystem research, a field that continues to increase in importance (e.g., Ellenberg *et al.* 1986). By allowing huge sets of data to be analyzed quantitatively from a great number of different angles, computer ecology has fundamentally altered the aims and methods and indeed the thinking of vegetation ecologists.

Here we would like to point out that the development of theories should be viewed in close relation with the quantification and formalization of methods in vegetation science. The lack of balance between theory and descriptive knowledge in vegetation science has been increasingly noticed and discussed (e.g., Austin 1980; Prentice & van der Maarel 1987; preface). The Working Group for Data Processing of the International Association for Vegetation Science founded in 1969 (van der Maarel, 1971; van der Maarel & Mucina 1989), organized in 1985 the Uppsala symposium 'Theory and Models in Vegetation Science'. The working group was renamed to 'Working Group for Theoretical Vegetation Science', which, we think, reflects how automatic data processing has contributed directly to theoretical discussions in vegetation science.

Here we present a structural analysis of the phytosociological procedures in the form of an outline which shows the different analytical steps in phytosociology and community ecology. Such an outline is helpful for developing strategies of analysis and to open up new perspectives for Central European community ecology. Allen *et al.* (1984) and Austin (1987) have presented similar descriptions of vegetation analysis and shown the weak points of data treatment and interpretation.

A detailed analysis of the history of Central European vegetation science is still lacking, despite several historical accounts (Becking 1957; Kuoch 1957; Whittaker 1962; Shimwell 1971; Westhoff & van der Maarel 1973; van der Maarel 1975; McIntosh 1985; Trepl 1987). What is needed is an analysis of phytosociology which would include the history of ideas and their sociological background starting from its roots in Humboldt and going up to the recent discussions revolving around the 'New Ecology' (Trepl 1987). Such an historical account from our present point of view would be valuable in that it could illuminate some of the many theoretical approaches that were not pursued further (e.g. Gams 1918). Such a review could prove to be a useful perspective for evaluating new fields like multivariate statistics, remote sensing, systems and chaos theory, and fractal geometry.

Special importance should also be given to detailed historical case studies about the 'emergence' and phytosociological acceptance of single community types. Wiegleb & Herr (1984) carried out such a study for aquatic communities and uncovered several problems of the intuitive disposition of traditional phytosociology with its bias against theory. This disposition and bias easily result in misleading statements of the ecology of plant species and of plant communities.

The practical training of phytosociologists and geobotanists in Central Europe continues to reflect the lack of clearly stated assumptions. The whole process of analysis from project planning to the display and categorization of the results is far from being well-defined. We also lack standards for a critical examination of the process. The discontent of numerous phytosociologists about sampling techniques, ordering of the relevé tables and methods of class membership identification in the European phytosociological reference system stems from phenomena of syntaxonomy and not from the process of investigation as a whole. This limited perspective precludes a constructive criticism of the results of phytosociological research and its application to landscape management and planning.

Even today training in phytosociology is often characterized by the relationship between a master and his disciples. The latter are trained in the craftsmanship of their profession while watching and imitating their master. They assimilate the necessary skills, a special view of the phytosociological universe, and in this way they gain sympathetic understanding (Tüxen 1951, p. 5). And, even more drastically, Tüxen (1951, p. 9): 'If at all, it is extremely difficult to learn them (these techniques) from written instructions alone. It is much easier to learn them in personal and practical intercourse. What is needed is a genuine talent, a feeling for relations in the phytosociological table and great experience in sociology, synecology and syngenesis of plant communities as well as in plant geography added to the reliable handling of the necessary techniques by well-trained auxiliary staff.' Such training has two sides. While working outdoors one must combine theoretical knowledge with observational sensitivity. Furthermore, an introduction inimical to theory prevents openness towards differing approaches and developments and, even more important, a critical reflection on one's own view.

Historical outline

Plant communities were 'discovered' at the turn of the century when the 'classificatory program' (Trepl 1987, p. 208) rapidly gained ground. At that time the development of theories in vegetation science and of more or less repeatable and systematic methods of sampling and data interpretation reached a peak (cf. McIntosh 1985; Trepl 1987; Noy-Meir & van der Maarel 1987 for a historical survey). Initially, phytosociology attempted to discover and describe all plant communities in existence. Trepl (1987) explains the phenomenon of the newly discovered community in terms of the assumption made by early phytosociologists that both the habitat and the interrelation of plants are linked to the species composition of a particular location. Thus plant communities become phenomena which can be investigated as more or less 'natural' units. The method that has come to be used for such descriptions is the floristic approach, which is based on species composition. Such a method allows more differentiated classification than an approach based on physiognomy and the environment. This floristic approach also necessitated a new methodology for data collection. This methodological tool was the sampling quadrat. Samples were taken from a defined plot and repeated on different locations. The sample quadrat as a systematic tool was emphasized by Pound and Clements (White 1985). Geobotanists dealt with landscapes and plant distribution and tried to explain coarse-grained patterns of their distribution. Other botanists concentrated on single organisms and species, their adaptations and their physiology. Between these alternatives a new concern with plant communities arose. To define communities as the interesting phenomena means to cut a piece out of a landscape and focus not on individual plants or species but on the species composition of a small area, the quadrat. Subsequent classification of these samples was based on the assumption that finite and ultimate units can be effectively described. In particular phytosociologists expected that if all associations were described it would be possible to contribute significantly to the solution of agricultural problems and landscape planning and to have a strong basis for further ecological research.

Different schools emerged and developed different hierarchical systems of communities based on different phenomena of interest. To mention just two examples, Du Rietz's system of 'sociations' (Du Rietz 1921) is based on the constancy and dominance of species and 'Grundformen'. Braun-Blanquet (1928) classified vegetation on the basis of the whole species composition of the sample sites. As a result quantitative characteristics could be only of marginal interest. The complexity of the phenomena could otherwise not be coped with without automatic data processing (Gams 1954). All these approaches share the assumption that discrete natural communities do exist. Whittaker (1962) discussed the basic assumptions of the different schools and gives a review of the theoretical consequences and the practical applications resulting from them.

To overcome the confusion in phytosociological terminology between different schools a series of sessions at subsequent International Botanical Congresses (3rd to 7th) tried to define the fundamental terms of phytosociology at the beginning of this century. However, although the fundamentals were defined, there was no conformity between the schools, as the definitions were reduced to a common, very general denominator and were interpreted differently, depending on the school. Even within the schools definitions were ambiguous. This is illustrated by the foundation of a working group 'Phytosociology' within the German Floristisch-Soziologische Arbeitsgemeinschaft (Dierschke 1985). This group, among other things, is supposed to decide which of the communities described during the last 75 years are to be accepted as associations. The disputes among the individual schools of phytosociology '... exclusively refer to the theoretical interpretative constructions and to sociological and systematic questions in particular' as Walter points out in his preface to Ellenberg's 'Tasks and Methods of Vegetation Science' (1956).

But there was a strong objection on the part of the leaders of the different schools to formalizing the classification process too much. Tüxen's words (quoted above) refer to the 'table work'. Du Rietz (1921, p. 215) similarly described the process of classification as '... simply walking around, looking at vegetation and acquiring clarity about the existing units in nature and their natural borders.' Braun-Blanquet used a similar intuitive, holistic classification.

In the traditional European schools of phytosociology the scientist can fall back on a familiar system and knows in advance how to sample, how to classify the data and how to compartimentalize the results. A couple of rules of thumb suffice to cope with special conditions, to determine the plot size, to estimate cover and to display the results. What matters above all is the sensitivity and experience of the observer (Tüxen 1951; Du Rietz 1921 p. 214). This simple and consistent methodology has certainly contributed to the rapid and successful spreading of phytosociology in Central Europe.

It were only Knapp (1948) and Ellenberg (1956) who gave an explicit description of the methods of sampling and tablework, respectively (See also Westhoff & van der Maarel 1973). The proper classification of the single stands and their assignment to associations, as described by Ellenberg, are performed with the complete set of relevés during the table work. This may be called a quasi-statistical, polythetic, divisive classification procedure (Moore *et al.* 1970). This procedure relies on some insight in the ultimate result of the classification as a guideline for the separation between typical and atypical relevés. Still, the discussion in classical phytosociology focusses on the separation of associations and on questions about their existence.

The delay in providing a precise and explicit description of the phytosociological method reflects the low esteem in formalization. Reference to the old masters was generally felt sufficient to justify the methods used. Horvat (1930, see Tüxen 1974 p. 3) stated bluntly: 'After the publication of Braun-Blanquet's sociology it is superfluous to talk about methodology. It will do to point out some questions which may seem important to the understanding of studies.' A similar practice is common for sampling methods. All this is centered around experience and what Du Rietz (1921) called the 'sociological view'. This sceptical attitude towards discussions regarding sampling methods and the reluctance to accept criticism emerged from the holistic comprehension of nature and the tradition of natural history. Here formalization and mathematization amounted to a restraint on the world of experience (e.g. Du Rietz 1921 p. 214; Oberdorfer 1972 p. 290). The very limited possibilities of processing the large number of seemingly incoherent data gathered in quantitative sampling made it easy to discount new methods.

In Central Europe the school of Zürich-Montpellier gradually became the most influential in phytosociology. With its ideal and hierarchical system of associations and higher units it had a definite organizing principle and offered the perspective and the hope of '... meeting the natural conditions up to a certain degree.' (Raabe 1957 p. 251). The challenging task was to ascertain a complete descriptive knowledge of vegetation. Complete knowledge about all variants and subvariants of the associations was assumed to be a prerequisite for synecological studies (Tüxen 1938). This goal contributed to the endless discussions about the 'true' classifications and the work done turned out to be mostly an occupation with the ideal hierarchy of plant communities. This occupation called 'synsystematics' became an end in itself (for a caustic criticism of this tendency see Gams 1954).

If we accept Hackings (1983) definition of phenomena dealt with by a particular discipline, plant communities are generally recognizable observations worthy of being mentioned, which occur under defined circumstances. Phenomena are not discovered but created. They are created by the common accord of a scientific community on basic assumptions and the definition of constraining conditions. The theorem of the ugly duckling states that without a reference system, everything is equal. A derived corollary would be that 'from a logical point of view, there is no such thing as a natural kind' (Watanabe 1985, p. 452). Thus, the search for a complete and natural classification as proposed by Tüxen (1938) will never find an end.

Schools of phytosociology fit into a general scheme in which the nature of their results reflects the assumptions embodied in their procedures of sampling and analysis (Fig. 1). With the choice of different phenomena of interest, transformations etc. and different emphasis on the stages of the analysis one obtains different types of vegetation analyses, as described above. Noy-Meir *et al.* (1975) have already pointed out the relationship between data standardization and the classical phytosociological schools. It therefore becomes important to have a formal description of the method in use, both in order to compare the results with those of different approaches and, on the other hand, to select an appropriate approach out of the variety of available ones.

The matrix concept in phytosociology

The erosion of traditional schools in the Anglo-American world started at the beginning of the fifties. The discussion was animated by the rediscovery of the 'individualistic concept' (Gleason

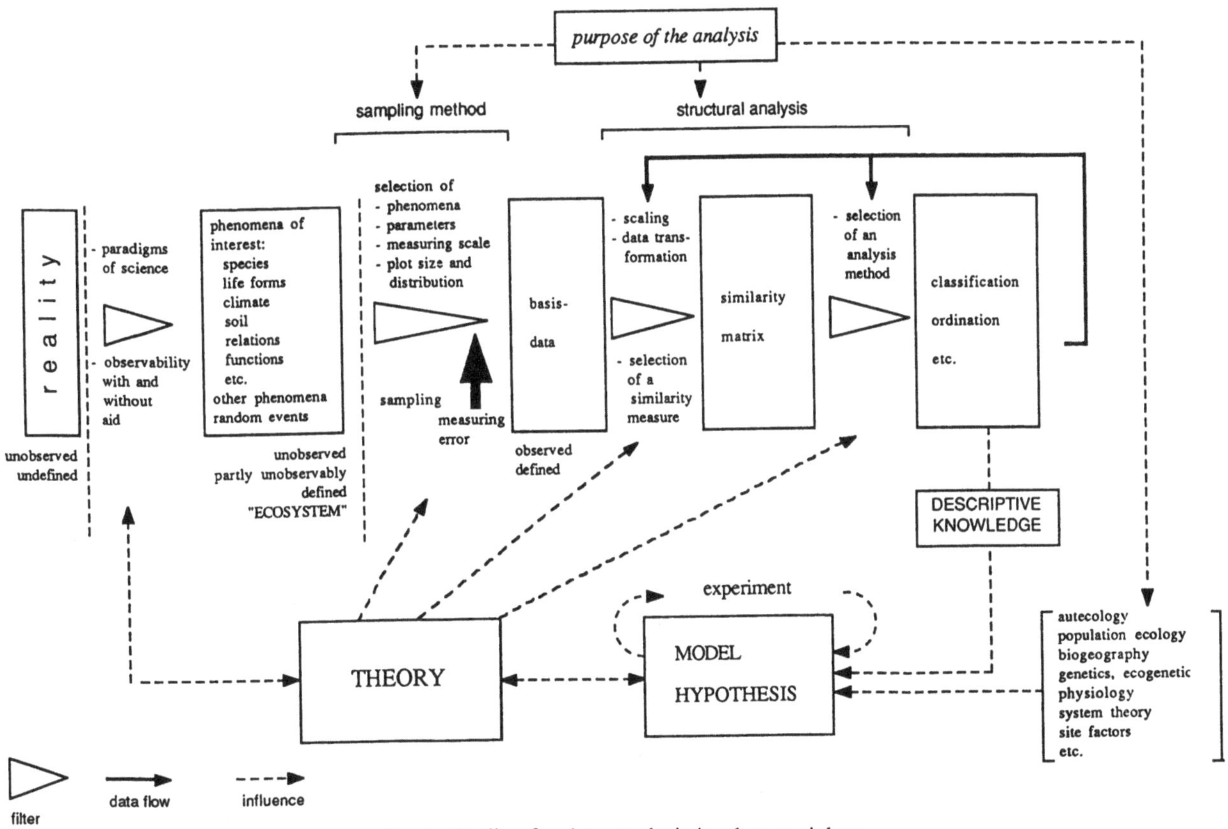

Fig. 1. Outline for data analysis in phytosociology.

1926, 1939), by the introduction of direct gradient analysis (Whittaker 1956) and by the work of the Wisconsin school of phytosociology. Now this discussion has calmed down and the different methods of analysis are used pragmatically and often in a mutually complementary way. As a result, Anglo-American vegetation science is very much matrix-orientated.

In Central European phytosociology the relevé tables can be understood as matrices. Contrary to the relevé table of the traditionalists, a matrix is formal, neutral as to contents and allows a great variety of applications and procedures. Moreover it does not conflict with the assumptions about the structure of fundamental units. What makes the difference is the fact that a matrix centers around samples (spot checks) and not around plant communities (cf. Fig. 1) or some kind of continuum. The traditional relevé table can be seen as a convenient way to display data rather than an analytical tool. There is no 'true' table (Wildi 1977) but a relevé table is one among many ways of presenting and discussing two-dimensional patterns of the data.

The matrix concept as a basis for the description and analysis of vegetation is closely connected with the idea of multidimensionality. As early as 1918, Gams pointed out: 'A polyphyletic system such as that of synusia has... as many dimensions as there are independently varying factors that determine affinity.... As a result the natural or polyphyletic system of synusia is not portrayable.' Raabe (1957) makes similar remarks.

It was not before the resources of multivariate statistics and automatic data processing came into existence that multidimensional phenomena could be easily described and analysed on all levels of integration and with different spatial and temporal scales. One of the most outstanding novelties in this respect is the generation of hypotheses (Lambert & Dale 1964; Noy-Meir & van der Maarel 1987). The use of new methods to develop and test new and old theories in vegetational science has now become a subject for discussion (Austin 1980) but it is still marginal in Central Europe.

A structuralistic outline of analysis in phytosociology

Our scheme (Fig. 1) marks the different stages of data collection and analysis as blocks and the operations during the analysis as arrows. Following Allen *et al.* (1984) the triangles represent filters which modify the signals in their transition from one stage to the other. Depending on the parameters of the filters chosen and the preceding data structure the quality of the impact differs from one stage to the other. The dotted arrows represent the flow of information that governs the process. Here we would like to make a clear distinction between two fields:

1. the purpose of the investigation which is obviously and directly connected with socio-cultural and economic influences and
2. the formulation of theories, models and ideas.

It was not until Kuhn (1967) that the connections between the two fields and the ideological background of the investigation were given major attention in epistemological and historical research (cf. Specht 1972; Weingart 1976). The point of departure is 'reality' regardless of the observer as against 'reality' as a function of the observer. We shall not go into the philosophic implications of reality independent of the observer as vegetation science deals with the reality dependent on the observer. This reality is determined by perceivability and the view of the world conditioned by theories.

In any case, the first step of analysis is observation, in other words, sampling, measuring and so on. This procedure results in a simplified representation of the observable part of reality and therefore is subject to errors such as wrong identification of species, mistakes of measurement and estimation of species quantities and variables of habitat as well as the implications of the distortions due to sampling problems. The data thus gained by observation are the basis for an implicit or explicit multivariate analysis whose results may in turn be used as a foundation for further lines of analysis, an iterative and repetitive process apt to provide a description of the community composition (cf Westhoff & van der

Maarel 1973). This can lead to hypotheses when combined with external information on autecology and population ecology, community ecology, biogeography, genetics, physiology, knowledge of factors involved in the respective stand, mathematical models etc.

These hypotheses can in turn be checked and falsified against the existing models of phytosociology. The aim is to exceed the mere representation of the individual object under examination, which may lead the way to new theories in vegetation science. For this procedure there is no single method that can best describe and explain the totality of patterns.

If one wants to compare two research projects it is necessary to exactly define their differences which is facilitated by using explicit concepts and a high degree of formalization. This is possible because all the individual steps of the procedure are separated from each other. Feoli (1984) used a set of basic data as an example of how to proceed from simple transformations to derived sets of data which he called 'conditioned spaces' falling back on the concept of the matrices. Here we see the crucial importance of scaling (Allen *et al.* 1984) and the significance of a clear distinction of all the single stages of the analysis. Methods can only complement each other and results can only be compared if the same or at least a comparable scale is applied and work is done on the same level of integration (Allen & Starr 1982).

Examples

We now want to discuss a few examples in order to demonstrate how different paths can be followed in an analysis, and how the filters selected in the different stages are brought into accordance.

(1) Wildi (1977) investigated an excentric bog complex in Switzerland with a regular grid and measured soil and water variables considered relevant for bogs. The quantitative species data were analyzed with PCA, with a subjective classification of the ordination results and with a discriminant analysis. The results obtained were matched with the measured environmental data. The environmental data were evaluated with factor- and regression analyses to find out their relationship.

(2) Grabherr (1982) worked with transect data from the Alps. The samples were taken rectangularly to the wind edges with differing slope and altitude, with a point quadrat method to obtain quantitative data. Using a PCA on the percent cover values he interpreted the data graphically to produce hypotheses about the relation between vegetation and environmental factors. The percent cover values were subsequently replaced by the estimation intervals of the Braun-Blanquet cover/abundance scale and by presence/absence values. With each transformation different structures of the data were filtered out. The analysis of the original percent cover data showed a gradient which was obviously determined by the degree of slope and aspect. One might say that the question asked was: How well do different plant species endure the environment? The analyses of the binary data showed an easily interpretable gradient of slope and altitude. The question asked here was rather: What type of environment can the plant species endure? The altitude thus became the most prominent factor. The Braun-Blanquet weighting showed gradients in both slope, aspect and altitude.

(3) In an investigation of phytoplankton communities from a freshwater lake Allen *et al.* (1977, 1984) concentrated on changes in this composition of algae in the course of one year. They attempted to account for the weekly changes in species composition by using a set of environmental variables. That is, they were interested in how changes in species composition tracked changes in environmental factors. Theoretical considerations suggested several particular transformations (filters), chosen so as to make the static observations dynamic. For example, if environmental conditions change, the plant communities often follow with a particular time lag. The transformation used was a type of smoothing of the time-series data matrix. The appearance or disappearance of a species from one week to the

other contributes to change and this species is included in the analysis. If a species is present in two successive weeks, it does not contribute to change and is scored zero. The results showed that the change of the species composition will be accelerated when there is a drastic change in the environment. The change of the species composition will slow down when the change of the environment is less drastic.

(4) Feoli & Lagonegro (1982) analyzed the community composition of beech forests in the Apennines on the basis of species presence. Data were collected from the literature. The dataset was then analyzed in various steps: A subsample was taken by a stratified random sampling procedure and the species groups important for the pattern in the matrix were chosen with the help of a non-hierarchical classification procedure. As a second step, those groups were classified hierarchically on the basis of the species groups. The next step was to find relationships between the vegetation units obtained with the help of an analysis of concentration. The use of presence/absence data as well as the use of species groups in the second step of analysis represent rather coarse filters and so address a fairly coarse level of community pattern. This level was adequate for the large geographical area covered.

(5) Box (1981) developed a global vegetation model of plant life-forms and their dependence on macroclimatic conditions. His sampling method was to collect information about the vegetation from the literature, preferring places where the climatic diagrams of Walter & Lieth (1960–1967) were available. The structural analysis is not explicitly formalized but it could be described as a subjective classification with a dominance transformation. The result of his classification were formations as vegetational units. He then developed a predictive model for those units. His use of formations instead of associations is in accordance with his global level of observation. Thus he worked at a much larger scale than Feoli & Lagonegro (1982).

These examples lead to the following conclusions. First, investigations which focus on species have a local character, typical of the pioneer investigations in plant sociology. Examples 1 and 2 reflect this interest in small-scale variability of vegetation in space and its relation with environmental factors. Example 3 concentrates on change in time of species and conditions. Examples 4 and 5 deal with a larger scale of observation: species groups and life-forms respectively. The nature of the questions asked and the level of investigation make it crucial to use different transformations and different analyses if consistent interpretations are to be obtained.

In these examples there is a strong emphasis on methods but there is no common standard methodology. Depending on the level and scale of observation and the eventual starting hypothesis each data set requires a special adaptation of the methods used. Adequate methods can only be found if a high degree of differentiation and formalization is reached and if their mathematical implications are evident and stated. These are the commensurable conditions. It would be a necessary condition for progress in community ecology to know the ecological and theoretical implications of the methods.

Consequences

What are the consequences and further implications (Specht 1972) of a formalization of the analyses in vegetation science subsequent to the establishment of numerical methods? We here take 'consequences' to mean both the direct impact on investigation resulting from changing questions and perspectives, and the indirect effects on the social and the scientific context of vegetation science and their influences for the use of numerical methods. The consequences of formalization can be summarized as follows:

1) Numerical methods require that the aims and problems of a study be explicitly and precisely formulated before data treatment, including the sampling design, structural analysis, and statistical evaluation.
2) As soon as the outline and the exact formulation of the problems have been fixed through algorithms, the latter can be executed con-

sistently. Thus they must be universally comprehensible and can be reproduced. The theories and conclusions can be communicated to and checked by others.

3) Numerical methods can be automatized to a considerable extent and large quantities of data can be processed effectively. These developments allow time saved on mechanical procedures to be used for project design and the interpretation of results. These are great advantages as the variation of vegetation can only be comprehended by way of a large number of samples.
4) Numerical methods enable biologists to work quantitatively and to apply sophisticated (e.g. stratified random) sampling strategies.
5) Numerical methods also provide opportunities to play with the data collected, i.e. to subject these data to varying filters and to screen them in order to discover information which can then be interpreted under the application of different criteria. Such tools complement traditional intuitive approaches. The criticism by Central European phytosociologists that numerical methods exclude experience and intuition dissolves in view of the multitude of algorithms and of possible transformations made possible and convenient through automated data processing.
6) Thus numerical methods can facilitate interdisciplinary communication and cooperation. They provide a common language, applicable in many different disciplines, as proved by the fact that many have been developed in other fields.

Two important areas of research methodology in vegetation science remain neglected and therefore constitute bottlenecks: sampling methods and sampling designs on the one hand and theoretical concepts on the other hand. The concepts are integrated as an organizing compartment. This compartment will determine – consciously or unconsciously – processes of decision and attempts at interpretation apart from the mere technicality of the methods used. In this context the recent claims requiring the formulation of appropriate theories and their evaluation are of crucial importance (Austin 1980, 1987; Wiegleb 1986; Prentice & van der Maarel 1987). As Austin put it in 1987: 'What is currently needed is for vegetation scientists to recognize that their studies imply a certain procedural model or paradigm. Further, the component procedures of this model can be used to test specific aspects of ecological theory.'

The traditional technique of data collection dominated by the ruling dogmas of phytosociologists still prevailing in Central Europe may be satisfactory for gaining an overview of the vegetation on a regional scale. Nevertheless, they are not particularly suited to the available tools of numerical methods (Wildi 1986; Podani 1987). The data not only consist of rough guesses of species performances but also are biased by subjectively chosen samples. These biases emerge in conspicuous author-clusters and author-gradients when data from different authors are simultaneously analysed with cluster analysis or ordination (Roth 1979). If the structures associated with various authors are spatially separated (Feoli & Lagonegro 1982), they generate uncertainties about the reliability of the data. These uncertainties also emerge when samples of the same area from different authors (Tüxen 1972) are compared. Methods of sampling in phytosociology must be reconsidered fundamentally in order not to block new perspectives in community ecology.

Appropriate results cannot be expected if the input-data are not tuned to the analytical outline used and considered as independent from any theory and context (the assumption being: 'We have to collect hard, objective data'). If the only consequence of the application of numerical methods were the recognition that new sampling methods must be introduced, this would already be a certain asset. What is necessary is a reformulation of the goals and methods of sampling. This is of importance not only in order to meet scientific standards but also to enhance the practical implications for planning, cartography, monitoring, prediction and evaluation.

New results and insights will not emerge if the classical, manual and often intuitive synopses and relevé tables are merely replaced by numerical classification procedures. In most cases data structures already known will only be reproduced which seem to support the traditional methods. Mere formalization of existing classification procedures can not exclude weaknesses and ambiguities that already exist implicitly in these approaches.

Further implications

The spreading use of electronic data processing in ecology is contributing to a far-reaching standardization of data, descriptions and procedures. While we have stressed the positive aspects of this process, such standardization could also potentially blind us to certain phenomena that are not easily standardized. Legitimate doubts still exist as to whether communities can and should be grasped only by way of census and quantifications. Such a far-reaching formalization and compartimentalization of communities involves the dangerous illusion of what we might call total ascertainability.

The increasing dependence on technical devices also tends to increase the time spent in front of the terminal to the detriment of time spent outdoors in observation. The world stored in computers is quite different from the world we walk around in. It provides many possibilities of dealing with large data sets from a complex world but we should also be alert to the risk that ecologists be alienated from ecology itself. The current methodological revolution will only be successful if we manage to focus our discussion on the biological meaning of the techniques and criteria applied. The ecology of the world should not be confused with the ecology of a man-made subsystem, the ecology of the computer. Comfortable instant program packages which reduce the technical expense and allow for more time for ecological work have the nasty side effect of providing results as long as the input data meet the formal input requirements. The practice of working and training, however, shows over and again the dangers of blindly crunching with numbers. The impressive display of graphic results often lacks content and understanding. thus the question of how to train ecologists is also broached.

The consequences and the new demands on university programs go deep and institutions are not at all prepared to cope with them. The problem does not reside so much in the handling of hardware and software, it concerns above all the specific implications of investigation processes and the methods applied in the area of theory and epistemology (Sattler 1986). To our mind, this is one of the crucial defects of our present situation, apart from the necessary extension of training in statistics, sampling design, and data processing.

Acknowledgements

We thank Dr D. Waller, Dr G. Wiegleb and anonymous reviewers for helpful comments on the paper. Mr A. Bemmerlein and Dr D. Waller helped us with the English translation.

References

Allen, T.F.H., Bartell, S.M. & Koonce, J.F. 1977. Multiple stable configuration in ordination of phytoplankton community change rates. Ecology 58: 1076–1084.

Allen, T.F.H. & Starr, T.B. 1982. Hierarchy: perspectives for ecological complexity. University Press, Chicago, London, 310 pp.

Allen, T.H.F., O'Neill, R.V. & Hoekstra, T.W. 1984. Interlevel relations in ecological research and management: some working principles from hierarchy theory. USDA Forest Service, General Technical Report RM-110: 1–11.

Allen, T.H.F., Sadowsky, D.A. & Woodhead, N. 1984. Data transformation as a scaling operation in ordination of plankton. Vegetatio 56: 14–160.

Austin, M.P. 1980. Searching for a model for use in vegetation analysis. Vegetatio 42: 11–22.

Austin, M.P. 1987. Models for the analysis of species' response to environmental gradients. Vegetatio 69: 35–45.

Becking, R.W. 1957. The Zürich-Montpellier School of Phytosociology. Bot. Review. 23: 411–488.

Box, E.O. 1981. Macroclimate and plant forms: An introduction to predictive modeling in phytogeography. Junk, The Hague.

Braun-Blanquet, J. 1928. Pflanzensoziologie. Grundzüge der Vegetationskunde. 1st ed. Springer, Berlin.

Braun-Blanquet, J. 1964. Pflanzensoziologie. 3rd ed. Springer, Berlin, Wien, New York.
Dierschke, H. 1985. Aufgaben pflanzensoziologischer Forschung in Mitteleuropa. Zur Gründung eines Arbeitskreises für Pflanzensoziologie innerhalb der floristisch-soziologischen Arbeitsgemeinschaft. Tuexenia 5: 561–563.
Du Rietz, G.E. 1921. Zur methodologischen Grundlage der modernen Pflanzensoziologie. Thesis, Uppsala University.
Ellenberg, H. 1956. Aufgaben und Methoden der Vegetationskunde. In: Walter, H. (ed.), Einführung in die Phytologie. Ulmer, Stuttgart.
Ellenberg, H., Mayer, R. & Schauermann, J. 1986. Ökosystemforschung – Ergebnisse des Sollingprojektes 1966–1986. Ulmer, Stuttgart.
Feoli, E. 1984. Some aspects of classification and ordination of vegetation data in perspective. Stud. Geobot. 4: 7–21.
Feoli, E. & Lagonegro, M. 1982. Syntaxonomical analysis of beech woods in the Apennines (Italy) using the program package IAHOPA. Vegetatio 50: 129–173.
Gams, H. 1918. Prinzipienfragen der Vegetationsforschung: Ein Beitrag zur Begriffsklärung und Methodik der Biocoenologie (Diss. Univ. Zürich). Vierteljahresschr. Naturforsch. Ges. Zürich 63: 1–205.
Gams, H. 1954. Vegationssystematik als Endziel oder Verständigungsmittel? Veröff. Geobot. Inst. Rübel 29: 35–40.
Gleason, H.A. 1926. The individualistic concept of the plant association. Bull. Torrey Bot. Club 53: 7–26.
Gleason, H.A. 1939. The individualistic concept of the plant association. Amer. Midl. Nat. 21: 92–110.
Grabherr, G. 1982. Die Analyse alpiner Pflanzengesellschaften mit Hilfe numerischer Ordinations- und Klassifikationsverfahren. Stapfia 10: 149–160.
Hacking, I. 1983. Representing and intervening. Cambridge Univ. Press, Cambridge.
Knapp, R. 1948. Einführung in die Pflanzensoziologie. Heft 1 & 2. Arbeitsmethoden der Pflanzensoziologie und Eigenschaften der Pflanzengesellschaften. Ulmer, Stuttgart.
Kuhn, T.S. 1967. Die Struktur wissenschaftlicher Revolutionen. Suhrkamp Wissenschaft, Frankfurt.
Kuoch, R. 1957. Vegetationskundliche Schulen. Mitt. Schweiz. Anst. Forstl. Versuchsw. 32: 419–446.
Lambert, M.J. & Dale, M.B. 1964. The use of statistics in phytosociology. Adv. Ecol. Res. 2: 59–99.
McIntosh, R. 1985. The background of ecology. Concept and theory. Cambridge Univ. Press, Cambridge.
Moore, J.J., Fitzsimmons, P., Lambe, E. & White, J. 1970. A comparison and evaluation of some phytosociological techniques. Vegetatio 20: 1–20.
Noy-Meir, I., Walker, D. & Williams, W.T. 1975. Data transformations in ecological ordinations. II. On the meaning of data standardization. J. Ecol. 63: 779–800.
Noy-Meir, I. & van der Maarel, E. 1987. Relations between community theory and community analysis in vegetation science: some historical perspectives. Vegetatio 69: 5–15.
Oberdorfer, E. 1972. Diskussionsbeitrag zu: P. Schönfelder, Systematisch-arealkundliche Gesichtspunkte bei der Erfassung historisch-geographischer Kausalitäten der Vegetation, erläutert am Beispiel des *Seslerio-Caricetum sempervirentis* in den Ostalpen. In: Tüxen, R. (ed.), Grundfragen und Methoden in der Pflanzensoziologie. Dr W. Junk, The Hague.
Podani, J. 1987. Computerized sampling in vegetation studies. Coenoses 2: 9–18.
Prentice, I.C. & van der Maarel, E. (eds). 1987. Theory and models in vegetation science. Vegetatio 69: 1–222.
Raabe, E.-W. 1957. Zur Systematik in der Pflanzensoziologie. Vegetatio 7: 271–277.
Roth, C. 1979. Soziologisch-ökologische Untersuchungen im Grenzbereich *Fagus sylvatica* L. / *Pinus sylvestris* L. in der nördlichen Schweiz. Veröff. Geobot. Inst. Rübel 66.
Sattler, R. 1986. Biophilosophy. Analytic and holistic perspectives. Springer, Berlin.
Shimwell, D.W. 1971. The description and classification of vegetation. Univ. of Washington Press, Seattle.
Specht, R. 1972. Innovation und Folgelast. Beispiele aus der neueren Philosophie- und Wissenschaftsgeschichte. Friedrich Frommann Verlag Günther Holzboog Kg, Stuttgart-Bad Cannstatt.
Trepl, L. 1987. Geschichte der Ökologie. Athenäum, Frankfurt.
Tüxen, R. 1938. Vorwort des Herausgebers. Mitt. Flor. Soz. Arbeitsgem. Niedersachsen 4: 1–2.
Tüxen, R. 1951. Erfahrungsgrundlagen für die pflanzensoziologische Kartierung des westdeutschen Grünlandes. Angew. Pflanzensoz. 4: 1–28.
Tüxen, R. 1972. Kritische Bemerkungen zur Interpretation pflanzensoziologischer Tabellen. In: Tüxen, R. (ed.), Grundfragen und Methoden in der Pflanzensoziologie. Junk, The Hague pp. 168–182.
Tüxen, R. 1974. Einleitung. Die Pflanzengesellschaften Nordwest-deutschlands: 1–32. 2. Aufl., Cramer, Lehre.
van der Maarel, E. 1971. Bericht über die Arbeitsgruppe für Datenverarbeitung der Internationalen Vereinigung für Vegetationskunde, 1969–1970. Vegetatio 22: 269–273.
van der Maarel, E. 1975. The Braun-Blanquet approach in perspective. Vegetatio 30: 213–219.
van der Maarel, E. & Mucina, L. 1989. Twenty years of numerical syntaxonomy. Vegetatio 81: 1–15.
Walter, H. & Lieth, H. 1960–67. Klimadiagramm-Weltatlas. Gustav Fischer Verlag, Stuttgart.
Watanabe, S. 1985. Pattern recognition: human and mechanical. Wiley, New York.
Weingart, P. 1976. Wissensproduktion und soziale Kultur. Suhrkamp, Frankfurt.
Westhoff, V. & van der Maarel, E. 1973. The Braun-Blanquet approach. In Whittaker, R.H. (ed.) Handbook of vegetation science, vol. 5. Classification and ordination, pp. 617–726, Junk, The Hague.
White, J. 1985. The census of plants in vegetation. In: White, J. (ed.), The population structure of vegetation, pp. 33–88. Junk, Dordrecht.

Whittaker, R.H. 1956. Vegetation of the Great Smoky Mountains. Ecol. Monogr. 26: 1–80.
Whittaker, R.H. 1962. Classification of natural communities. Bot. Rev. 28: 1–239.
Wiegleb, G. 1986. Grenzen und Möglichkeiten der Datenanalyse in der Pflanzenökologie. Tuexenia 6: 365–377.
Wiegleb, G. & Herr, W. 1984. Zur Entwicklung vegetationskundlicher Begriffsbildung am Beispiel der Fließwasservegetation Mitteleuropas. Tuexenia 4: 303–325.
Wildi, O. 1977. Beschreibung exzentrischer Hochmoore mit Hilfe quantitative Methoden. Veröff. Geobot. Inst. Rübel 60.
Wildi, O. 1986. Analyse vegetationskundlicher Daten: Theorie und Einsatz statistischer Methoden. Veröff. Geobot. Inst. Rübel 90: 1–226.

Vegetatio **81**: 29–39, 1989

Influences of the individualistic concept of vegetation on syntaxonomy

J. Moravec
Botanical Institute of the Czechoslovak Academy of Sciences, CS-252 43 Průhonice near Prague, Czechoslovakia

Accepted 8.9.1988

Keywords: Concept of vegetation, Phytocoenosis, Species individuality, Syntaxonomy, Vegetation continuum

Abstract

The individualistic concept of vegetation is based on three principles: (1) vegetation continuum, (2) ecological and chorological individuality of species and (3) multidimensional variability of vegetation. All principles reflect phenomena existing in nature. The conclusions arising from these principles are these: (1) plant communities (phytocoenoses) are not natural objects (wholes) as such but merely mixtures of plant individuals coexisting on the same site as the result of migration and environmental selection, (2) the classification of vegetation is quite arbitrary if not pseudoscientific, and (3) vegetation phenomena can be reduced to phenomena of plant individuals.

The integrated concept of vegetation considers the phytocoenoses as real functional systems (wholes) where plant populations are integrated both by the environment and by interactions among and within plant populations. The principles of the individualistic concept, but not the above conclusions arising from them, are compatible with the integrated concept. Phytocoenoses limit the individualistic performance of species through restriction of ecological amplitudes and shifting of ecological optima (as compared with physiological amplitudes and optima). The individualistic behaviour of species does not exclude the formation of coenological groups of species which form a basis for the classification of vegetation according to floristic-coenological criteria. Under natural conditions, plant communities are usually interconnected by transition zones to form a vegetation continuum. Phenomena of the vegetation continuum do not cause greater difficulties in syntaxonomy than those of polymorphic taxonomic groups in plant taxonomy.

Introduction

The individualistic concept of vegetation, having been neglected for about 30 yr, started to influence syntaxonomy in the 1950s. This concept was rediscovered by some American authors who criticized and refused the classification of vegetation, especially that of the Zürich-Montpellier School, as fully arbitrary and claimed ordination as the only method appropriate to an investigation of vegetation, i.e. vegetation as a continuum (see e.g. Whittaker 1962).

In fact, there was enough in the practice of this school to be criticized: (1) as soon as the classification system of vegetation advanced geographically, the majority of character-species appeared to be of regional validity only and no principal difference between character- and differential-species could be seen any longer (see also Gams 1941), (2) the diagnostic value of species was frequently changed according to deviating experiences of individual authors working in different regions; this induced thus frequent changes of the vegetation units, (3) floristic criteria were one-sidedly emphasized, although other criteria (ecological, chorological and syndynamical) were also applied tacitly, (4) no clear description of the method of tabular synthesis of phytosociological data was available at that time. Ellenberg (1954) correctly described this situation as the 'crisis of the character-species-doctrine'.

In the 1960s the rapid development of computers began to influence syntaxonomy. Computer-assisted ordination techniques were developed and proved to be an efficient tool in gradient analysis. The application of numerical classification techniques followed only slightly later. No wonder that the numerical techniques were welcomed for their objectivity. However, it was soon found that different numerical techniques gave different results for identical data, while all claimed to be objective. An explanation of the procedure of vegetation classification according to the Zürich-Montpellier School ('the tabular synthesis') was published relatively late, by Scamoni (1955) and Ellenberg (1956). It was approached by computer algorithm only at the end of the 1960s (see van der Maarel *et al.* 1976 for a survey).

Since long, also in phytosociology, ordination and classification of vegetation are regarded rather as complementary procedures which do not exclude each other (see e.g. Westhoff & van der Maarel 1973; van der Maarel 1975). The objections raised against classification from the standpoint of the individualistic concept (recently e.g. Whittaker 1978: p.5) still seem to be valid, and have especially been emphasized from the viewpoint of syndynamics. So, one really wonders if something may be wrong with the classification of vegetation. Meanwhile, specialists of other branches of biology regard syntaxonomy with suspicion and young scientists are discouraged to apply it (see also Wiegleb 1986).

The different concepts of vegetation

The individualistic concept originated partly as a reply to Clements's organismal concept and both concepts gave origin to an integrated concept.

The organismal concept

This concept, proposed by Clements (without using this term) is the oldest one. Clements formulated it perhaps as early as 1905, but published it in a complete and clear form later, in his monograph on plant succession (Clements 1916). According to this concept, the climax formation represents an adult complex organism and succession towards the climax is similar to the ontogeny. This concept received strong criticism in Europe (Gams 1918; Du Rietz 1919 (see 1921); Tansley 1920) and was not followed in its original version.

The designation 'organismal (or organismic) concept' has been used in a broader sense and later even in a different sense. Westhoff (1970: p. 14) considered as adherents to this concept all authors who regarded the plant community (phytocoenosis) as an organized system. Mirkin (1986a; see also Mirkin 1986b: p. 8) derived this concept from the definition of the association as accepted by the International Botanical Congress in Brussels 1910, as a basic unit for the classification system of vegetation comparable with the Linnaean species.

The individualistic concept

This term was introduced by Gleason (1917: p. 464) in his reply to Clements's (1916) book, but the first steps were apparently taken by Ramenskij as early as 1910 and independently of Clements's standpoint. Gleason (l.c.) conceived the plant community as a mixture of independent plant individuals and formulated the principle of reductionism in the following sentences: 'The development and maintenance of vegetation is... merely the resultant of the development and maintenance of the component individuals... the phenomena of vegetation depend completely upon the phenomena of the individual'. Thus Gleason's approach can be regarded as individualistic-reductionistic.

A more complete version of the individualistic concept (without using this term) was published independently by Ramenskij (1924 ex. 1971) in the following principles:

1. The principle of vegetation continuum (Pravilo nepreryvnosti rastitel'nogo pokrova) is based on the graphs showing the overlapping bell-shaped curves of the quantitative performance of populations of

more abundant species along an environmental transect in the field. Similar graphs were published by Whittaker (1956, 1967) 30 yr later, and also used as a basis for the theory of vegetation continuum.

2. The principle of ecological individuality of species (Pravilo ekologičeskoj individual'nosti rastitel'nych vidov) says that every species follows its own law of quantitative distribution depending upon environmental conditions..., that not two species even have exactly identical curves of quantitative distribution... and that the quantitative distribution of every species reflects the physiological constitution of each of these.

3. The principle of multidimensional variability of vegetation (Pravilo mnogomernosti rastitel'nogo pokrova) expresses the fact that the variability of vegetation does not follow one single ecological gradient only, but a 'multidimensional space' of gradients of ecological factors the varied combinations of which constitute the variability of 'ecotopes'.

In these three principles Ramenskij defined the concept of the real vegetation continuum, i.e. of the real complex of plant communities joined together with transition zones in accordance with the continuous changes of ecological factors. However, the author notes that the above principles are valid also when various plant communities of an extended territory are studied. In a complete and voluminous phytocoenological material (set of vegetation relevés) an indefinite variability of more or less distinct plant groupings is manifested as well as the fact that every detailed and quantitative relevé shows its individuality and that no two relevés are exactly the same. Therefore the relevés can be arranged into continuous and varied ecological series which join them in a single complex system.

In these lines Ramenskij formulated the concept of the 'syntaxonomic vegetation continuum' (without using this term), and consequently proposed the use of ordination ('Ordnung' – Ramenskij 1930; see also Goodall 1954: pp. 168–169) for the processing of vegetation data. Thus he laid the foundation of gradient analysis as recognized by Whittaker (1967: p. 210). Despite his views, Ramenskij (1924) wrote quite distinctly that the principle of ecological individuality of species does not contradict the existence of internally integrated coenoses and his approach can be designated as individualistic-integrated.

Gleason (1926) published the mature version of the individualistic concept independently of Ramenskij as a reply to Clements organismal concept. He touched again upon the nature of the plant community ('plant association'). This version can be summarized in two principles:

1. The ecological and chorological individuality of species is formulated in Gleason's often cited phrase '... every species of plant is a law unto itself, the distribution of which in space depends upon its individual peculiarities of migration and environmental requirements' (l.c.: p. 26).

2. The plant community represents merely a 'juxtaposition of individuals of the same or different species' (l.c.: p. 25) depending 'solely on the coincidence of environmental selection and migration' (l.c.: p. 26) 'and therefore is not an organic entity of itself' (Gleason 1917: p. 473).

In the first paragraph Gleason formulated the principle of individuality of species nearly in the same words as Ramenskij did. However, in the second paragraph he considerably differs from Ramenskij's view. (In the same year Lenoble 1926 expressed ideas similar to those of Gleason, apparently independently.)

The organismal and individualistic concepts are often treated as alternative approaches, however, they ought to be treated as polar concepts between which a gradient of other concepts occurs.

The community-unit concept

Whittaker (1967: p. 209) expressed the idea that basic vegetation units exist in nature as real entities and do not represent merely abstract categories (logical classes) of our knowledge: the 'community-unit theory'. This concept seems to be closer to the organismal then to the individualistic concept. It had been formulated already much earlier, e.g. by Du Rietz (1921: p. 15) and Alechin (1925), but had not become widely accepted in vegetation science. At present the term 'community-unit theory' or 'community-unit approach' is often used for all approaches adopting the classification of plant communities into abstract vegetation units (types) on the basis of common features (see Werger, 1974: p. 99), as is the case in the integrated concept.

The integrated concept

This concept originated from a reply to the failures both of the organismal and the individualistic concept of vegetation and has been accepted mainly by phytocoenologists of the Zürich-Montpellier approach (Westhoff 1972). It emerged from a paper by Tansley (1920) which refuted Clements' identification of a plant community as a complex organism. Tansley regarded the plant community as an organic entity integrated through a dynamic balance of competition between individuals, both of the same and of different species. Therefore, he adopted the term 'quasi-organism' (which was used for human society), which he later (1935) replaced by the term 'ecosystem' after he realized that the functional whole included not only the plant community but also other living components and that part of the environment with which the living components were in interactions. As early as in 1920, Tansley stated (l.c.: p. 125) that plant communities 'are certainly entities, in the sense that they behave in many respects as wholes and therefore have to be studied as wholes'. Thus the integrated concept denied the individualistic-reductionistic idea that plant communities are merely mixtures of plant individuals selected from the immigrants by the environment.

Westhoff (1972: p. 5) summarized the integrated concept as follows:

1. Phytocoenoses are functional wholes, which represent more than merely a sum of individual components selected by the abiotic environment.

2. The integration of phytocoenoses is the result of interactions between their components and is manifested primarily by their capability of self-regulation.

The principles of the individualistic concept as formulated by Ramenskij (1924) are compatible with the integrated concept which can be extended as formulated in the abstract.

Implications of the individualistic concept for syntaxonomy

Ramenskij (1924) and Gleason (1926) denied the possibility of an objective classification of vegetation, but their opinion received little attention. The American authors from the 1950s who formulated the concept of vegetation continuum independently of Ramenskij were more successful in arguing from the position of the individualistic concept against the classification of vegetation as quite arbitrary or even pseudoscientific (see Whittaker 1962: p. 95, p. 124, p. 127, pp. 158–160); their arguments were repeated even in the Handbook of vegetation science (Whittaker 1973: p. 327: 'Species do not fit naturally into groupings that correspond to community types... Community-types are not natural but arbitrary units...' and without changes in Whittaker 1978: p. 5) and are sometimes considered to be still valid. These authors based their arguments (1) on the fact that no two stands of plants are exactly the same, (2) on the individualistic behaviour of species, and (3) on the phenomena of vegetation continua. Although these arguments may be real, the objections derived from them against the classification of vegetation are not adequate. Recently similar objections based on the individualistic concept have been raised against succession as an ecosystemic process (see Drury & Nisbet 1973; Horn 1976; Connell & Slatyer 1977; Noble & Slatyer 1980).

Uniqueness of stands

Neither the general theory of classification nor special classifications in the natural sciences (botany, zoology, soil science, etc.) postulate that objects assigned to the same classification unit (logical class) must be exactly identical. The only condition is that they must have common properties or characters by means of which the unit can be defined and upon which it can be decided which objects belongs to it (see Fig. 1). In botany, it is well known that there are no two plant individuals that are exactly identical

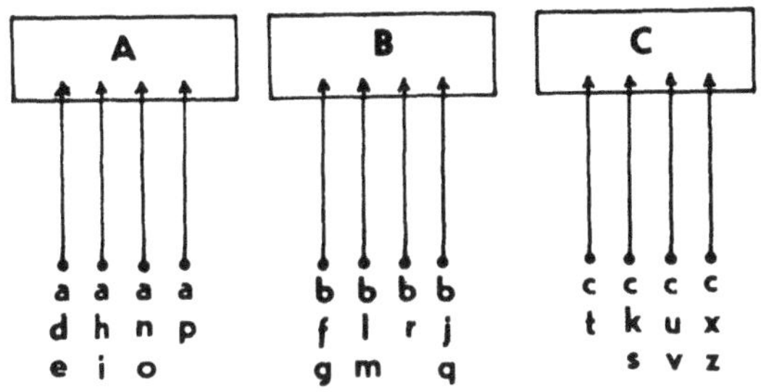

Fig. 1. Schematic illustration of the principle of delimitation of logical classes.

and, still nobody ever expressed doubts about the adequacy of a classification of the plant kingdom.

Individuality of species

The majority of plant species are independent of each other as well as of plant communities in which they occur (they can live outside communities as well, for instance in gardens). Only a limited number of species display dependence relations (e.g. specialized parasites) or directly exclude each other (e.g. through allelopathy). The vast majority does not possess mechanisms precluding the formation of groups of species on similar sites owing to their similar ecological requirements. Therefore, the principle of ecological and chorological individuality of species is not incompatible with the existence of coenological groups of species which are the basis of the floristic-phytocoenological approach in the classification of vegetation. These groups are characterized by clustered boundaries of occurrence in narrow transition zones, marking community boundaries, as can be seen along transects across complexes of plant communities (e.g. Dierssen & Dierssen 1984: p. 150; Shipley & Keddy 1987). Often, a certain species group shows a simultaneous disappearance and thus a marked limitation in one direction along the ecological gradient but a diversified successive disappearance and thus continuous transition into the opposite direction.

Rather distinct species groups were detected in the tabular synthesis of vegetation relevés. However, also in this case the principle of ecological and chorological individuality of species remained valid. The coenological groups of species are often geographically limited and outside the territory of their validity the same species can enter into different groups of species. Hegg (1965) proved the existence of syntaxonomically relevant coenological groups of species statistically on the basis of their joint occurrence in small sample plots.

Vegetation continuum and syntaxonomic continuum

An example of coenological groups of species within a syntaxonomic continuum as well as a fixation of certain species to certain vegetation units (syntaxa) can be seen in Fig. 2 showing the constancy of certain species in associations of natural mesophilous deciduous forests (the Carpinion, Tilio-Acerion and Fagion) in the Czech Socialist Republic. The figure has been constructed on the basis of a synoptic table (Moravec *et al.* 1982: pp. 18–33; 1989) made of 2019 relevés sampled by 27 authors between 1925 and 1981. These associations form a multidimensional 'field' with a main complex altitudinal gradient combined by a less distinct gradient of soil nutrient status (and reaction) completed with special variation regarding ecotope (ravines, scree). The influence of this field of ecological gradients is combined with the influence of chorological phenomena (vicarious associations of the Carpathian province, the influence of the Pannonian region, northern influences, local migration streams etc.). The arrangement of associations within alliances in Fig. 2 cannot fully represent this multidimensionality which is schematically outlined in Fig. 3. Of course, any rearrangement of associations in Fig. 2 will change the shape of the curves, but their peaks will remain in the same associations.

The upper part of Fig. 2 contains the main tree species the curves of which do not enable a delimitation of the individual syntaxa (except for *Ulmus montana* – a diagnostic species of the Tilio-Acerion). However, the curves of selected herb species show a distinct link with particular associations as well as their coherence in coenological groups of species. It may even occur that a particular species occurs in two associations belonging to different al-

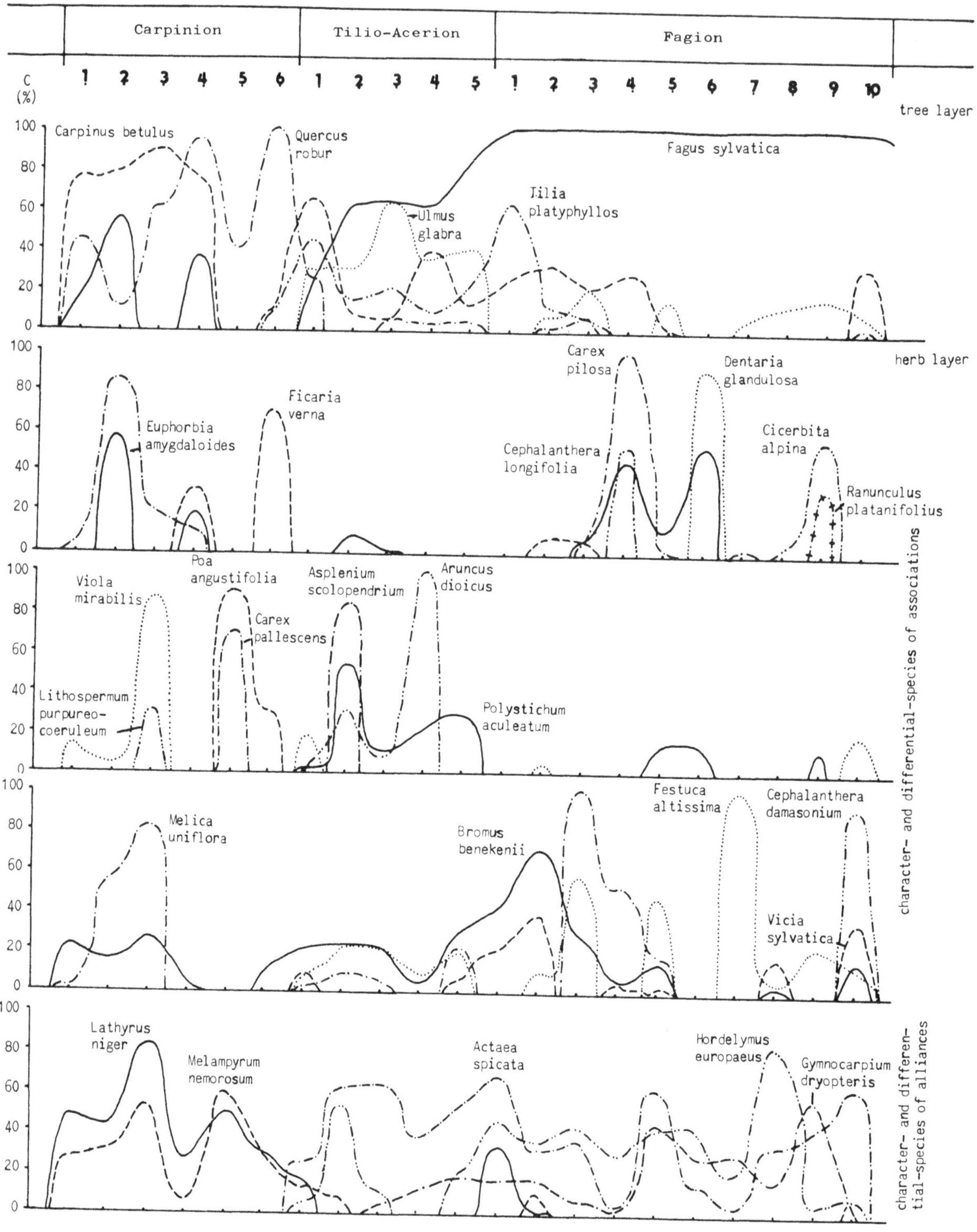

Fig. 2. Curves of a quantitative distribution (in constancy percentage) of selected species in associations of mesophilous deciduous forests in the Czech Socialist Republic.

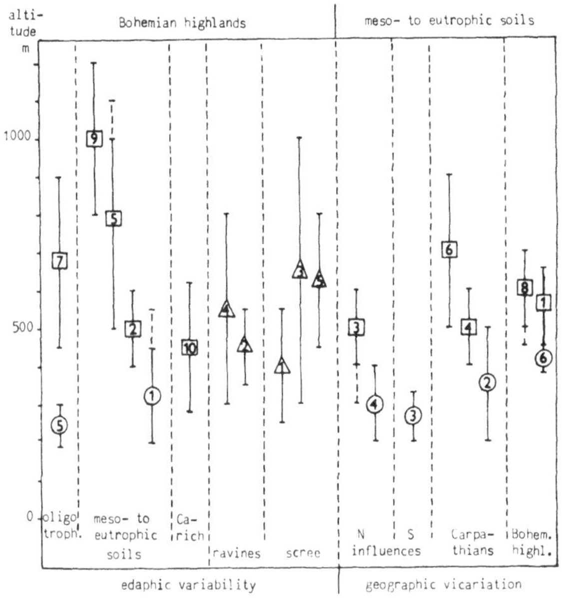

Fig. 3. Schematized 'field' of ecological and geographical variability of mesophilous deciduous forests of the Czech Socialist Republic (○ – *Carpinion*: 1 – *Melampyro nemorosi-Carpinetum*, 2 – *Carici pilosae-Carpinetum*, 3 – *Primulo veris-Carpinetum*, 4 – *Tilio-Carpinetum*, 5 – *Tilio-Betuletum*, 6 – *Stellario-Tilietum*; △ – *Tilio-Acerion*: 1 – *Aceri-Carpinetum*, 2 – *Scolopendrio-Fraxinetum*, 3 – *Lunario-Aceretum*, 4 – *Arunco-Aceretum*, 5 – *Mercuriali-Fraxinetum*; □ – *Fagion*: 1 – *Tilio platyphylli-Fagetum*, 2 – *Tilio cordatae-Fagetum*, 3 – *Melico-Fagetum*, 4 – *Carici pilosae-Fagetum*, 5 – *Dentario enneaphylli-Fagetum*, 6 – *Dentario glandulosae-Fagetum*, 7 – *Festuco-Fagetum*, 8 – *Violo reichenbachianae-Fagetum*, 9 – *Aceri-Fagetum*, 10 – *Cephalanthero-Fagetum*.

liances and co-occurs with different groups of species respectively (e.g. *Carex pilosa*). The bottom part shows species that are present in all associations of a certain alliance (with varying constancies) and which are therefore used as diagnostic species of the alliances.

Here two contradictory views should be discussed: Ramenskij (1924) and later Whittaker (1967) stated that the bell-shaped curves of species performance, overlapping continuously along environmental gradients, permit merely an arbitrary classification of vegetation with arbitrary boundaries of community-types (Whittaker 1973: p. 327; 1978: p. 5). On the other hand, the results of Hegg (1965) as well as Fig. 2 suggest that this classification may have a real basis. The explanation can be found in the fourth principle of Ramenskij's version of the individualistic concept – the principle of quantity (Pravilo količestvennosti). This principle infers that the quantitative representation of abundant species is much more important in the characterization of a habitat than the presence or absence of rare species. Thus Ramenskij's conclusions are based only on abundant (grassland) species with broad ecological amplitudes. Later, Ramenskij (1952) changed his opinion and admitted the insufficiency of dominants for the delimitation of associations. He admitted also that associations and higher classification units reflect objective phenomena existing in nature.

The same is true of Whittaker's above-mentioned statement, which was based on the bell-shaped curves of woody species only. Mueller-Dombois & Ellenberg (1974: pp. 328 – 333) analyzed Whittaker's results and noted that the bell-shaped form of species performance curves and their scattered distribution in the graphs were dependent on the procedures of data sampling and processing applied. – Thus Whittaker corroborated unconsciously the old experience of European phytocoenologists and foresters that the classification of the forest vegetation cannot be based on woody species only.

The characteristics of a real vegetation continuum – viz. broad transition zones between plant communities rather than sharp boundaries (except for man-induced situations) really exist. Still, communities can usually be distinguished by the relatively larger extent of uniform areas as compared with the smaller extent of transition zones. Similar phenomena can be found in the syntaxonomic continuum. For the above reasons Tüxen (1955) identified the fundamental units of the classification system (i.e. associations) as logical types of the natural sciences (naturwissenschaftliche Typen). These types represent a maximum correlative concentration of properties and characters of certain natural objects (see also von Glahn 1965), combined with the joint absence of properties and characters specific of other types (see Fig. 4).

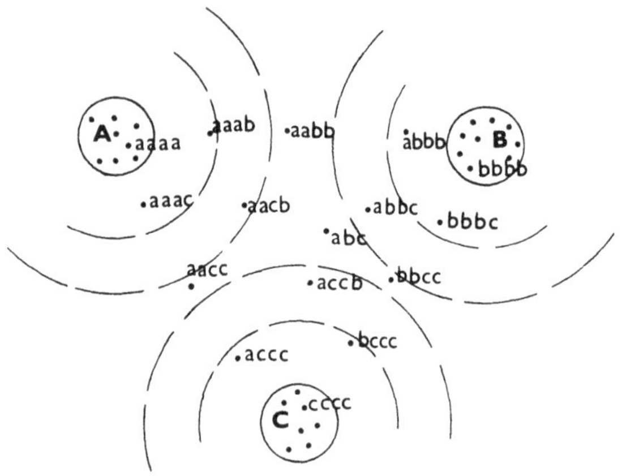

Fig. 4. Schematic illustration of the principle of delimitation of logical types.

Formal aspects of syntaxonomy

In syntaxonomy, real stands of vegetation are the objects of study, and relevés of such stands are the elements (samples) of the syntaxonomic treatment. The plant species recorded in the relevés, represent syntaxonomic characters. Syntaxa (formalized vegetation units) are defined by means of the presence or dominance of certain species or groups of intercorrelated species (diagnostic species), as well as by the remaining species and ecological and/or geographical features. The diagnostic species involved in the intentional definition of a syntaxon seldom reach a constancy of 100% in the set of relevés assigned to that syntaxon. This means that the group of diagnostic species is not always represented in full in any single relevé. In fact, we may find a continuous series of relevés with a successively decreasing number of diagnostic species, from those with the highest number, which represent the nucleus of the syntaxon, towards poorly characterized relevés at the syntaxon's margin, which is not sharp. The margin is determined subjectively as a minimal fraction of the diagnostic species group (usually the successive increase of diagnostic species of neighbouring syntaxa is also taken into account). Stands included within this limit are then regarded as elements of the syntaxon in the sense of a logical class. In this way, syntaxa defined as logical types are further interpreted as logical classes which can be arranged hierarchically within the system. This procedure is in full agreement with the principles of logical typology generalized by Hempel & Oppenheim (1936). According to Tüxen (1955) and Whittaker (1962: p. 116) the 'nuclei' and not the margins are of primary importance in syntaxonomic evaluation of differences between syntaxa. (See further Westhoff & van der Maarel 1973).

The occurrence of continuous transitions between classification units can be encountered in plant taxonomy as well, especially in polymorphic taxonomic groups where more or less distinct types are connected by continuous transitions (e.g. in some species of *Crataegus* and *Quercus*). However, this is never considered a basic obstacle in the classification of plants.

Subjectivity of the phytocoenological classification system

The predicates 'subjective' and 'objective' are usually seen as alternatives, meaning that either of these is valid in a particular case. However, in natural sciences, these predicates should be taken as extremes, between which a gradient occurs from more subjective (and less objective) to less subjective (and more objective). Our syntaxonomical knowledge is supposed to advance along this gradient from the subjective to the objective end, but it will not necessarily reach the latter. This also applies to many other classification systems in the natural sciences (e.g., for the systems of plants, animals, soils, etc.). None of these systems is absolutely objective because each is based on a comparative method, i.e., on an evaluation of differences between natural objects. Every individual scientist dealing with the creation or revision of a classification system will be found somewhere between the subjective and the objective endpoints of the gradient – an experienced and accurate scientist will move towards the objective end, a less experienced and less accurate scientist may be towards the subjective ends. This implies that classification systems in the natural sciences may be subjective in the beginning when initiated by a few individual scientists, but will become more and more objective as the science concerned advances.

This objectivation of classification systems proceeds by means of repeated critical revisions by later authors. Poore (1956) described this procedure in phytosociology as 'successive approximation'. However, there will always remain subjectivity for instance in the delimitation of the extent of classification units: we now have and we will keep both narrowly deliminated syntaxa and plant taxa besides broadly delimited ones. In plant taxonomy, again, this element of subjectivity is not regarded as a fundamental obstacle to systematics and the same attitude should be adopted in syntaxonomy.

Numerical syntaxonomy greatly contributed to the objectivation of the synthetical elaboration of relevés. However, the subjectivity has not been eliminated in this way but merely shifted towards the preparatory phase (transformation of original data, choice of similarity or dissimilarity indices, choice of numerical techniques, etc.).

There is a pair of alternatives related to that of subjectivity versus objectivity in classification, viz. artificial versus natural classification of vegetation. Again, these predicates indicated endpoints along a continuous gradient. Principally, no classification system (of plants, animals, soils, plant communities, etc.) exists in nature as such; each is the product of human endeavour. Whether a system is more artificial or more natural depends upon the criteria applied to the classification. When a classification is based on properties and characters of the natural objects, the classification system can never be completely artificial and the more substantial the properties and characters of the natural objects, the more natural is the classification system. Thus, the phytocoenological classification system is probably more natural than the original Linnaean ('sexual') classification system of plants, but less natural than the phylogenetic system of plants.

The nature of plant communities

The nature of plant communities is still not well understood (see e.g. Noy-Meir & van der Maarel 1987 for a historical review of this problem). Most authors consider plant communities simply as assemblages of plants living together with some degree of permanence (see Greig-Smith 1986: p. 19). However, from the point of view of integration, different kinds of plant assemblages exist in nature (see e.g. Alechin 1926; Neuhäusl 1963; Burrichter 1964) for the distinction of which suitable field methods are still lacking. Theoretically, we can distinguish the following three kinds of plant assemblages, which are not sharply delimited from each other:

1. Unstable incidental plant assemblages not capable of maintaining the identity of their species composition (even within one vegetation period). Usually, there are not considered plant communities.
2. Environmentally stabilized plant groupings (Standortsgemeinschaften – Burrichter 1964) lacking interactions between plants owing to an extreme environment.
3. Phytocoenoses or plant communities in the strict sense (Alechin 1926), integrated and stabilized by a regular framework of interactions within and between the component plant populations balanced by the environment. Even in this ,e we may expect different degrees of interactions when comparing an agro-phytocoenosis with a climax phytocoenosis.

The individualistic-reductionistic concept of Gleason seems to be valid for the first two cases, the integrated concept (including the individualistic-integrated concept of Ramenskij) seems to be valid for the phytocoenosis. (Cf. the definition by Westhoff & van der Maarel 1973: a system of interacting plant populations).

Recently, American authors have denied the existence of plant communities as integrated wholes according to Gleason's reductionistic idea (see e.g. McIntosh in West *et al.* 1981: pp. 10–23 for more details). The criterion for a whole is seen in the presence of 'emergent' properties that cannot be derived from the properties of individual components. In fact, there is little agreement on the question as to which properties of plant communities should be considered emergent and which properties only represent a sum of properties of constituent plant individuals ('collective properties'). A good discussion of this problem is found in West *et al.* (1981).

In my opinion, one of the most important emergent properties of phytocoenoses is their homeostasis, e.g. the capability of self-regulation, both the regulation of numbers of individuals of species in accordance with the carrying capacity of the environment, and the regulation of the species composition by exclusion of certain species which have the capability to grow in the environment concerned.

The second important property of phytocoenoses as wholes consists in their complex response to changes in ecological factors, the influence of which is transmitted through the framework of interactions between the constituents to species populations not directly influenced by these factors (see e.g., Greig-Smith 1986: p.29).

Moreover, phytocoenoses act as wholes while limiting the performance of certain constituent species. This can be seen in the narrowing of ecological amplitude and the shifting of ecological optima as compared with the respective physiological amplitudes and optima, as known from forestry, horticulture and ecological experiments (Ellenberg 1953; Ellenberg & Snoy 1957; Olsen 1923).

That phytocoenoses are functional wholes does not mean that they are deterministic wholes where the species composition should be unequivocally determined by the environment or where the presence of species B should be unequivocally determined by the presence of species A (which is only true for some specialized parasites). The actual fluctuation of the species composition around a certain mean, within a phytocoenose as well as between phytocoenoses of the same syntaxon, shows that phytocoenoses are rather stochastic wholes the constituents of which create a 'working system'.

References

Alechin, W. W. 1925. Ist die Pflanzenassoziation eine Abstraktion oder eine Realität? Bot. Jahrb. 60 Beibl. 135: 17–25.

Alechin, W. W. 1926. Was ist eine Pflanzengesellschaft? Ihr Wesen und ihr Wert als Ausdruck des sozialen Lebens der Pflanzen. Repert. Sp. Nov. Reg. Veget. Beih. 37: 1–50.

Burrichter, E. 1964. Wesen und Grundlagen der Pflanzengesellschaften. Abh. Landesmus. Naturk. Münster 26/3: 1–16.

Clements, F. E. 1905. Research methods in ecology. Lincoln. (not seen).

Clements, F. E. 1916. Plant succession. Carnegie Inst. Washington Publ. 242, Washington.

Connell, J. H. & Slatyer, R. O. 1977. Mechanisms of succession in natural communities and their role in community stability and organisation. Amer. Nat. 111: 1119–1144.

Dierssen, B. & Dierssen, K. 1984. Vegetation und Flora der Schwarzwaldmoore. Beih. Veröff. Naturschutz Landschaftspfl. Baden-Württemberg 39: 1–512.

Drury, W. H. & Nisbet, I. C. T. 1973. Succession. J. Arnold Arbor. 54: 331–368.

Du Rietz, G. E. 1921. Zur methodologischen Grundlage der modernen Plfanzensoziologie. Thesis Uppsala. A. Holzhausen, Wien.

Ellenberg, H. 1953. Physiologisches und ökologisches Verhalten derselben Arten. Ber. Deutsch. Bot. Ges. 65: 350–361.

Ellenberg, H. 1954. Zur Entwicklung der Vegetationssystematik in Mitteleuropa. Angew. Pflanzensoz. Wien Festschr. E. Aichinger 1: 134–143.

Ellenberg, H. 1956. Grundlagen der Vegetationsgliederung. I. Aufgaben und Methoden der Vegetationskunde. E. Ulmer, Stuttgart.

Ellenberg, H. & Snoy, M. L. 1957. Physiologisches und ökologisches Verhalten von Ackerunkräutern gegenüber der Bodenfeuchtigkeit. Mitt. Staatsinst. Allg. Bot. Hamburg 11: 47–87.

Gams, H. 1918. Prinzipienfragen der Vegetationsforschung. Vierteljahrschr. Naturf. Ges. Zürich 63: 293–493.

Gams, H. 1941. Über neue Beiträge zur Vegetationssystematik unter besonderer Berücksichtigung des floristischen Systems von Braun-Blanquet. Bot. Arch. 42: 201–238.

Gleason, H. A. 1917. The structure and development of the plant association. Bull Torrey Bot. Club 44: 463–481.

Gleason, H. A. 1926. The individualistic concept of the plant association. Bull. Torrey Bot. Club 53: 7–26.

Goodall, D. W. 1954. Vegetational classification and vegetational continua. Angew. Pflanzensoz. Wien Festschr. E. Aichinger 1: 168–182.

Greigh-Smith, P. 1986. Chaos or order – organization. In: Kikawa, J. & Anderson, D. J. (eds), Community ecology, pp. 19–29. Blackwell Melbourne.

Hegg, O. 1965. Untersuchungen zur Pflanzensoziologie und Ökologie im Naturschutzgebiet Hochgant (Berner Voralpen). Beitr. Geobot. Landesaufn. Schweiz. 46: 1–188.

Hempel, C. G. & Oppenheim, P. 1936. Der Typusbegriff im Lichte der neuen Logik. Leiden.

Horn, H. S. 1976. Succession. In: May, R. M. (ed.), Theoretical ecology. pp. 187–204. W.B. Saunders, Philadelphia.

Lenoble, F. 1926. A propos des associations végétales. Bull. Soc. Bot. France 73: 873–893.

Mirkin, B. M. 1986a. Čto takoje rastitel'nyje soobščestva. Moskva.

Mirkin, B. M. 1986b. Paradigm change and vegetation classification in Soviet phytosociology. Vegetatio 68: 131–138.

Moravec, J. 1989. Organismální, individualistické a integrované pojetí vegetace (Organismal, individualistic and integrated concepts of vegation). Preslia 61: (in print).

Moravec, J., Husová, M., Neuhäusl, R. & Neuhäuslová-Novotná, Z. 1982. Die Assoziationen mesophiler und hygrophiler Laubwälder in der Tschechischen Sozialistischen Republik. Vegetace ČSSR, Ser. A, 12. Academia, Praha.

Mueller-Dombois, D. & Ellenberg, H. 1974. Aims and methods

of vegetation ecology. J. Wiley & Sons. New York.

Neuhäusl, R. 1963. Allgemeine Fragen der phytozönologischen Terminologie. Preslia 35: 302–315.

Noble, I. R. & Slayter, R. O. 1980. The use of vital attributes to predict successional changes in plant communities subjected to recurrent disturbances. Vegetatio 43: 5–21.

Noy-Meir, I. & van der Maarel, E. 1987. Relations between community theory and community analysis in vegetation science: some historical perspectives. Vegetatio 69: 5–15.

Olsen, C. 1923. Studies in hydrogen-ion concentration of the soil and its significance to the vegetation, especially to the natural distribution of plants. Compt. Rend. Trav. Lab. Carlsberg 77: 1–160.

Poore, M. E. D. 1956. The use of phytosociological methods in ecological investigations. IV. General discussion of phytosociological problems. J. Ecol. 44: 28–50.

Ramenskij, L. G. 1910. O sravnitel'nom metode ekologičeskogo izučenija rastitel'nych soobščestv. Dnevnik XII Sjezda Russ. Jestestvoispyt. i Vračej Otd. II, 9: 389–390. (not seen)

Ramenskij, L. G. 1924. Osnovnyje zakonomernosti rastitel'nogo pokrova i ich izučenie. Vestn. Opytn. Dela Voronež pp. 37–73.

Ramenskij, L. G. 1930. Zur Methodik der vergleichenden Bearbeitung und Ordnung von Pflanzenlisten und anderen Objekten, die durch mehrere, verschiedenartig wirkende Faktoren bestimmt werden. Beitr. Biol. Pfl. 18: 296–304.

Ramenskij, L. G. 1952. O nekotorych principial'nych položenijach sovremennoj geobotaniki. Bot. Žurn. 37/2: 181–201.

Ramenskij, L. G. 1971. Izbrannyje raboty, pp. 5–33. Nauka, Leningrad.

Scamoni, A. 1955. Einführung in die praktische Vegetationskunde. Deutsch. Verl. Wiss., Berlin.

Shipley, B. & Keddy, P. A. 1987. The individualistic and community-unit concepts as falsifiable hypotheses. Vegetatio 69: 47–55.

Tansley, A. G. 1920. The classification of vegetation and the concept of development. J. Ecol. 8: 118–149.

Tansley, A. G. 1935. The use and abuse of vegetational concepts and terms. Ecology 16: 284–307.

Tüxen, R. 1955. Das System der nordwestdeutschen Pflanzengesellschaften. Mitt. Flor.-Soz. Arbeitsgem., N.F. 5: 155–176.

van der Maarel, E. 1975. The Braun-Blanquet approach in perspective. Vegetatio 30: 213–219.

van der Maarel, E., Orlóci, L. & Pignatti, S. 1976. Data-processing in phytosociology. Junk, The Hague.

von Glahn, H. 1965. Der Begriff des Vegetationstyps im Rahmen eines allgemeinen naturwissenschaftlichen Typenbegriffes. Ber. Geobot. Inst. ETH Zürich 36: 14–27.

Werger, M. J. A. 1974. The place of the Zürich-Montpellier method in vegetation science. Fol. Geobot. Phytotax. 9: 99–109.

West, D. C., Shugart, H. H. & Botkin, D. B. 1981. Forest succession. Springer-Verlag, New York.

Westhoff, V. 1970. Vegetation study as a branch of biological science. In: Venema, H. J., Doing, H. & Zonneveld, I. S. (eds) Vegetatiekunde als synthetische wetenschap. Miscell. Papers Landbouwhogesch. Wageningen 5: 11–30.

Westhoff, V. 1972. Die Stellung der Pflanzensoziologie im Rahmen der biologischen Wissenschaften. In: van der Maarel, E. & Tüxen, R. (eds), Grundfragen und Methoden in der Pflanzensoziologie, pp. 1–12. Junk. Den Haag.

Westhoff, V & van der Maarel, E. 1973. The Braun-Blanquet approach. In: Whittaker, R. H. (ed.), Ordination and classification of communities, pp. 617–726. Junk, The Hague.

Whittaker, R. H. 1956. Vegetation of the Great Smoky Mountains. Ecol. Monogr. 26: 1–80.

Whittaker, R. H. 1962. Classification of natural communities. Bot. Rev. 28: 1–239.

Whittaker, R. H. 1967. Gradient analysis of vegetation. Biol. Rev. 42: 207–264.

Whittaker, R. H. (ed.) 1973. Ordination and classification of communities. Junk, The Hague.

Whittaker, R. H. (ed.) 1978. Classification of plant communities. Junk, The Hague.

Wiegleb, G. 1986. Grenzen und Möglichkeiten der Datenanalyse in der Pflanzenökologie. Tuexenia 6: 365–377.

Vegetatio **81**: 41–60, 1989

Similarity measures for structured data: a general framework and some applications to vegetation data

M. B. Dale
Dept. di Biologia, Università di Trieste, Via Valerio 30-32, Trieste I-34100, Italy; Permanent address: C.S.I.R.O., Division of Tropical Crops and Pastures, 306 Carmody Rd., St. Lucia, 4067, Qld. Australia

Accepted 8.9.1988

Keywords: Classification, Frequency distribution, Graph, Levenshtein distance, Profile, Richness, String, Succession, Transect, Tree, Weighting

Abstract

Although there are many measures of similarity existing in the phytosociological literature, these almost all apply to data for which the describing attributes have only single values. In many cases, however, there can be a richer structure in the attribute values, either directly from the nature of the attributes or derived from relationships between the stands. In this paper, I first examine a range of possible sources of such structure in phytosociological data, and then propose a similarity measure sufficiently general to be applicable to all the variant types. Finally I present some examples of applying such measures to frequency data from tropical grasslands and to successional data from subtropical rain forest.

Introduction

Many methods of classification and ordination employ pairwise similarity measures which are used to organise stands into clusters or display them in some axis configuration. While it is sometimes possible to observe similarity values between items directly (Lehmann 1972), in most cases the similarities are calculated from various attribute values associated with the stands; in phytosociology the attribute values are mostly species performance estimates, although other descriptors have occasionally been employed. The choice of attributes is obviously critical in determining patterns in the data; by comparison the choice of similarity measure to be calculated from the attribute values, while still important, is of minor importance. It is desirable, then, that we have available similarity measures suitable for whatever attribute values we desire to record. One important feature of such similarity calculations is that the attribute values used have been restricted, in almost all cases, to be single valued. Yet there are many interesting cases where such single values are inappropriate, the desirable attribute values being structured or multiple valued. These composite structured or multiple values can derive from the nature of the attributes themselves; in the case of logical dependency, for example, the abundance of a species cannot be measured if the species is absent. Composite values may also be obtained from relationships between stands, by redefining the basic unit being classified to be some aggregate of primary sampling units. This might be desirable when studying successional sequences or spatially arranged transects, where it is desired to examine the sequences or transects, not the elements of which they are composed.

With values reflecting intrinsic structure in the attributes and/or extrinsic structure in the stands, the common similarity measures are inapplicable and some new forms are required. Such new forms should not prohibit the use of single valued attributes so we must seek a more general class of measures which can encompass all the varying types of attributes. Such a general class of measures would also have the advantage of providing a common context within which coefficients could be compared and evaluated, thus avoiding the need for extensive empirical studies which presently provide us with information on the properties of similarity measures.

In this paper I examine the sources of multiple values and structure in attributes, and then propose a general approach to the problems of measuring similarity for structured data. This proposal does provide a means by which the same kind of similarity can be measured for all attribute types. Thus all contributions to the overall similarity can be made compatible and any incompatibilities which are desired are made explicit. The proposed similarity measure is then applied to two exemplary phytosociological data sets. Finally some problems and some alternative similarity measures which might also be applicable are noted for further study.

Sources of structure

Intrinsic structure

Intrinsic structure in the attributes can be derived from many sources. First, the attributes may have states related in some functional or evolutionary way, as in the character state trees of taxonomists (see e.g. Estabrook & Meacham 1979). In phytosociology, composite cover-abundance coding leads to a complex relationship between the various scale values (Dale 1989). The relationships between states need explication in order that changes in cover and density can be in some sense equivalenced.

Next, the states of an attribute may be nonexclusive; this would occur for example, if we were to record several flower colours existing within the same stand for a single species. Lance (1970) has presented some variations, his pseudo- and quasi-ordered types, which strongly suggest that nonexclusive values are found where a class description of a multivariate situation is being attempted. Further, the attributes themselves may be related through logical dependency as examined by Dale (1968, see also Ben-Bassat & Zaidenberg 1984). Phytosociological examples include the use of quantitative data since this necessarily involves the preexistence of presence/absence values, or in Lambert & Dale's (1964) suggestions that quantitative values be represented by a series of cumulatively coded binary values, popularised as pseudospecies by Hill (1979).

Attributes may record not single values but frequency distributions; examples would include the use of girth or height classes. Dale & Dale (1986) have shown that incorporating the ordering relationship between classes into an analysis can modify the results obtained. Frequency distributions can themselves be multidimensional, as with transition matrices, and the simple approach adopted by Dale & Dale then becomes impossible to use. For unordered classes the method of Legendre & Chodorowski (1977) may be relevant. Note too that usually the frequencies are determined for intervals rather than for simple values.

Several restricted forms of this kind of ordered interval structure can occur. The ordered categorical Braun-Blanquet codes are probably the most commonly used measures of species performance, yet each code represents an interval, not a single value (Dale 1989). Further, a single attribute can be assigned a range of values. For example a range of leaf sizes might be recorded as 10–25 cm^2 or assigned to several adjacent categories of size class; Blackburn (1980) has considered a possible measure for this case. We can extend this to cases where, for any attribute, several values may be observed and collected to form a *bag*; unlike a set a bag may have repeated identical elements. A phytosociological example is Jensén & van der Maarel's (1980) use of samples within lakes, if it were desired to classify the lakes. The sample could be used to characterise a frequency distribution but this simply places us in the situation described above. Sibson's (1969) information radius or Matusita's (1977) affinities are examples of similarity measures which might be useful in this case, although the problem of measuring distance between distributions has exercised statisticians for a long time.

Sometimes attributes are naturally structured

mathematically. Angular measures have a two-dimensional nature requiring representation in a complex plane. Thus aspect, a commonly employed environmental measure, requires first that some slope exists, a logical dependency, and further that 360° and 0° be regarded as equal values. Such measures can be recoded by employing two correlated variables in place of the single aspect value, but the two introduced variables need to be treated together as a single entity, i.e. as a multiple value. Slope, also a commonly employed environmental factor, poses less problems for it is usually restricted in range so that, while in principle it requires the identity of 360° and 0°, in practice this is of little importance because of the restricted range of values encountered. For vegetation, any attempt to characterise the three-dimensional distribution of foliage in a stand will also be likely to produce angular data and in three rather than two dimensions; Wallbrecher (1976) discusses 3-dimensional angular measures which might be used if slope and aspect were to be combined in a single measure. Attempts to measure shape of objects or texture of surfaces often result in attributes with comparable problems. See for example the shape descriptors employed by Dale *et al.* (1970a).

Finally, for intrinsic structure, the way in which we measure can pose difficulties. The scale of measure may be bounded. In many cases phytosociological data are censored in some manner, if only because we cannot measure how much a species is NOT present, but boundedness can arise from our measuring procedures in other ways. If we estimate percent cover by recording shoot frequency, i.e. by counting the number of subsamples containing some part of the species, we may find that the upper limit of 100% can be attained in several, nonequivalent, ways. For example there could be a dense populatio of small individuals or a single large leaf. Even assuming we recorded this difference, the presently available similarity measures would only permit us to use the 100% value. Borg & Staufenbiel (1986) have discussed some problems with distance measures for such scales while Hohn & Nuhfer (1980) have discussed possible, asymmetric, correlation measures for such kinds of data. Also, if our observations are made using visual estimates and the observers differ in their interpretation of class boundaries, we must allow for possible disagreements in our similarity calculation. Brook & Stirling (1984) have discussed this problem.

In all these cases the notion of a single value for each attribute is not acceptable, so that the usual similarity measures are inapplicable. Note that the existence of correlation between attributes is not here regarded as introducing structure unless it is a logically necessary correlation. It may of course sometimes be desirable to remove the effects of empirically observed correlation between attributes but this is a different problem (see e.g. Ito *et al.* 1984).

Extrinsic structure

These are not the only sources of structured descriptions which can be found in phytosociology. We can extend the notion of an attribute to more complex structures which derive from spatial and temporal relationships between stands, an extrinsic structure. Thus in successional studies it may be desirable to classify, or ordinate, not so much the descriptions of the 'stands at specific times', but the temporal sequences themselves. This would be useful in palynology for example, although there are complications if the sequences are truncated or fragmented by gaps. (c.f. Hogeweg & Hesper 1984). Following Dale *et al.* (1970b), Williams *et al.* (1969) attempted classifications of successional sequences while Norris & Dale (1971) later classified soil profile descriptions, both using transition matrices. Unfortunately this approach has some major difficulties, to be discussed later, and loses some of the information contained in the sequences. For continuous time records Bellacicco (1977) has suggested decomposition using spectral analysis.

Somewhat analogous data can be obtained when stands are spatially arranged in clumps. Such patterns are often 2-dimensional arrays of records rather than the linear sequences of the temporal studies, although linear arrangements are not unknown; for example Mäkirinta's (1978) scheme could be regarded as a circular linear structure. The use of physiognomic data would necessitate the comparison of shapes which is usually accomplished by reducing them to collections of spatially related simple polygons (e.g. Bykat 1979). Spatial studies of this kind can be regarded as studies of landscape patterns, of catenary variation and zonations.

All these various forms of structured data are potentially relevant to phytosociological studies, so it is important to discover what kinds of similarity measures can be used with them. In addition it would also be desirable to have some relationship between similarity measures employed with structured data and other measures used with single valued attributes, so that the two may be sensibly combined (c.f. Lerman & Peter 1985). It may also be desirable to calculate similarities differently for different subsets of variables, even where they are of the same kind, and then combine these with varying weights. Little & Ross (1985) present some examples of such weighting in soil profile studies. It seems, to me at least, to be important that in general we measure the same kind of similarity, however that is to be defined, for all the different sets of attributes which we seek to combine into a single measure of similarity. If this is not to be done then at least the choice should be conciously made. Of course, in some cases it may be preferable to use one part of the similarity as a constraint as suggested by Dale & Walker (1970), rather than include it in the overall assessment.

In this paper I shall concentrate on structure deriving from frequency distributions and from relationships between stands, in particular temporal sequences, and will examine the use of the minimum mutation or Levenshtein distance (Levenshtein 1965) to measure similarity in these cases. This particular measure can be adapted conceptually to cover most of the structured attribute types noted previously, although I shall not attempt such a complete coverage here. This flexibility of application seems highly desirable, enabling us to combine a wide variety of attribute contributions to similarity within a single kind of measure. The Levenshtein measure can provide a basis for formulating many of the commonly used similarity measures for single valued data, giving the comparable measure sought. I do not claim it to be the only, nor necessarily the best, such measure.

Levenshtein distance: strings and changes

Consider first two sequences of observations of a succession, with specific vegetation types denoted by one of 6 letters of the alphabet, A, B, C, D, E, F. If one succession runs AABBBBBCCD while the other runs ABBCEECF, how can a measure of similarity for these two strings be derived. Let us start with the child's game of converting one word into another by a series of letter changes. – The game is called 'doublets' and was apparently invented by Lewis Carroll although I have no reference. Interestingly enough Carroll invented another word game, 'syzygies' which is effectively seriation! – For example we may wish to change the word 'SEEM' into the word 'POUT'. One possible sequence of words is: SEEM SEAM BEAM BEAR BOAR BOOR POOR POUR POUT. However we can get a shorter series by using: SEEM SEER BEER BOER BOOR POOR POUR POUT and still shorter with: SEEM SEER PEER PEUR POUR POUT.

The rules of the game are fairly simple. They are:

1. Change only (substitute) one letter at a time.
2. All the words must be meaningful in whichever language you are using. This makes our third, shortest solution unacceptable, as it contains a french word.
3. The number of changes must be minimal.

We can regard this minimal number of changes as the distance between the two words, and this is, in effect, what the Levenshtein distances do. They count the minimal number of changes needed to convert one string to another. Our phytosociological 'words' I will take to be sequences of symbols, each symbol representing a different vegetation type. We need not restrict ourselves to allowing substitution of one type by another as the only permissible change. Indeed if our observation periods are of different lengths, as they may be in pollen diagrams for example, then we must be able to add, delete or merge some symbols and leave gaps or we can never change one sequence to the other. What this means is that, first, a set of permissible changes must be defined; for example we might allow only insertion and deletion. Next, the weight to be given to each of these changes has to be assigned and finally the minimal number of changes which convert one string into the other must be identified, which gives one of the *Levenshtein distances*. If we use equal weights for all symbols, for our example successional strings the procedure goes as follows:

Delete A; Delete 3 B's; Insert 2 E's; Delete D; Insert F total = 8 changes

We can get a similarity rather than a distance if we count the symbols in the shortest common subsequence (a subsequence has symbols in correct order but does not maintain adjacency whereas a string maintains adjacency). For our example the longest common subsequence is ABBCC, of 5 symbols. The shortest common supersequence is defined as the shortest string subsuming both of the observed strings as sequences; for our example it is AABBBBBCEECDF, of 13 symbols. Our similarity could then be taken to be 5/13. This particular similarity is analogous to the Bray-Curtis and Jaccard measures both of which are essentially ratios of (cardinalities of) set intersection and set union measures.

The general principle of minimising the number of changes needed to convert one string to another can be used in a wide variety of cases. A general reference is Sankoff & Kruskal (1983) which also shows how the distance between strings can be calculated, using a dynamic programming algorithm. However van Rijsbergen (1986) suggested a very similar, if not identical approach when he espouses the minimal program to change one specification to another, in information retrieval. Here I shall only indicate ways in which the simple form shown above can be changed to reflect specific features of similarity which might be of importance.

1. We can allot different weights to operations involving different symbols. Thus the substitution 'AC' might be regarded as more costly than substituting 'CF'. Costs of insertion and deletion of a symbol could differ and we could weight classes of symbols too; thus the replacement of any herb by any tree might be regarded as less costly than replacement of a tree by a herb. One problem here is simply how to obtain the necessary weights, which can be very numerous if a large number of types are identified. This form of weighting differs from the proposals of Feoli & Lagonegro (1983). They differentially weighted contributions to the overall measure derived from elements of similarity (matching) as well as dissimilarity (mismatching). If we had two sequences which match in some fragments, but with large sections not matching then it is necessary to introduce such weighting of both matches and mismatches.

2. We can weight operations on the same symbol differently depending on the surrounding context of other symbols. The context will most often be some adjacent symbol(s) but can even be randomly defined.

3. We can weight operations on the same symbol differently depending on their position in the string. Thus adding symbols to the beginning or end of a string could be made easier than inserting them in the middle of the string (c.f. Lu & Fu 1978). This is one means of coping with sequences whose lengths are very different.

4. The set of operations permitted can be extended. As an example, in addition to insertion, deletion and substitution, we might allow pairs of symbols to be transposed as a single operation; thus **ht th** would only be a single change apart (see Lowrance & Wagner 1975); this sort of change would be relevant to errors in typescripts for example. Alternatively the operations can be restricted; for example only substitution might be allowed as in the game described above. The Levenshtein formulation of squared Euclidean distance requires such a restriction, with weights derived from the squared differences of the (numeric) symbols. Using transpositions of adjacent symbols is characteristic of permutation measures of rank correlation (Kendall 1938). With rankings, if we allow transpositions between non-adjacent symbols we obtain a distance measure first used by Cayley (1849), while if we assume that the ordering is only partial then we obtain a Hausdorf extension to Cayley's measure (c.f. Critchlow 1985). Such Hausdorf measures will be discussed elsewhere (Dale 1989).

5. We can regard the string as being slightly elastic, so that the same symbol can appear in runs of varying length without any great penalty; methods for calculating distances in this case are felicitously known as 'time-warping'. In a spatial transect along a gradient and in successional sequences the *rate* of change need not be constant and then time warping

would be valuable. While related to usual deletion and insertion operations, such compression and expansion reflect the underlying continuous nature of the processes being examined. Alternatively, we could compress all runs of the same symbol to a single representative and compare only the symbol content; this is like using presence/absence data instead of abundance data since we forgo information of the magnitude of sequences involving only a single symbol. We could even use only the global frequencies of different symbols, the Parikh vector, effectively foregoing all sequence information, if this is sufficiently discriminatory.

6. Instead of finding the overall minimal number of changes we can constrain the solution in some way, insisting on some synchronisation of the two strings being compared. The constraints can be in the form of limits to the positions of possible substitutions, of limits to the number of consecutive changes of some kind which are permitted or of limits to the overall numbers of particular kinds of changes. We can also allow gaps in strings with penalty either dependent on length of gap or independent of it. This latter would allow for gaps in the record to be acknowledged without too severe a penalty due to differences in length of strings (Fredman 1984).

7. In the children's game we insist that the words be meaningful. We can equally phytosociologically value the strings produced as intermediates, and place constrains on them, to be 'meaningful'; sequences without meaning are simply prohibited.

8. We can extend the notion of minimal change from pairs of strings to sets of strings (Findler & van Leeuwen 1979; Lemone 1982), to tree structures such as dendrograms (Lu 1984; Waterman & Smith 1978), to arrays (Moore 1979) and more generally to directed graphs, by adding extra operations relevant to these structures. In phytosociological studies 2-dimensional arrays of symbols, representing areas of landscape, would seem most pertinent. However with these more complex structures the computational demands are increased and their widespread use will almost certainly require the availability of highly parallel computers.

9. In some circumstances it may be permissible to allow 'wild' characters which match all others, or more general pattern matching. Thus in our word example we could insist that only certain classes of syllabic replacement were permissible; soft vowels for hard, voiced syllables for unvoiced or glottal consonants for laryngeal or labial; phytosociological equivalents would be life form or growth form classes. Whatever merits these might have lie presently in the study of counterpoint or language not phytosociology.

These possibilities are sufficient to show the great flexibility of the Levenshtein approach, although at present we lack any great experience with most of these alternatives in phytosociological studies. Note though that for special purposes, such as identification, other methods, such as the redundant hash coding approach of Reuhkala *et al.* (1979), may be more efficacious. In the examples which follow only the simplest, equal weighted insert/delete version of the Levenshtein measure are used, and this seems reasonably effective. Some of the variations will be examined in a companion paper.

The sources of strings

So far I have said nothing about the nature of the symbols used to form the sequences except to state that they are vegetation types. (This may not be the only possible phytosociological interpretation, but it is the only one I shall consider here). Where do the symbols come from and what is the significance of changes in the symbol set? I shall briefly examine two possibilities, one for frequency data, the other for successional data. However the general approach can be adapted for most of the sources of structure mentioned earlier.

With frequency data we have records of the numbers of times a particular species has some property in subsamples of a stand. If this is a single number then we already have available suitable similarity measures, such as information coefficients, and there is no necessity to invoke string conversion at all. But perhaps we have recorded frequency by species and size class; our data are now 2-way frequency distributions with ordered categories. Dale & Dale (1986) have suggested a simple method of dealing with data of this kind, but the Levenshtein approach of Werman *et al.* (1985) offers more possibilities. We first allocate to each species and size class combination a unique symbol; thus species 1 size class 1 might be assigned A, species 1 size class 2 assigned B, species 1 size class 3 assigned C and so on. We now examine the first stand and if species 1 in size class 1 occurs n times then we write n A's, if it occurs in size class 2 m times then we append m B's, if it occurs in size class 3 p times then we append p C's and so on for all size classes, and all species. If the species-size class combination has a zero frequency the relevant symbol is omitted. This process is applicable to n-dimensional contingency tables (Orlóci & Stofella 1986). Our Levenshtein distance can now be defined in terms of deletion, insertion or any other operation which we deem pertinent to transforming the strings. We might, for example, decide to make the deletion of a symbol of low weight and insertion of high weight, except when the symbol is already present in the target string.

If we have only a single frequency measure for each species, and are using the simple Levenshtein metric with equal weights for insertion and deletion, we in fact obtain the Bray-Curtis coefficient. However if we allow substitution with differential weighting, we can allow for the ordering of categories in frequency distributions, and further can allow multidimensional tables! We simply make a change from $A \rightarrow B$ less costly than a change from $A \rightarrow C$, but perhaps equal to the cost of the change $C \rightarrow B$. We can also, if desired, give different weights to changes in size class for different species and still others to changes between species with the same, or different, size classes. Indeed the problem would seem to be how to determine appropriate weighting for all the possibilities open to us in the more complicated versions. The frequency coding proposed here produces strings of varying length reflecting the presence or absence of species in stands. Thus it is likely that, unless some compensatory action is taken, our similarity measure will be sensitive to absences; this is indeed the case with the Bray-Curtis coefficient and the conceptually identical similarity ratio. There are various modifications which can be used to lessen the importance of variation in sequence length, which is here reflecting species richness and abundance; for example a fixed gap-weighting option is a suitable choice, though positional weighting might also be attractive.

If we examine the kind of data which is usually available for a study of succession, it consists of observations on a number of stands at several different times. At each time every stand would be described by the presence and/or absence of species, or some similar data set. If we have a priori classes for vegetation then these records can simply be used to allot the stands to their appropriate class. Each class can be identified by some unique symbol and the sequence of observations on each stand then provides the string of symbols needed. If no a priori classification is available then we have two choices. Either we can classify the 'stand at time' data to form classes or we can attempt to use the data without prior classification by defining a suitable Levenshtein measure. We could count the cost of inserting, deleting or substituting suites of species simultaneously. If the species performance values s_j lie in the range $0-1$, then we could perhaps take a deletion as costing s_j, an insertion as costing $|s_j$ and the substitution of species j by species k as costing $|s_j - s_k|$. More complex formulae, perhaps involving Kullback's (1959) J-divergences for example, could be devised if this seemed necessary.

The classification alternative is easily accomplished numerically, but there are problems. Firstly, the number of types (classes) has to be determined, which requires the use of effective stopping rules. Although such rules have been developed (c.f. Dale 1988) they are not widely available, so that some study of the effects of changing the number of types seems desirable. Secondly it seems extremely likely that a stand could have affinities for 2 (or more) types while it was in a transitional stage, so the allocation of a single symbol is a somewhat dubious procedure. Residual elements of a previous type as well as the initial elements of a succeeding type may be present in a plot. Some form of fuzzy classification would seem appropriate with stands belonging to types with varying degree, but we then have each stand described by its degree of belonging to each type, a form of data little different to the original species description. – There are analogies here with problems in continuous speech recognition and cursive script recognition (see e.g. Sakoe 1979); in these problems there is a possibility of incorrectly identified symbols being present in the string. We may have merged symbols which should be separate, rather like kerning 'a' and 'e' to form 'ae' or have separated parts of symbols which should be merged to form a single symbol, as in the erroneous replacement of 'u' by 'ii'. In sampling the processes of succession we are likely to make these kinds of mistakes. Analogous situations arise with spatial sampling and Levenshtein distances can be used in such cases but alternative approaches, for example the probabilistic finite state machines of Boinovič & Srihari (1982) might be more appropriate. – Just as the frequency transformation produced sequences of variable length, so might successional studies. However with successional studies the data are really sampled from a continuous time process, with variable rates of change. Thus the definition of a Levenshtein measure suitable for the successional case will probably not be identical with one suitable for frequency data. For example, time warping options might be more attractive than gap-length weighting because we expect variation in rate of development.

Data and analyses

I have used two sets of data to illustrate the use of Levenshtein measures. One consists of a rather simple set of ecological data, due to Bowman & Wilson (1986). The data themselves were coded cover estimates for species from two areas in the Adelaide River flood plain, in northern Australia. Each area was sampled so as to include a range of the variation in vegetation. In Table 1, the order of plots and species from the original paper is shown, obtained using a reciprocal averaging (correspondence analysis) (Hayashi 1956; Hill 1973) seriation. Table 2 shows data after the conversion of the cover codes for the species to numeric symbols, which were then classified. In this case we assign to each species a unique code number, shown in Table 1, and encode the frequency as discussed earlier. Pairwise similarities are then calculated allowing only insertion and deletion with equal weights for both operations, but the distances were normalised by dividing by the sum of the lengths of the two sequences. The classification method used was a sequential agglomerative hierarchical non-overlapping (SAHN) grouping method (Sokal & Sneath 1963). The transposed data were also converted and classified and Table 3 shows the two-way table resulting from these two analyses.

The other data set is taken from Williams *et al.*'s (1969) study of succession in subtropical rain forest, which has previously been examined using transition matrices. A total of 10 stands were recorded on each of 12 unequally spaced occasions, with some 118 species occurring in the presence/absence data. The 120 'stand-times' were first classified using a Bray-Curtis

Table 1. North Australian wetland vegetation: Adelaide River data (Bowman & Wilson 1986).

Original gradient order: – stands

30 34 28 31 24 26 36 37 41 29 27 32 33 40...

20 6 18 12 39 19 11 23 25 22 21 35 38...

17 5 8 13 4 16 14 2 15 17 3 10 9

Original gradient order: – species

1. Cyperus rotundus	2. Abelmoschus ficulneus	3. Ipomoea coptica
4. Cynodon arcuatus	5. Merremia hederacea	6. Alysicarpus vaginalis
7. Panicum cambodiencee	8. Abelmoschus moschatus	9. Melochia corchorifolia
10. Waltheria indica	11. Ludwigia octovalvis	12. Poaceae spp. 1
13. Heliotropium crispatum	14. Euphorbia vachellii	15. Echinochloa colona
16. Phyla nodiflora	17. Paspalum scrobiculatum	18. Echinochloa elliptica
19. Poaceae spp. 2	20. Phyllanthus spec.	21. Goodenia purpurescens
22. Cardiospermum halicacabum	23. Sesbania spec.	24. Heliotropium indicum
25. Dentalla dioeca	26. Ipomoea aquatica	27. Oryza spec.
28. Cassia obtusifolia	29. Eleocharis spec.	30. Pseudoraphis spinescens
31. Ludwigia adscendens	32. Polygonum attenuatum	33. Aeschynomene indica

Table 2. Adelaide River data coded as sequences, for stands.

Stand	Sequence
1.	27 27 27 27 27 29 29 31
2.	9 27 27 27 27 27 29 30 31 32
3.	15 26 27 27 30 31 31 31 31
4.	10 23 27 27 27 27 27 31
5.	10 15 23 23 23 24 24 27 27 27 27 29 29 31
6.	9 16 16 16 16 18 18 23 23 24 29 33
7.	18 23 27 27 29 29 30 30 30 30 31
8.	10 22 23 23 24 27 27 28 31
9.	24 26 27 27 31 32 32 32 32 32 32
10.	24 27 29 29 30 30 30 30 30 31 31
11.	9 15 16 16 16 18 18 18 24 26 29 30 33
12.	9 16 16 16 16 16 18 22 24 26 29 30
13.	10 15 18 23 24 27 27 27 27 27 31
14.	9 24 26 27 27 27 27 29 30 30 30 31
15.	27 27 27 27 30 30 31 32 32 32
16.	10 10 22 22 23 26 27 27 27 29 29 29 30 30 31
17.	9 23 23 23 26 27 27 27 31
18.	9 16 16 16 16 16 18 18 22 22 24 29
19.	6 7 9 16 16 16 16 16 18 18 19 23 24 26 27 29 33
20.	6 6 12 13 15 16 16 16 16 18 24 27 29 33 33
21.	19 24 27 27 27 29
22.	7 9 14 15 16 16 16 17 24 27 27 27 27
23.	6 7 9 14 16 16 16 16 17 22 22 22 24 24 27 27 27 33
24.	4 4 4 7 9 14 16 17 23
25.	7 16 16 16 16 17 24 27 27 27 29
26.	1 1 1 1 10 14 15 16 16 16 23
27.	5 5 7 7 7 7 7 7 8 9 9 14 15
28.	1 3 4 7 9 10 14 15 23
29.	1 3 4 7 9 10 10 10 10 16 16 16 23 24
30.	1 1 1 1 3 5 7 9 10 14 23
31.	1 1 3 4 4 4 4 9 10 10 15 23
32.	7 7 7 7 9 9 10 11 14 18 18 23
33.	1 7 8 9 9 16 16 16 16 23
34.	1 1 1 2 4 4 4 4 9 14 15 23
35.	7 9 14 14 15 18 24 24 25 27 27
36.	4 4 4 4 7 9 14 16 16 18 23 24 33
37.	4 4 4 4 5 7 9 9 14 18 18 23 23 26
38.	7 9 14 14 15 18 20 21 27 27 27 27
39.	9 16 16 23 24 29
40.	9 14 15 16 16 16 16 16 17 27 29
41.	4 4 4 7 7 7 9 14 18 27 27

(1957) similarity measure and SAHN clustering. Sneath's (1985) test on dendrogram levels yielded a significance difference from multivariate hypersphericity, using the Lilliefors (1967) statistic ($p<0.003$). – I have assumed that the similarity measure is equivalent to the complement of a simple matching coefficient and the SAHN method to weighted-pair-group analysis. The intragroup similarity was assumed to be 0.5. The technique does not seem overly sensitive to the assumptions, which are in any case regarded only as indicative. – As to the numbers of groups, both Mojena's (1977) test and the Ratkowsky-Lance (1978) test indicated 7 groups, which form the vegetation types. However all these tests should be regarded as indicative since their power is largely unknown. With these data I have therefore explored the effects of varying the number of vegetation types over the range 2–8.

Each of the 10 sequences is then redescribed using a type code to indicate which type existed on the stand at each time. Pairwise similarities for these sequences were then calculated using a Levenshtein distance permitting only insertion and deletion, with equal weighting to both operations, normalised as above. These similarities, which now include formation both on the composition and on the temporal patterns of the sequences, were then classified using a SAHN algorithm. I have also examined the use of transition matrices but because this approach has known problems, the quantitative diversity information analysis of Dale & Anderson (1972) was also used. – The problems arise because the transition matrix information measure does not distinguish between identical transition matrices and matrices which share no transitions at all! The program recognises this condition and prevents the disparate items from joining in a cluster, but there may remain no acceptable fusions. The system is then termed dissociated and classification must stop. – Both the transition matrix method and the diversity information method provide measures of the heterogeneity within groups and the change in heterogeneity caused by fusing items, but I have only presented the latter for the transition matrix results. Although it seems likely that the two successional series identified by Williams *et al.* (loc. cit.) are distinct, at least over the period of the observations, the small number of sequences prevents any useful significance tests. In any case substantive phytosociological interpretation is not intended here.

Results

For the first data set, the results are not surprising. Sneath's (1985) test was applied, and the results sug-

Table 3. Adelaide River data. Two-way table resulting from SAHN classification of Levenshtein distances. Note that the species classification used data coded by stand frequency, while the stand classification uses data coded by species frequency.

Group labels	a	b	c	d	e	f
Species	22	11	.13	1112	11222233	.112223
	01	23	693	13457890453	68467901	2172582
Plots						
A						
30	..	..	...	41.11.1.111	....	1......
34	..	..	...	3.4...1.111	....	....
28	..	..	...	111.1.11111	....	....
31	..	..	...	214...12.11	....	....
26	..	..	...	4......1111	3.......	....
35	..	..	...	....1.1.11.	.12.2...	....1..
38	11	..	...	....1.1.21.	.1..41..	....
24	..	..	...	..3.11..1.1	1.......	..1....
36	..	..	..1	..411.1.1.1	2111....	....
37	..	..	...	..4112..1.2	.2.1....	....
41	..	..	...	..3.3.1.1..	.1..2...	....
27	..	..	...	...2612.11.	....	....
32	..	..	...	....4.211.1	.2......	.1.....
B						
29	..	..	...	111.1.14..1	3.1.....	....
33	..	..	...	1...112...1	4.......	....
39	..	..	...	......1...1	2.1..1..	....
40	..	..	...	......1.11.	5...11..	..1....
20	..	11	2.2	.........1.	411.11..	....
6	..	..	..1	......1...2	421..1..	....
18	..	..	...	......1....	521..1..	...2...
12	..	..	...	......1....	5111.11.	...1...
19	..	..	111	....1.1...1	521111..	....
11	..	..	..1	......1..1.	3311.11.	....
23	..	..	1.1	....1.1.1..	4.2..3..	..13...
25	..	..	...	....1......	4.1.31..	..1....
22	..	..	...	....1.1.11.	3.1.41..	..1....
C						
17	..	..	...	......1...3	...13..1	....
5	..	..	...	.......1.13	..2.42.1	....
8	..	..	...	.......1..2	..1.2..1	...2.1.
13	..	..	...	.......1.11	..1.5..1	....
4	..	..	...	.......1..1	....5..1	....
21	..	..	.1.	....	..1.31..	....
16	..	..	...	.......2..1	...13321	...2...
14	..	..	...	......1....	..114131	....
2	..	..	...	......1....	5111	1
15	..	..	...	....	....4.21	3
1	..	..	...	....	....52.1	....
7	..	..	...	1	.1..2241	....
3	..	..	...	.........1.	...12.14	....
10	..	..	...	....	..1.1252	....
9	..	..	...	....	..112..1	6

gest that the data cannot be considered multivariate hyperspherical ($p < 0.0001$), although it is not clear that such a declaration necessarily means that the data are actually clustered. Interestingly, Sneath attempts to estimate a kind of intrinsic dimensionality from the dendrogram levels, by assuming a model where only some of the species are variable, the remainder being regarded as 'absent throughout' or 'present throughout' the stands. For these data the estimate is 10 and if this model is acceptable then the data could be regarded as a single 10-dimensional cluster. Since the original authors seem to regard the data as exemplifying a single environmental gradient there are some interesting questions raised by this result regarding the relationship between 'species dimensionality' and environmental dimensionality. Obviously they cannot be regarded as equivalent.

Since the data are, or the full data at least, likely to be clustered, then using the Mojena (1977) test for number of groups suggests that three are appropriate and these can be easily seen in the table. The groupings obtained obviously reflect the original sequence; with a few aberrations the groups reflect the three parts of the gradient shown separated in Table 1. This particular example is of course a simple case, for the distance measure is in fact exactly equivalent to the Bray-Curtis (1957) measure, though both motivated and calculated somewhat differently. However some basic properties of such a coefficient are revealed. With these data, variation in the length of the strings, which is due to a combination of species richness and abundance variation, does not seem to have any significant impact with the stand classification. In the species classification it appears important only in the identification of groups composed of rare species only, e.g. groups a, b and g.

For the second data set, although the sequences are all of fixed length, the number of types is being varied and the interpretation more complex. The results of the analyses using different numbers of vegetation types are presented in Appendix 1, with seven tables, each table referring to one particular number of initial vegetation types. In the Levenshtein-SAHN results only in the 2- and 3-type results do the Mojena tests indicate the existence of groups, in both cases suggesting that two groups of sequences exist. With only ten sequences Sneath's tests are weak, but they do not indicate any marked clustering anywhere. Sneath argued for use of fractional powers of the levels of fusion and, using these transformed values in Mojena's test, most analyses suggest division into two groups of sequences. What is less attractive is that the clusters so formed are not constant in membership. With 2, 7 and 8 types a consistent result separating sequences 2, 3 and 5 is apparent, but for the other numbers of types the situation is less satisfactory. Individually the results look sensible for the sequences but they are rather different among themselves, 3 and 4 types agreeing with each other, 5 and 6 disagreeing with all others. They do in general involve sequences 2, 3 and 5, either singly or in combination, but the variation in underlying definition of vegetation types does cause variations in the precise content of the clusters. But such sensitivity may be desirable, if it reflects added precision when more types are recognised.

The transition matrix analyses proves very disappointing. If two vegetation types only are distinguished then a complete hierarchy is obtained, suggesting two groups; with 3 types only the final fusion is rendered impossible, and the others dissociate even earlier with increasing numbers of types. This reflects the short lengths of the sequences which allow very few transitions to be found in each, as well as the occurrence of runs of identical symbols. Thus it would seem that the analysis of transition matrices is really useful only with relatively long series and where it is the variation in frequencies of transitions rather than their existence which is of greatest interest. I have not shown details of the tests for the diversity analysis, but it seems in all cases to derive two clusters with some analyses suggesting a possible third. But again the cluster composition is not constant as the number of vegetation types changes. With 8 types, sequences 2, 3 and 5 are distinguished as a cluster, with 5, 6 or 7 types, the diversity result is identical with the 5-type Levershtein results, but for 2, 3 and 4 types the results differ among themselves and from the Levenshtein results. It is also interesting to examine the final heterogeneity (the Level values) in the diversity analyses as the number of vegetation types decreases. Starting at 301.84 for 8 types,the fall seems to follow a logarithmic form;

there is only a slight indication of irregularity with the 2-type result. The gaps between the final pair of groups (the Changes) show a different pattern falling slowly to 4-types and then much faster in the final two analyses. While in accord with intuition, this does suggest some stability in the results while the number of groups remains fairly high. Thus it probably does not matter too much if we misidentify the number of groups by one or two either way. However, it is clear that large errors in estimating the number of groups can lead to very different assessments of the clarity of separation of the sequence groups.

In summary, the transition matrix analysis largely fails, while both the diversity and Levenshtein approaches suggest that two, or perhaps three, groups of somewhat variable composition are present; the composition reflects changes in the definition of vegetation types. Can we then identify one approach as preferable to the other?

Since 7 types was the preferred solution in the initial classificatory analysis, we may take that result as an example. The diversity analysis identifies the sequence clusters (1, 4, 5, 6) and (2, 3, 7, 8, 9, 10). The only difficulty with this result is the placement of sequence 5; this is only sequence with vegetation type 4, as can be seen from the Appendix, and certainly is not well placed. Indeed it would seem to be positioned solely on the occurrence of types 2 and 7 in the succession. Such reliance on small features is hardly comforting! In contrast the Levenshtein measure places sequences (2, 3, 5) together reflecting the common occurrence of vegetation type 1. If we accept three groups then the two others formed, (1, 4, 6) and (7, 8, 9, 10) are equivalent to the diversity groups, less 5. Thus the Levenshtein result would seem to be more appropriate in its placement of sequence 5 and no worse for the remainder. It must be remembered that we are using only the simplest form of Levenshtein measure and selection of a more appropriate measure could still further improve the results.

Discussion

What can we conclude from these studies. First, even the simplest Levenshtein distances can be useful, though they do need some care with interpretation. They require the user to consider the semantics of similarity in order to derive an appropriate measure of similarity from knowledge of the subject matter. This was the approach adopted by Austin & Belbin (1982; see also Lerman 1977) in suggesting asymmetric measures of similarity for species. Asymmetric measures have also been examined by Ozawa (1983) who has suggested various situations where asymmetry might be appropriate for stand comparisons. His objective was to combine pairwise measures of interstand differences with specific properties of a single stand, such as the local density, or neighbourhood linearity; his definitions of similarity clearly relate to his objectives. Other examples of using semantic information to guide formation of similarity measures include Mountford's (1962) use of the species-area curve parameters to define a similarity measure, and Faith's (1985) discussion of measuring genetic similarity.

Levenshtein measures have a wide range of applicability. Indeed Dale (in press), has suggested that most common similarity measures are Levenshtein in basic form. The breadth of the range is exemplified by van Rijsbergen (1986) who has suggested an extended version where distance is measured as the shortest computer program for changing one item into another.

Not all distances, though, are Levenshtein variants. Thus the Bartels *et al.* (1970) distance, which is based on the notion of 'between-ness', reflects topological relationships and is not, I believe, a Levenshtein metric, though it may still be a Hausdorf metric. It is likely that neither Lewis *et al.*'s (1967, c.f. Dale *et al.* 1984) synonymity measures based on common context, nor the predicate calculus formulae adopted by Vesely (1981) are Levenshtein measures. If so then they represent other aspects of similarity not captured in the Levenshtein frame.

Earlier we noted the problematic effects of variation in sequence length. Although with the data used here it did not seem overly important, with other data this may not be the case. Where variation in the string lengths solely reflects variations in abundance of symbols there is no problem; this is the variation which we seek to utilise. But in some cases the species richness varies considerably and stands may share no common species, while the species which are present in them can be abundant. Barson & Dale (this vol.) with aquatic floristic data and Dale, Coutts & Dale (in press) with data from sclerophyllous forest have both identified such properties in their data. The distance between stands then tends to reflect only the length of the sequences, and this is unhelpful phytosociologically. The problem is common with many similarity measures, as might be expected since most are variant forms of the Levenshtein distance anyway (c.f. Dale 1987b). I should note here that Austin & Belbin (1982) previously argued that stands and species require different similarity measures although I would regard the difference as more one of degree of impact of the richness problem than a difference in kind. Finally, as an example of appropriate and extendable treatment of frequency data the Levenshtein measure can be regarded as successful in the present application. However the use of the Levenshtein formulation has, I believe, clarified the nature of the richness problem somewhat.

What can be done to allow the user to control the effects? The use of gap-weighting or time warping have already be noted. Another possiblity is to use a subset of the species only, specifically those which are in some way 'interesting'. Dale *et al.* (1986) have examined various ways of accomplishing such selection, although their results do not suggest that any one method is always successful. We might also use descriptors other than species so that variation in length due to differences in richness are minimised or abolished. Dale & Clifford (1976) have attempted, successfully, to employ higher taxa, but other workers have been tempted to the use of structural descriptors. This has been supported recently by Orlóci & Stofella (1986) although the physiognomic and structural approaches have, of course, a long history. Unfortunately even this approach does not always work well. For example, Barson & Dale (this vol.) found with aquatic vegetation that use of life forms did not avoid variations in richness. Thus while choosing different descriptors can often be of assistance, we need other means of handling the problem.

One possibility is to modify the Levenshtein measures themselves. For example, Coggins (1983) has made a number of specific suggestions designed to make the basic Levenshtein distance measure more sensitive to various other aspects of string and sequence similarity. He first considered the possibilities of extensive weighting schemes but these pose difficulties because of the number of weights which must be given; if there are p symbols in the alphabet of the strings then we require $(p+1)^2$ weights for insertion, deletion and substitution. Coggins (see also Ozawa 1983) therefore suggested that several different coefficients be combined to improve the resolution. Thus he defined:

$$d_{ab} = \text{len}(a) + \text{len}(b) - 2\, q_{ab} \qquad (1)$$

where len(x) is the length of x and q_{ab} is the length of the longest common subsequence

$$D_{ab} = \text{len}(a) + \text{len}(b) - 2\, s_{ab} \qquad (2)$$

where s_{ab} is the length of the longest common substring

$$d_{ab} = \text{char}(a) + \text{char}(b) - 2\, \text{charand}(a, b) \quad (3)$$

where char(x) is the number of different symbols in x and charand(x, y) is the number of symbols common to x and y. He then suggested various combinations of these basic coefficients, for example

$$D_c = d_{ab} * D_{ab} + d_{ab} \qquad (4)$$

in order to render the overall coefficient more sensitive. Note that all the components are of the form '#(union) – #(intersection)' where # indicates cardinality, as with the Bray-Curtis coefficient itself (and various others), and would all probably reflect variation in total length of the strings. I would suspect, therefore, that it might be more profitable to first examine the operations used and the weights involved, and also possible normalisation techniques, before opting for these complications.

A third solution seems possible. There is more general class of distances which might be effective in treating the richness problem, the Hausdorf distances. These were mentioned earlier in discussing partial ranking. Bednarek & Ulam (1979) present a general method of calculating such distances in the following manner. Given an item x, define an operation $f(x)$ which returns a set of items associated with, and including, x. Now define $f(f(x))$ or $f^2(x)$ as the union set resulting if the same operation is applied to all members of the set $f(x)$, and more generally $f^n(x)$ as the operation applied n times. Now consider two items A, B and determine the value of n where (B)C $f^n(A)$ and m where (A)C $f^m(B)$ and define $d(A,B) = \max(n,m)$. This value is the (integer) Hausdorf distance between A and B. Obviously the result depends on the nature of the function used; one possibility would be that $f(x)$ returns all items which are one 'rule-application' away from x. Bednarek & Smith (1980) have shown how this can be operationalised in a taxonomic study and perhaps the taxonomic hierarchy could be used to define the operation. Alternatively the species can be partially ordered in some manner (c.f. Ito *et al.* 1984) and this order used to determine what are 'neighbouring' species. Perhaps some functional relationships might also be used for this purpose, such as ability to fix nitrogen or use of the C_4 pathway. There are also possibilities of deriving Hausdorf functions from relationships between stands. Nakamura & Iwai (1982) propose something similar when they admit 'properties of high order'. If x is like y then the properties of y are regarded as properties of x, but at a second order, and if z is like y then properties of z are regarded as third order properties of x, and so on for higher orders. This involves a certain circularity, for it is not clear exactly what should determine the nature of 'likeness'; clearly not the properties under study. However Hausdorf distances would seem to offer some ways of overcoming the richness problem and deserve further study.

In conclusion it seems that the Levenshtein approach using the minimal mutation distance has a wide range of applications, and serves as a general basis for many coefficients. This does not mean that it is necessary to calculate every coefficient using the standard Levenshtein algorithm; many coefficients can be more simply calculated in other ways and approximations to the, computationally expensive, full dynamic programming algorithm have been proposed, (e.g. by Ukkonen 1985). In the more complex versions the costs of computation can be quite significant although, or those activities with sufficient commercial strength, dedicated hardware might be available as it is for speech processing. But whatever the computational problems it seems that the Levenshtein measures do form a large family within which the properties of the various coefficients can be compared and assessed. Such assessment is vitally needed if users have to have rational grounds for selecting particular coefficients of similarity, related to the process extant in the situation they are studying. However the usefulness of the minimal mutation approach should not be overrated either. Other means of measuring similarities may be better suited to some problems. The reason for this is simply that the minimal mutation distance, for all its flexibility and power, is only one measure of similarity, and may not capture some relevant features of relationships. Thus it is not easy to relate analogy with a minimum mutation approach, and analogy is a fundamental mode of reasoning.

Acknowledgements

To Pauline Hogeweg for discussion, Enrico Feoli and Mario Lagonegro for their encouragement, and *lis stries* for their unfailing enthusiasm and support.

References

Austin, M. P. & Belbin, L. 1982. A new approach to the species classification problem in floristic analysis. Aust. J. Ecol. 7: 75–89.

Bartels, P. H., Bahr, G. F., Calhoun, D. W. & Wied, G. L. 1970. Cell recognition by neighbourhood grouping techniques in Ticas. Acta Cytol. 14: 313–324.

Bednarek, A. R. & Smith, T. F. 1980. A taxonomic distance applicable to paleontology. Math. BioSci. 50: 285–295.

Bednarek, A. R. & Ulam, S. M. 1979. An integer valued metric for patterns. Fundamentals of computation theory, pp. 52–57, Academic-Verlag, Berlin.

Bellacicco, A. 1977. Clustering time varying data. In: Barra, J. R., Brodeau, F., Romier, G. & van Cutsen, B. (eds), Recent developments in statistics, pp. 739–748, North Holland, New York.

Ben-Bassat, M. & Zaidenberg, L. 1984. Contextual template matching: a distance measure for patterns with hierarchically dependent features. IEEE Trans Patt. Anal. Mach. Intell. PAMI 6: 201–211.

Blackburn, D. T. 1980. A generalized distance metric for the analysis of variable taxa. Bot. Gaz. 141: 325–335.

Borg, I. & Staufenbiel, T. 1986. The MBR metric. J. Math. Psychol. 30: 81–84.

Bowman, D. M. J. S. & Wilson, B. A. 1986. Wetland vegetation pattern on the Adelaide River floodplain, Northern Territory, Australia. Proc. Roy. Soc. Qld. 97: 69–77.

Boinovič, R. & Srihari, S. N. 1982. A string correction algorithm for cursive script recognition. IEEE Trans. Patt. Anal. Mach. Intell. PAMI 4: 655–663.

Bray, J. R. & Curtis, J. T. 1957. An ordination of the upland forest communities of southern Wisconsin. Ecol. Monogr. 27: 325–349.

Brook, R. J. & Stirling, W. D. 1984. Agreement between observers when the categories are not specified. Brit. J. Math. Statist. Psychol. 37: 271–282.

Bykat, A. 1979. On polygon similarity. Inform. Process. Lett. 9: 23–25.

Cayley, A. 1849. A note on the theory of permutations. Phil. Mag. 34: 527–529.

Critchlow, D. 1985. Metric methods for analyzing partially ranked data. Springer-Verlag, New York.

Coggins, J. M. 1983. Dissimilarity measures for clustering strings. In: Sankoff, D. & Kruskal, J. B. (eds), Time warps, string edits and macromolecules: the theory and practice of sequence comparison. pp. 311–321. Addison Wesley, London.

Dale, M. B. 1968. On property structure, numerical taxonomy and data handling. In: Heywood, V. H. (ed), Modern methods in plant taxonomy, pp. 185–197. Academic Press, London.

Dale, M. B. 1989. Dissimilarity for partially ranked data and its application to cover-abundance data. Vegetatio (in press).

Dale, M. B. 1988. Knowing when to stop: cluster concept-concept cluster. Coenoses 3: 11–32.

Dale, M. B. in press. Mutational and nonmutational similarity measures: a preliminary examination. Coenosis.

Dale, M. B. & Anderson, D. J. 1972. Qualitative and quantitative information analysis. J. Ecol. 60: 639–653.

Dale, M. B. & Clifford, H. T. 1976. The effectiveness of higher taxonomic ranks for vegetation analysis. Austral. J. Ecol. 1: 37–62.

Dale, M. B., Clifford, H. T. & Ross, D. R. 1984. Species, equivalence and morphological redescription: a Stradbroke Island vegetation study. In: Coleman, R. J., Covacevich, J. & Davie, P. (eds), Focus on Stradbroke: New information on North Stradbroke Island and surrounding areas, 1974–1984. Boolarong Publ., Brisband & Stradbroke Island Management Organization, Amity Point.

Dale, M. B. & Dale, P. E. R. 1986. Similarity and structured attribute in ecological classification. Abstr. Botan. 10: 17–34.

Dale, M. B., Ferrari, C., Beatrice, M. & Venanzoni, R. 1986. A comparison of some methods of selecting species in vegetation analysis. Coenoses 1: 35–52.

Dale, M. B., Groves, R. H., Hull, V. J. & O'Callaghan, J. F. 1970. Computer assisted description of leaf shapes. New Phytol. 70: 437–442.

Dale, M. B., MacNaughton-Smith, P., Williams, W. T. & Lance, G. N. 1970. Numerical classification of sequences. Austr. Comput. J. 2: 9–13.

Dale, M. B. & Walker, D. 1970. Information analysis of pollen diagrams. Pollen et Spores 2: 21–37.

Estabrook, G. F. & Meacham, C. A. 1979. How to determine the compatability of undirected character state trees. Math. Bio-Sci. 46: 251–256.

Faith, D. P. 1985. A model of immunological distance in systematics. J. Theor. Biol. 114: 511–526.

Feoli, E. & Lagonegro, M. 1983. A resemblance function based on probability: applications to field and simulated data. Vegetatio 53: 3–9.

Findler, N. V. & van Leeuwen, J. 1979. A family of similarity measures between strings. IEEE Trans. Patt. Anal. Mach. Intel. PAMI 1: 116–118.

Fredman, M. L. 1984. Algorithms for computing evolutionary similarity measures with length independent gap penalties. Bull. Math. Biol. 46: 553–566.

Hayashi, C. 1956. Theory and examples of quantification II. Proc. Inst. Statist. Math. 4: 19–30. (in Japanese).

Hill, M. O. 1973. Reciprocal averaging: an eigenvector method of ordination. J. Ecol. 61: 237–249.

Hill, M. O. 1979. TWINSPAN, a FORTRAN program for analysing multivariate data in an ordered two-way table by classification of individuals and attributes. Dept. Ecology & Systematics, Cornell Univ. Ithaca, New York.

Hogeweg, P. & Hesper, B. 1984. The alignment of sets of sequences and the construction of phylogenetic trees: an integrated method. J. Mol. Evol. 20: 175–184.

Hohn, M. E. & Nuhfer, E. B. 1980. Asymmetric measures of association, classed data and multivariate analysis. Math. Geol. 12: 235–246.

Ito, T., Kodama, Y. & Toyoda, J. 1984. A similarity measure between patterns with non-independent attributes. IEEE Trans. Patt. Anal. Mach. Intell. PAMI 6: 111–115.

Jensén, S. & van der Maarel, E. 1980. Numerical approaches to wetland classification with special reference to macrophyte communities. Vegetatio 42: 117–128.

Kendall, M. G. 1938. A new measure of rank correlation. Biometrika 30: 81–93.

Kullback, S. 1959. Information theory and statistics. Wiley, New York.

Lambert, J. M. & Dale, M. B. 1964. The use of statistics in phytosociology. Adv. Ecol. Res. 2: 59–66.

Lance, G. N. 1970. Mixed and discontinuous data. In: Anderssen, R. S. & Osborne, M. R. (eds), Data representation, pp. 102–107. Univ. Queensland Press, St. Lucia, Qld.

Legendre, P. & Chodorowski, A. 1977. A generalization of Jaccard's association coefficient for Q analysis of multi-state ecological data matrices. Ekol. Polska 25: 297–308.

Lehmann, D. R. 1972. Judged similarity and brand-switching data as similarity measures. J. Marketing Res. 9: 331–334.

Lemone, K. A. 1982. Similarity measures between strings extended to sets of strings. IEEE Trans. Patt. Anal. Mach. Intel. PAMI 4: 345–347.

Lerman, I.-C. 1977. Formal analysis of a general notion of proximity between variables. In: Barra, J. R., Brodeau, F., Romier, G. & van Cutsen, B. (eds), Recent developments in statistics, pp. 787–795. North Holland, New York.

Lerman, I.-C. & Peter, P. 1985. Elaboration et logiciel d'un indice de similarité entre objets d'un type quelconque. IRISA, Rennes, Publ. Intern. 262, 72 pp.

Levenshtein, V. I. 1965. Binary codes capable of correcting deletions, insertions and reversals. Dokl. Akad. Nauk SSR. 163: 825–828. (in Russian).

Lewis, P. A. W., Baxendale, P. B. & Bennett, J. L. 1967. Statistical discrimination of the Synonymy/Antonymy relationship between words. Assoc. Comput. Mach. J. 14: 20–44.

Lilliefors, H. W. 1967. On the Kolmogorov-Smirnov test for normality with mean and variance unknown. Amer. Statist. Assoc. J. 62: 399–402.

Little, I. P. & Ross, D. R. 1985. The Levenshtein metric: a new means for soil classification tested by data from a sand-podzol chronosequence and evaluated by discriminant analysis. Aust. J. Soil Res. 23: 115–130.

Lowrance, R. & Wagner, R. A. 1975. An extension to the string-to-string correction problem. J. A. C. M. 22: 177–183.

Lu, S.-Y. 1984. A tree matching algorithm based on node splitting and merging. IEEE Trans. Patt. Anal. Machine Intell. PAMI 6: 249–256.

Lu, S.-Y. & Fu, K.-S. 1978. A sentence-to-sentence clustering procedure for pattern analysis. IEEE Trans. Systems, Man & Cybernetics SMC 8: 381–389.

Mäkirinta, U. 1978. Die pflanzensoziologische Gliederung der Wasservegetation im See Kukkia, Südfinnland. Acta Univ. Ouluens. Ser. A 75, Biol. 5.

Matusita, K. 1977. Cluster analysis and affinity of distributions. In: Barra, J. R., Brodeau, F., Romier, G. & van Cutsen, B. (eds), Recent developmnts in statistics. pp. 537–544. North Holland, New York.

Mojena, R. 1977. Hierarchical grouping methods and stopping rules: an evaluation. Comput. J. 20: 359–363.

Moore, R. K. 1979. A dynamic programming algorithm for the distance between two finite areas. IEEE Trans. Patt. Anal. Machine Intell. PAMI 1: 86–88.

Mountford, M. D. 1962. An index of similarity and its application to classificatory problems. In: Murphy, P. W. (ed.), Progress in soil zoology, pp. 43–50. Butterworth, London.

Nakamura, K. & Iwai, S. 1982. A representation of analogical inference by fuzzy sets and its application to information retrieval system. In: Gupta, M. M. & Sanchez, E. (eds), Fuzzy information and decision processes, pp. 373–368. North Holland.

Norris, J. M. & Dale, M. B. 1971. Transition matrix approach to numerical classification of soil profiles. Proc. Soil. Sci. Soc. Amer. 35: 487–491.

Orlóci, L. & Stofella, S. K. 1986. A taxon-free numerical approach to the study of plant communities. Ann. Arid Zone 25: 111–131.

Ozawa, K. 1983. CLASSIC: a hierarchical clustering algorithm based on asymmetric similarities. Patt. Recog. 16: 201–211.

Ratkowsky, D. A. & Lance, G. N. 1978. A criterion for determining the number of groups in a classification. Austral. Comp. J. 10: 115–117.

Reuhkala, E., Jalanko, M. & Kohonen, T. 1979. A redundant hash addressing method adapted for the postprocessing and error-correction of computer-recognized speech. ICASSP 79: IEEE Internatl. Conf. Acoustics, Speech and Signal Processing. pp. 591–594.

Sakoe, H. 1979. Two-level DP-matching – a dynamic programming-based pattern matching algorithm for connected voice recognition. IEEE Trans. Acoustics, Speech and Signal Processing ASSP 27: 588–595.

Sankoff, D. & Kruskal, J. B. 1983. Time warps, string edits and macromolecules: the theory and practice of sequence comparison. Addision Wesley, London.

Sibson, R. 1969. Information radius. Z. Wahrsch. Verw. Geb. 14: 149–160.

Sneath, P. H. A. 1985. Testing levels on a dendrogram for multivariate hypersphericity. Comput. GeoSci. 11: 767–785.

Sokal, R. R. & Sneath, P. H. A. 1963. Principles of numerical taxonomy. San Francisco, Freeman.

Ukkonen, E. 1985. Algorithms for approximate string matching. Inform. Control. 64: 100–118.

van Rijsbergen, C. J. 1986. A non-classical logic for information retrieval. Comput. J. 29: 481–485.

Vesely, A. 1981. Logically oriented cluster analysis. Kybernetika 17: 82–92.

Wallbrecher, E. 1976. Ein Cluster-Verfahren zur richtungsstatistischen Analyse tektonischer Daten. Geol. Rdsch. 67: 840–857.

Waterman, M. S. & Smith, T. F. 1978. On the similarity of dendrograms. J. Theor. Biol. 73: 789–800.

Werman, M., Pelg, S. & Rosenfeld, A. 1985. A distance metric for multidimensional histograms. Comput. Vision Graph. Image Process. 32: 328–336.

Williams, W. T., Lance, G. N., Webb, L. J., Tracey, J. G. & Dale, M. B. 1969. Studies in the numerical analysis of complex rain forest communities. III. The analysis of successional data. J. Ecol. 57: 515–535.

Appendix 1

Mt. Glorious data a) (Williams *et al.* 1969) 8 vegetation types. Sequences and classifications show the 10 sequences of 12 vegetation types, a SAHN classification of the Levenshtein distance, with Mojena's critical value. The transition matrix classification, and the diversity analysis follow. This last shows both the dissimilarity of the groups joining (Change) and the heterogeneity of the group formed (Level).

8 group

Sequences	Fusing ⇒	Group	Level
		Levenshtein + SAHN	
7 7 2 2 2 8 5 5 5 8 5 5	7 & 9	11	0.167
7 1 1 1 1 8 8 8 8 8 3 3	10 & 11	12	0.219
1 1 1 1 1 8 3 3 3 3 3 3	2 & 3	13	0.417
6 7 2 2 2 8 6 6 6 6 6 6	4 & 6	14	0.417
1 1 1 1 2 8 4 4 4 4 4 4	1 & 8	15	0.500
7 7 1 1 2 8 7 7 6 6 6 6	12 & 15	16	0.560
7 7 2 2 2 8 8 8 8 8 6 6	5 & 13	17	0.625
7 7 2 2 2 8 3 3 3 3 3 3	14 & 16	18	0.839
7 7 2 2 2 8 8 8 8 8 8 3	17 & 18	19	1.105
7 8 2 2 2 8 8 8 8 8 3 3			

Mojena critical distance = 1.24

Transition matrix

Fusing ⇒	Group	Change
4 & 7	11	2.01
8 & 10	12	3.39

Dissociated system

Information analysis

Fusing ⇒	Group	Change	Level
7 & 9	11	2.14	38.52
2 & 3	12	4.99	35.31
10 & 11	13	5.87	63.84
1 & 8	14	8.32	45.57
4 & 6	15	8.57	47.21
5 & 12	16	14.90	65.34
13 & 13	17	16.76	126.17
15 & 17	18	25.63	199.00
16 & 18	19	37.10	301.34

b) Mt. Glorious data, 7 vegetation types. Sequence and classifications. For details see a.

7 group

Sequences	Fusing ⇒	Group	Level
		Levenshtein + SAHN	
6 6 2 2 2 7 5 5 5 7 5 5	1 & 4	11	0.167
6 1 1 1 1 7 7 7 7 7 3 3	7 & 9	12	0.167
1 1 1 1 1 7 3 3 3 3 3 3	10 & 12	13	0.219
5 6 2 2 2 7 5 5 5 5 5 5	2 & 3	14	0.417
1 1 1 1 2 7 4 4 4 4 4 4	6 & 11	15	0.427
6 6 1 1 2 7 6 6 5 5 5 5	8 & 13	16	0.538
6 6 2 2 2 7 7 7 7 7 5 5	5 & 14	17	0.625
6 6 2 2 2 7 3 3 3 3 3 3	15 & 16	18	0.840
6 6 2 2 2 7 7 7 7 7 7 3	17 & 18	19	1.113
6 7 2 2 2 7 7 7 7 7 3 3			

Mojena critical distance = 1.21

Transition matrix

Fusing ⇒	Group	Change
1 & 4	11	1.96
11 & 7	12	3.32
8 & 10	13	3.39

Dissociated system

Information analysis

Fusing ⇒	Group	Change	Level
7 & 9	11	2.14	38.52
1 & 4	12	2.50	39.75
2 & 3	13	4.99	35.31
10 & 11	14	5.87	63.84
8 & 14	15	9.56	90.34
5 & 6	16	10.23	46.96
12 & 16	17	19.37	106.08
13 & 15	18	21.05	146.71
17 & 18	19	39.04	291.82

c) Mt. Glorious data, 6 vegetation types. Sequences and classifications. For details see a.

6 group

Sequences
5 5 2 2 2 6 4 4 4 6 4 4
5 1 1 1 1 6 6 6 6 6 3 3
1 1 1 1 1 6 3 3 3 3 3 3
4 5 2 2 2 6 4 4 4 4 4 4
1 1 1 1 2 6 4 4 4 4 4 4
5 5 1 1 2 6 5 5 4 4 4 4
5 5 2 2 2 6 6 6 6 6 4 4
5 5 2 2 2 6 3 3 3 3 3 3
5 5 2 2 2 6 6 6 6 6 6 3
5 6 2 2 2 6 6 6 6 6 3 3

Levenshtein + SAHN

Fusing ⇒	Group	Level
1 & 4	11	0.167
7 & 9	12	0.167
10 & 12	13	0.219
5 & 6	14	0.333
2 & 3	15	0.417
12 & 14	16	0.451
8 & 13	17	0.538
15 & 17	18	0.767
16 & 18	19	1.136

Mojena critical distance = 1.15

Transition matrix

Fusing ⇒	Group	Change
1 & 4	11	1.96
11 & 7	12	3.32
8 & 10	13	3.39

Dissociated systems

Information analysis

Fusing ⇒	Group	Change	Level
7 & 9	11	2.14	38.52
1 & 4	12	2.50	39.75
5 & 6	13	4.93	41.66
2 & 3	14	4.99	35.31
10 & 11	15	5.87	63.84
8 & 15	16	9.56	90.34
12 & 13	17	12.48	93.89
14 & 16	18	21.05	146.71
17 & 18	19	38.42	279.02

d) Mt. Glorious data, 5 vegetation types. Sequences and classificiations. For details see a.

5 group

Sequences
5 5 2 2 2 5 4 4 4 5 4 4
5 1 1 1 1 5 5 5 5 5 3 3
1 1 1 1 1 5 3 3 3 3 3 3
4 5 2 2 2 5 4 4 4 4 4 4
1 1 1 1 2 5 4 4 4 4 4 4
5 5 1 1 2 5 5 5 4 4 4 4
5 5 2 2 2 5 5 5 5 5 4 4
5 5 2 2 2 5 3 3 3 3 3 3
5 5 2 2 2 5 5 5 5 5 5 3

Levenshtein + SAHN

Fusing ⇒	Group	Level
9 & 10	11	0.083
1 & 4	12	0.167
7 & 11	13	0.188
5 & 6	14	0.333
2 & 3	15	0.417
12 & 14	16	0.451
8 & 15	17	0.521
13 & 17	18	0.801
16 & 18	19	0.976

Mojena critical distance = 1.08

Transition matrix

Fusing ⇒	Group	Change
1 & 4	11	0.58
8 & 10	12	0.63
5 & 6	13	1.50
2 & 3	14	1.50
1 & 7	15	2.32

Dissociated system

Information analysis

Fusing ⇒	Group	Change	Level
9 & 10	11	0.74	31.92
1 & 4	12	1.12	38.37
7 & 11	13	3.43	52.29
5 & 6	14	3.55	38.37
2 & 2	15	4.99	35.31
8 & 13	16	7.33	76.57
12 & 14	17	11.02	87.76
15 & 16	18	20.10	131.98
17 & 18	19	30.93	250.67

e) Mt. Glorious data, 4 vegetation types. Sequences and classifications. For details see a.

4 group

Sequences
4 4 1 1 1 4 3 3 3 4 3 3
4 1 1 1 1 4 4 4 4 4 2 2
1 1 1 1 1 4 2 2 2 2 2 2
3 4 1 1 1 4 3 3 3 3 3 3
1 1 1 1 1 4 3 3 3 3 3 3
4 4 1 1 1 4 4 4 3 3 3 3
4 4 1 1 1 4 4 4 4 4 3 3
4 4 1 1 1 4 2 2 2 2 2 2
4 4 1 1 1 4 4 4 4 4 4 2

Levenshtein + SAHN

Fusing ⇒	Group	Level
2 & 10	11	0.083
9 & 11	12	0.135
1 & 4	13	0.167
3 & 8	14	0.167
6 & 7	15	0.167
5 & 13	16	0.219
12 & 15	17	0.380
16 & 17	18	0.800
14 & 18	19	0.899

Mojena critical distance = 0.93

Transition matrix

Fusing ⇒	Group	Change
2 & 10	11	0.05
6 & 7	12	0.09
1 & 4	13	0.58
11 & 8	14	0.66
13 & 5	15	1.63
14 & 3	16	1.64
15 & 12	17	3.31

Dissociated system

Information analysis

Fusing ⇒	Group	Change	Level
2 & 10	11	0.16	34.64
6 & 7	12	0.78	36.12
9 & 11	13	0.99	49.87
1 & 4	14	1.12	38.37
3 & 8	15	1.73	31.45
5 & 14	16	2.64	53.79
12 & 16	17	7.01	96.92
13 & 15	18	9.92	91.4
17 & 18	19	30.13	218.28

f) Mt. Glorious data, 3 vegetation types. Sequences and classifications. For details see a.

3 group

Sequences
3 3 1 1 1 3 3 3 3 3 3 3
3 1 1 1 1 3 3 3 3 3 2 2
1 1 1 1 1 3 2 2 2 2 2 2
3 4 1 1 1 3 3 3 3 3 3 3
1 1 1 1 1 3 3 3 3 3 3 3
3 3 1 1 1 3 3 3 3 3 3 3
3 3 1 1 1 3 3 3 3 3 3 3
3 3 1 1 1 3 2 2 2 2 2 2
3 3 1 1 1 3 3 3 3 3 3 2
3 3 1 1 1 3 3 3 3 3 2 2

Levenshtein + SAHN

Fusing ⇒	Group	Level
1 & 40	11	0.0
6 & 7	12	0.0
11 & 12	13	0.0
2 & 10	14	0.083
9 & 13	15	0.130
3 & 8	16	0.167
5 & 15	17	0.287
14 & 17	18	0.327
16 & 18	19	0.905

Mojena critical distance = 0.74

Transition matrix

Fusing ⇒	Group	Change
6 & 7	11	0.0
4 & 11	12	0.0
1 & 12	13	0.0
2 & 10	14	0.05
14 & 8	15	0.66
13 & 5	16	0.89
15 & 3	17	1.64
16 & 9	19	1.85

Dissociated system

Information analysis

Fusing ⇒	Group	Change	Level
1 & 4	11	0.0	22.74
6 & 7	12	0.0	22.74
11 & 12	13	0.0	45.48
2 & 10	14	0.16	34.64
9 & 14	15	0.99	48.87
5 & 13	16	1.54	57.10
3 & 6	17	1.73	31.45
15 & 16	18	5.46	112.42
17 & 18	19	19.58	163.45

g) Mt. Glorious data, 2 vegetation types. Sequences and classifications. For details see a.

2 group

Sequences
2 2 1 1 1 2 2 2 2 2 2 2
2 1 1 1 1 2 2 2 2 2 2 2
1 1 1 1 1 2 2 2 2 2 2 2
2 2 1 1 1 2 2 2 2 2 2 2
1 1 1 1 1 2 2 2 2 2 2 2
2 2 1 1 1 2 2 2 2 2 2 2
2 2 1 1 1 2 2 2 2 2 2 2
2 2 1 1 1 2 2 2 2 2 2 2
2 2 1 1 1 2 2 2 2 2 2 2
2 2 1 1 1 2 2 2 2 2 2 2

Levenshtein + SAHN

Fusing ⇒	Group	Level
1 & 4	11	0.0
3 & 5	12	0.0
6 & 7	13	0.0
8 & 9	13	0.0
10 & 11	15	0.0
13 & 14	16	0.0
15 & 16	17	0.0
2 & 12	18	0.104
17 & 18	19	0.312

Mojena critical distance = 0.24

Transition matrix

Fusing ⇒	Group	Change
9 & 10	11	0.0
8 & 7	12	0.0
6 & 4	13	0.0
13 & 8	14	0.0
1 & 5	15	0.0
15 & 3	16	0.0
3 & 12	17	0.0
16 & 2	18	0.06
18 & 17	19	1.75

Information analysis

Fusing ⇒	Group	Change	Level
1 & 4	11	0.0	22.74
3 & 5	12	0.0	20.16
6 & 7	13	0.0	22.74
8 & 9	14	0.0	22.74
10 & 11	15	0.0	34.11
13 & 14	16	0.0	45.48
15 & 16	17	0.0	79.58
2 & 17	18	0.25	92.17
12 & 18	19	2.88	115.21

Vegetatio **81**: 61–77, 1989.

New combinatorial clustering methods

János Podani
Department of Plant Taxonomy and Ecology, L. Eötvös University, Kun B. tér 2, Budapest, H-1083, Hungary; and Research Institute of Ecology and Botany, Hungarian Academy of Sciences, Vácrátót, H-2163, Hungary

Accepted 12.12.1988

Keywords: Cluster homogeneity, Dendrogram, Flexible method, Hierarchical classification, Rock grassland, Ultrametric

Abstract

Sixteen clustering methods are compatible with the general recurrence equation of combinatorial SAHN (sequential, agglomerative, hierarchical and nonoverlapping) classificatory strategies. These are subdivided into two classes: the d-SAHN methods seek for minimal between-cluster distances the h-SAHN strategies for maximal within-cluster homogeneity. The parameters and some basic features of all combinatorial methods are listed to allow comparisons between these two families of clustering procedures. Interest is centred on the h-SAHN techniques; the derivation of updating parameters is presented and the monotonicity properties are examined. Three new strategies are described, a weighted and an unweighted variant of the minimization of the increase of average distance within clusters and a homogeneity-optimizing flexible method. The performance of d- and h-SAHN techniques is compared using field data from the rock grassland communities of the Sashegy Nature Reserve, Budapest, Hungary.

Abbreviations: CP = Closest pair; RNN = Reciprocal nearest neighbor; SAHN = Sequential, agglomerative, hierarchical and nonoverlapping

Nomenclature of syntaxa follows Soó, R. 1964. Synopsis systematico-geobotanica florae vegetationisque Hungariae I. Akadémiai, Budapest.

Introduction

The sequential, agglomerative, hierarchical and nonoverlapping clustering techniques (the so-called SAHN methods, Sneath & Sokal 1973) are commonly used procedures of numerical classification in vegetation science (see Orlóci 1978; van der Maarel 1979; and Greig-Smith 1983, for review), including synsystematics (e.g., Orlóci & Stanek 1979; Mucina 1982; Moreno-Casasola & Espejel 1986). A family of these methods requires only a symmetric distance (dissimilarity, similarity, etc.) matrix **W** to be stored in computer memory during computations; the raw data may be released once this matrix has been calculated (*stored matrix* approach, Anderberg 1973). The original data are not needed because there is a *combinatorial* solution to recompute between-cluster measures using the information contained in **W** and in an array of cluster sizes. Lance &

Table 1. Parameters for combinatorial d-SAHN (1–8) and h-SAHN (9–16) clustering methods.

Clustering method	α_i	α_j	β	γ	λ_h	λ_i	λ_j
1. Single linkage (SL)	1/2	1/2	0	– 1/2	0	0	0
2. Complete linkage (CL)	1/2	1/2	0	1/2	0	0	0
3. Unweighted average (UPGMA)	$\frac{n_i}{n_h + n_i}$	$\frac{n_j}{n_h + n_j}$	0	0	0	0	0
4. Weighted average (WPGMA)	1/2	1/2	0	0	0	0	0
5. Centroid (UPGMC)	$\frac{n_i}{n_h + n_i}$	$\frac{n_j}{n_h + n_j}$	$-\frac{-n_i n_j}{(n_i + n_j)^2}$	0	0	0	0
6. Median (WPGMC)	1/2	1/2	– 1/4	0	0	0	0
7. β-Flexible (β-FLEX)	$(1-\beta)/2$	$(1-\beta)/2$	<1	0	0	0	0
8. (β, γ)-Flexible ((β, γ)-FLEX)	$(1-\beta)/2$	$(1-\beta)/2$	unrestricted	unrestricted	0	0	0
9. Minimum increase of sum of squares (MISSQ)	$\frac{n_h + n_i}{n.}$	$\frac{n_h + n_j}{n.}$	$-\frac{n_h}{n.}$	0	0	0	0
10. Minimum sum of squares of new cluster (MNSSQ)	$\frac{n_h + n_i}{n.}$	$\frac{n_h + n_j}{n.}$	$\frac{n_i + n_j}{n.}$	0	$-\frac{n_h}{n.}$	$-\frac{n_i}{n.}$	$-\frac{n_j}{n.}$
11. Minimum increase of variance (MIVAR)	$\left(\frac{n_h + n_i}{n.}\right)^2$	$\left(\frac{n_h + n_j}{n.}\right)^2$	$-\frac{n_h(n_i + n_j)}{n.^2}$	0	0	0	0
12. Minimum variance of new cluster (MNVAR)	$\left(\frac{n_h + n_i}{n.}\right)^2$	$\left(\frac{n_h + n_j}{n.}\right)^2$	$\left(\frac{n_i + n_j}{n.}\right)^2$	0	$-\left(\frac{n_h}{n.}\right)^2$	$-\left(\frac{n_i}{n.}\right)^2$	$-\left(\frac{n_j}{n.}\right)^2$

13. Minimum increase of weighted average distance (WMIDIS)	$\frac{b_{hi}}{b.}$	$\frac{b_{hj}}{b.}$	$\frac{b_{ij}}{b.} - \frac{1}{2}$	0	$-\frac{b_h + n_i n_j}{2b.}$	$-\frac{b. - 2b_i - 2n_h n_j}{4b.}$	$-\frac{b. - 2b_j - 2n_h n_i}{4b.}$
14. Minimum increase of unweighted average distance (UMIDIS)	$\frac{b_{hi}}{b.}$	$\frac{b_{hj}}{b.}$	$\frac{b_{ij}}{b.} - \frac{b_{ij}}{b_h + b_{ij}}$	0	*	*	*
15. Minimum average distance within new cluster (MNDIS)	$\frac{b_{hi}}{b.}$	$\frac{b_{hj}}{b.}$	$\frac{b_{ij}}{b.}$	0	$-\frac{b_h}{b.}$	$-\frac{b_i}{b.}$	$-\frac{b_j}{b.}$
16. λ-Flexible (λ-FLEX)	$(1-3\lambda)/3$	$(1-3\lambda)/3$	$(1-3\lambda)/3$	0	$\leqq 0$	$= \lambda_h$	$= \lambda_h$

$b_i = \binom{n_i}{2}$, n_i = number of objects in C_i, $n. = n_h + n_i + n_j$, * expression too long, see formula (14) in Appendix.

Williams (1966, 1967) presented the classical example for this approach; they suggested the recurrence formula

$$w_{h,ij} = \alpha_i w_{hi} + \alpha_j w_{hj} + \beta w_{ij} + + \gamma | w_{hi} - w_{hj} | \quad (1)$$

to update the values of **W** for the single linkage, complete linkage, group average, centroid and median (Gower 1967) clustering algorithms (Table 1). If clusters C_i and C_j are merged in a clustering cycle, then $w_{h,ij}$ gives the updated criterion value to be used in the next cycle for cluster $C_i \cup C_j$ with any other cluster C_h. The simple average (or weighted average linkage, WPGMA) method originally suggested by Sokal & Michener (1958) is also combinatorial. Wishart (1969) developed first the parameters for the incremental sum of squares agglomeration technique. The β-flexible strategy of Lance & Williams (1967) has no fixed parameters, the values of α_i, α_j, and β may be changed under certain conditions to provide transitions between extremely space-dilating and space-contracting algorithms (see also Sneath & Sokal 1973). Less constrained is the (β, γ)-flexible technique introduced by DuBien & Warde (1979) which in fact contains the single linkage, complete linkage, simple average and β-flexible strategies as special cases. Diday *et al.* (1982) presented the parameters for the minimization of the increase of variance, so that the number of methods formally compatible with equation (1) increased to ten.

In the meantime, it was revealed that further SAHN methods have combinatorial solutions. Jambu (1978) and Podani (1978, 1979) independently derived updating parameters for two techniques which had formerly been described by Anderberg (1973) as representatives of the *stored data* approach. These methods minimize the error sum of squares or the variance of the newly formed clusters. Podani (1978, 1979) also showed that the method of average linkage within the new group (Anderberg 1973, p. 139), in which cluster homogeneity is defined as the average of within-cluster similarities, is also combinatorial. These three methods differ from those compatible with equation (1) in that within-cluster measures are also used in recalculating between-cluster measures. The updated value, $w_{h,ij}$, is determined according to six values of **W**:

$$\begin{matrix} w_{hh} & w_{hi} & w_{hj} \\ & w_{ii} & w_{ij} \\ & & w_{jj} \end{matrix}$$

where w_{hh}, w_{ii} and w_{jj} (replaced by w_h, w_i, and w_j, for simplicity) denote either sum of squares, variance or average similarity within clusters C_h, C_i, and C_j, respectively. Jambu (1978, see also Jambu & Lebeaux 1983) suggested that the scope of the Lance-Williams formula be extended to these methods by adding three terms to equation (1) so that a more general recurrence relation is obtained:

$$w_{h,ij} = \alpha_i w_{hi} + \alpha_j w_{hj} + \beta w_{ij} + \gamma \mid w_{hi} - w_{hj} \mid + + \lambda_h w_h + \lambda_i w_i + \lambda_j w_j \quad (2)$$

(see Table 1 for parameters of the above three methods).

This paper shows that three more SAHN techniques fit into equation (2): the minimization of the increase of weighted and unweighted average distance within clusters, and a new flexible strategy. A classification of the 16 known combinatorial techniques is suggested considering whether inter-cluster distances or within-cluster homogeneities are optimized during the clustering process. Those concerned with cluster homogeneity are discussed and the derivation of their updating parameters is presented. Two tables summarize basic information on the combinatorial SAHN methods compatible with the general recurrence formula given by equation (2). 23 different algorithms of these methods are applied to a phytosociological data set from grassland communities, and the resulting dendrograms are assessed by a multiple comparison method.

Combinatorial h-SAHN methods

Distance- versus homogeneity-optimizing strategies

The SAHN procedures have been the subject of intensive research for many years, but an interesting aspect recognized by Lance & Williams (1967) remains largely overlooked. Those authors distinguished among three basic types of measures used in cluster analysis; these types will serve as the starting point in this paper for a more comprehensive categorization of combinatorial SAHN methods.

For a number of SAHN techniques inter-cluster distances (or dissimilarities, similarities, etc.) are defined (the (i,j)-measures in Lance & Williams' terminology). Distances are geometrically interpretable in a Euclidean space and compatible with all methods to be discussed; therefore they will be used in the sequel unless otherwise stated. Two clusters, C_i and C_j, are fused if their distance, $d(C_i, C_j)$, is minimal in the given clustering step. One entry in **W** is defined in two ways to ensure compatibility with equation (1):

$$w_{ij} = d(C_i, C_j) \text{ or } w_{ij} = d^2(C_i, C_j) .$$

There are no restrictions on within-cluster homogeneity; and the fusion levels indicated in the dendrogram are between-cluster distances containing no information on within-cluster structure. Typical examples are the group average, centroid and single linkage methods.

In other SAHN procedures the fusion criterion relies on some measure of within-cluster homogeneity, even if inter-object distances are calculated first in the analysis. Using an appropriate homogeneity measure, $h(C_i)$, the analysis may proceed in two different ways. One possibility is to maximize the homogeneity of the newly formed clusters, that is,

$$w_{ij} = h(C_i \cup C_j) .$$

This criterion corresponds with the (i)-measures of Lance & Williams (1967) although they considered such measures to have relevance in nonhierarchical clustering only. An example is the minimization of sum of squares in new clusters, a method already mentioned in connection with the general recurrence formula (2). The other optimization procedure involves minimization of the change of homogeneity upon the fusion of two clusters, so that one entry of **W** will have the following general form,

$$w_{ij} = h(C_i \cup C_j) - \rho_i h(C_i) - \rho_j h(C_j)\,.$$

where ρ_i and ρ_j are weights specific to each procedure. Such criteria were called the (ij, k)-measures by Lance and Williams, with k referring to the union of C_i and C_j. The incremental sum of squares technique which amalgamates clusters so as to minimize the increase of within-cluster sum of squares is an example. The increments w_{ij} are not used directly as fusion levels in the dendrogram; a more appropriate level is $h(C_i \cup C_j)$ so that the results of the alternative homogeneity-optimizing strategies become directly comparable.

I think that the distinction between the *d*istance- and *h*omogeneity-optimizing SAHN strategies is important and facilitates the discussion of combinatorial methods. I suggest the use of abbreviations d-SAHN and h-SAHN, respectively, to cover these two main groups of procedures. Within the second category, further distinction is made between the nh-SAHN and ch-SAHN techniques depending on whether the homogeneity of *n*ew clusters or the *c*hange of homogeneity is minimized. It is noted that the use of this terminology is not restricted to combinatorial methods; there are d-SAHN methods (e.g., the *U*-statistic clustering method proposed by d'Andrade 1978) and h-SAHN methods (e.g., those utilizing information theoretic criteria to measure cluster homogeneity, Lance & Williams 1967, Sneath & Sokal 1973) which do not satisfy the recurrence relation (2).

The close relationship among the h-SAHN combinatorial methods is that the derivation of updating parameters for equation (2) follows the same logic (see Appendix). This further supports the importance of an at least technical distinction between d-SAHN and h-SAHN combinatorial procedures.

Fusion criteria in combinatorial h-SAHN methods

Four definitions of within-cluster homogeneity have been proposed in association with combinatorial SAHN methods. These are the error sum of squares (*SSQ*), variance (*VAR*), and average distance (*DIS*) or similarity (*SIM*) within clusters. The latter measure is especially important in various approaches to phytosociological classification (e.g., Popma *et al.* 1983). The clustering process may proceed in two basically different ways: 1) maximization of the homogeneity of new clusters, and 2) minimization of the decrease of homogeneity. The combination of homogeneity measures and strategy types gives rise to 6 h-SAHN methods, one of them with weighted and unweighted variants. These include both widely used and less known methods of numerical classification, as well as two procedures for which the updating parameters are presented for the first time in this paper. This section gives a summary of fusion criteria; the derivation of parameters is presented in the Appendix. The parameters are shown in Table 1 while other useful information is summarized in Table 2 which allows for comparing the basic features of d-SAHN and h-SAHN combinatorial procedures. A practical importance of the subsequent discussion is that several publications do not specify exactly the fusion criterion actually used; reference to terms such as minimum variance clustering and Ward's method is a potential source of confusion.

Optimization of new within-cluster heterogeneity

Minimization of error sum of squares (dispersion) within the new cluster (MNSSQ)
Cluster homogeneity is expressed in terms of error sum of squares calculated from pairwise distances, d_{ij}, of objects. One element of the starting matrix is $d_{ij}^2/2$. At each stage of the analysis any C_r and C_s are fused provided that

$$w_{rs} = min\{SSQ(C_i \cup C_j)\colon\ 1 \leqq i < j \leqq n\}\,,$$

(Anderberg 1973: p. 148).

Table 2. Some properties of combinatorial SAHN clustering methods.

Clustering method	Initialization of **W**	Fusion level of $C_i \cup C_j$ * (and update of w_i)	$\alpha_i + \alpha_j + \beta + + \lambda_h + \lambda_i + \lambda_j$	Monotone fusion levels CP	Monotone fusion levels RNN	Results of CP and RNN agree
1. SL	d_{ij}	w_{ij}	1	yes	yes	yes
2. CL	d_{ij}	w_{ij}	1	yes	yes	yes
3. UPGMA	d_{ij}	w_{ij}	1	yes	yes	yes
4. WPGMA	d_{ij}	w_{ij}	1	yes	yes	yes
5. UPGMC	d_{ij}^2	$\sqrt{w_{ij}}$	<1	no	no	no
6. WPGMC	d_{ij}^2	$\sqrt{w_{ij}}$	.75	no	no	no
7. β-FLEX	d_{ij}	w_{ij}	1	yes	yes	yes
8. (β, γ)-FLEX	d_{ij}	w_{ij}	1	yes/no	yes/no	yes/no
9. MISSQ	$d_{ij}^2/2$	$w_{ij} + w_i + w_j$	1	yes	yes	yes
10. MNSSQ	$d_{ij}^2/2$	w_{ij} *	1	yes	yes	yes
11. MIVAR	$d_{ij}^2/4$	$w_{ij} + \frac{n_i}{n_i + n_j} w_i + \frac{n_j}{n_i + n_j} w_j$	$\leqq 1 \leqq$	no	no	no
12. MNVAR	$d_{ij}^2/4$	w_{ij} *	1	yes	yes	yes
13. WMIDIS	d_{ij}	$w_{ij} + \frac{1}{2} w_i + \frac{1}{2} w_j$ *	$\leqq 1 \leqq$	no	no	no
14. UMIDIS	d_{ij}	$w_{ij} + \frac{b_i}{b_i + b_j} w_i + \frac{b_j}{b_i + b_j} w_j$ *	$\leqq 1 \leqq$	yes?	no	no
15. MNDIS	d_{ij}	w_{ij} *	1	yes	no	no
16. λ-FLEX	d_{ij}	w_{ij} *	1	yes	no	no

Minimization of variance within the new cluster (MNVAR)
Cluster homogeneity is measured by the average contribution of objects to the total sum of squares of the cluster (i.e., variance). One element of the starting matrix is $w_{ij} = d_{ij}^2/4$. The condition for the fusion of two clusters C_r and C_s is that the variance of the new cluster be minimal:

$$w_{rs} = min\{VAR(C_i \cup C_j):\ 1 \leqq i < j \leqq n\}$$

(cf. Anderberg 1973: p. 148).

Optimization of average distance or (dis)similarity within the new cluster (MNDIS).

Originally, I suggested that cluster homogeneity be measured by the simple matching coefficient generalized to more than 2 objects (Podani 1978, 1979). In this case, homogeneity was defined as the number of agreements among objects divided by the possible number of agreements. This ratio is simply the average of all pairwise similarity coefficients within the cluster. However, the strategy equally applies to other types of similarity measures, as well as to dissimilarity and distance coefficients. If cluster homogeneity is defined as the average of pairwise similarities (*SIM*), then C_r and C_s are selected for fusion if

$$w_{rs} = max\{SIM(C_i \cup C_j):\ 1 \leqq i < j \leqq n\}\ .$$

For dissimilarities and distance, the criterion is

$$w_{rs} = min\{DIS(C_i \cup C_j):\ 1 \leqq i < j \leqq n\}\ .$$

This method, termed as average linkage within the new group, was considered formerly by Anderberg (1973: p. 139) as a representative of the stored matrix approach: only the distances have to be retained in computer memory during calculations even if the combinatorial algorithm is not used. However, the combinatorial procedure is much faster than the algorithm suggested by Anderberg.

Minimization of the increase of heterogeneity

Minimization of the increase of sum of squares (MISSQ)

This technique has been referred to under various and often misleading names (e.g., Ward's method, minimum variance (!) clustering, sum of squares agglomeration, and a better one: incremental sum of squares clustering) and belongs to the most widely used clustering algorithms. As clusters to be fused any C_r and C_s are chosen so that

$$w_{rs} = min\{SSQ(C_i \cup C_j) - SSQ(C_i) - SSQ(C_j):\ 1 \leqq i < j \leqq n\}\ ,$$

(see e.g., Anderberg 1973; Orlóci 1967; Wishart 1969).

Minimization of the increase of variance (MIVAR)

In this strategy clusters C_r and C_s are fused provided that

$$w_{rs} = min\{VAR(C_i \cup C_j) - \frac{n_i}{n_i + n_j} VAR(C_i) - \frac{n_j}{n_i + n_j} VAR(C_j):\ 1 \leqq i < j \leqq n\}\ ,$$

(Diday *et al.* 1982). Jambu & Lebeaux (1983) stated that the parameters for this method agree with those of MISSQ, but this is not the case. Diday *et al.* (1982, p. 89) listed first the correct parameters for this technique without showing the derivation of parameters which is presented in the Appendix.

Minimization of the increase of average within-cluster distances (MIDIS)

This is a new strategy with two alternative variants. The average within-cluster distances for clusters C_h and $C_i \cup C_j$ may be calculated in two different ways, i.e., with and without considering cluster sizes. Accepting the terminology of Sneath & Sokal (1973), these variants are termed as unweighted and weighted MIDIS, respectively, because when cluster sizes are neglected the smaller cluster receives greater weight. In this sense, these alternatives are analogous to the pair of group average and weighted average (UPGMA and WPGMA) methods as well as to the pair of the median and centroid strategies from the group of d-SAHN methods.

In the weighted case (WMIDIS), clusters C_r and C_s are amalgamated provided that

$$w_{rs} = min\{DIS(C_i \cup C_j) - \frac{1}{2} DIS(C_i) - \frac{1}{2} DIS(C_j):\ 1 \leqq i < j \leqq n\}\ .$$

The condition for the fusion of C_r and C_s in the unweighted strategy (UMIDIS) is as follows:

$$w_{rs} = min\{DIS(C_i \cup C_j) - \frac{b_i}{b_i + b_j} DIS(C_i) - \frac{b_j}{b_i + b_j} DIS(C_j): 1 \leqq i < j \leqq n\},$$

with $b_i = \binom{n_i}{2}$. For similarities, *max* should replace *min* in both formulae.

Ultrametric properties of combinatorial SAHN methods

A hierarchy produced by a clustering algorithm may be described in terms of a matrix **D** in which δ_{ij} is the lowest hierarchical level at which objects *i* and *j* belong to the same cluster. If any triplet δ_{ij}, δ_{hi}, and δ_{hj} of such values satisfies the relations

$$\delta_{ij} \leqq max\{\delta_{hi}, \delta_{hj}\},$$

$$\delta_{hi} \leqq max\{\delta_{ij}, \delta_{hj}\}, \quad and$$

$$\delta_{hj} \leqq max\{\delta_{ij}, \delta_{hi}\},$$

the output values are *ultrametric distances* (cf. Johnson 1967). In this case the fusion levels monotonically increase: the fusion level of any cluster C_h and $C_i \cup C_j$ cannot be lower than that of cluster C_i with C_j. Failure to satisfy the above relationships is manifested as "reversals" in the tree diagram. The methods compatible with formula (1) will produce monotone increasing fusion levels provided that

$$\alpha_i + \alpha_j + \beta \geqq 1, \tag{3a}$$

$$\alpha_i + \alpha_j \geqq 0, \tag{3b}$$

$$\gamma \geqq -min\{\alpha_i, \alpha_j\}, \tag{3c}$$

(see Milligan 1979; Batagelj 1981). Therefore, 5 d-SAHN strategies are monotonic, the exceptions being UPGMC, WPGMC and, depending on the choice of parameters, the (β, γ)-flexible method (Table 2).

The h-SAHN methods require separate and more thoroughful scrutiny for monotonicity. First, the nh-SAHN methods are considered. Diday (1983) suggested four necessary and sufficient conditions to ensure monotonicity of methods compatible with equation (2). These conditions include 3a–c and

$$\lambda_1, \lambda_2, \lambda_3 \geqq 0. \tag{3d}$$

From Table 2 it is easily seen that neither nh-SAHN methods discussed meet the requirement expressed by condition (3d). One should observe, however, that $\alpha_i + \alpha_j + \beta + \lambda_h + \lambda_i + \lambda_j = 1$ for MNSSQ, MNVAR and MNDIS, therefore it is worth examining if the constraints

$$\alpha_i + \alpha_j + \beta + \lambda_h + \lambda_i + \lambda_j \geqq 1, \tag{4a}$$

$$\lambda_h, \lambda_i, \lambda_j \leqq 0, \tag{4b}$$

$$\alpha_i, \alpha_j, \beta \geqq 0, \quad \text{and} \tag{4c}$$

$$\gamma = 0 \tag{4d}$$

are sufficient to prove the ultrametric feature of these methods. The hierarchical levels monotonically increase if it can be shown that $w_{h,ij} \geqq w_{ij}$ for every clustering step. The proof below utilizes some elements of Milligan's (1979) proof applied to d-SAHN procedures.

Proof. The assumed constraint (4a) may be rewritten as $\beta \geqq 1 - \alpha_i - \alpha_j - \lambda_h - \lambda_i - \lambda_j$. Substituting this constraint into (2) yields

$$w_{h,ij} \geqq \alpha_i w_{hi} + \alpha_j w_{hj} + (1 - \alpha_i - \alpha_j - \lambda_h - \lambda_i - \lambda_j) w_{ij} + \lambda_h w_h + \lambda_i w_i + \lambda_j w_j.$$

After rearrangement we have

$$w_{h,ij} \geqq w_{ij} + \alpha_i(w_{hi} - w_{ij}) + \alpha_j(w_{hj} - w_{ij}) + \lambda_h(w_h - w_{ij}) + \lambda_i(w_i - w_{ij}) + \lambda_j(w_j - w_{ij}). \tag{5}$$

Since in the first clustering step $w_h, w_i, w_j = 0$, it must be that $w_{ij}, w_{hi}, w_{hj} \geqq w_i, w_j, w_h$. The clustering procedure always selects the smallest value in **W**, so that $w_{ij} \leqq w_{hj}, w_{hi}$. Since the constraints (4b–c) require that the λ-s are non-positive and that α_i, α_j cannot be negative, the last five terms in (5) must be greater than or equal to zero and may be deleted from the inequality without losing its validity, and we end up with the desired inequality:

$$w_{h,ij} \geqq w_{ij} . \tag{6}$$

After the fusion, w_i is set equal to w_{ij} and row and column j of **W** are masked. It is apparent that w_i will not be greater than any off-diagonal value of **W**. According to (6), the off-diagonal values of **W** cannot decrease in the subsequent steps, therefore the newly computed fusion levels are never lower than the earlier values. Thus, the monotonicity of levels holds, no matter whether *SSQ*, *VAR* or *DIS* are used as the homogeneity measure.

In the proof above it was assumed that the classical paradigmatic SAHN clustering algorithm is employed, i.e., a single fusion is performed in every clustering step (*c*losest *p*air or CP algorithm). When reciprocal nearest neighbors are fused in each cycle to accelerate the analysis (*r*eciprocal *n*earest *n*eighbor or RNN algorithm, see e.g., Anderberg 1973; Murtagh 1983; Day & Edelsbrunner 1984), the constraints (4a–d) are insufficient to ensure monotonicity and the nature of the homogeneity measure will be of primary concern. It is enough to examine whether the reducibility condition.

$$w_{h,ij} \geqq min\{w_{hi}, w_{hj}\} \text{ for all } h$$

(Bruynooghe 1978), holds for all reciprocal nearest neighbors i and j. Since this condition is satisfied for MNSSQ and MNVAR, their results (and thus their ultrametric properties) are unaffected by the choice between the CP and RNN algorithm. However, for the MNDIS criterion the reducibility condition is not satisfied as the following simple example demonstrates. Let the matrix of Euclidean distances of four objects be given by

$$\mathbf{W} = \begin{matrix} 0 & 1 & 2+\varepsilon & 2+\varepsilon \\ & 0 & 2+\varepsilon & 2+\varepsilon \\ & & 0 & 2 \\ & & & 0 \end{matrix}$$

Object pairs 1–2 and 3–4 are reciprocal nearest neighbors if $0 < \varepsilon$, but $w_{1,34}$, $w_{2,34} = (6 + 2\varepsilon)/3 < 2 + \varepsilon$. As a consequence, the RNN algorithm does not exclude the possibility of reversals. Thus, whereas the ultrametric properties of d-SAHN methods remain the same for the CP and RNN algorithms (Gordon 1987; see Table 2), there is at least one counter-example among the h-SAHN strategies.

The ch-SAHN methods are less similar to one another in general properties than the nh-SAHN strategies. There are differences in the way of calculating the fusion levels and the sum of parameters is not a constant, except in MISSQ (Table 2). The increment of MISSQ is monotonic, because the parameters satisfy relation (3). Also, it is easy to see that $SSQ(C_i \cup C_j) = w_{ij} + w_i + w_j$ cannot be lower than w_i or w_j (i.e., the sums of squares are additive). The hierarchy produced by MISSQ is monotonic regardless whether the increments or the new sums of squares are indicated as fusion levels. This is not so with MIVAR, because from the inequality

$$\left(\frac{n_h + n_i}{n.}\right)^2 + \left(\frac{n_h + n_j}{n.}\right)^2 - \frac{n_h(n_i + n_j)}{n.^2} \geqq 1$$

it should follow that

$$n_h^2 \geqq n_h n_i + n_h n_j + 2 n_i n_j$$

which does not always hold, so the increments are not monotonic. The result of MIVAR is very strongly influenced by the algorithm employed. For example, the RNN algorithm produces a tree which might suggest the existence of some clusters even in random data, while the dendrogram of the CP algorithm exhibits extensive chaining (Fig. 1).

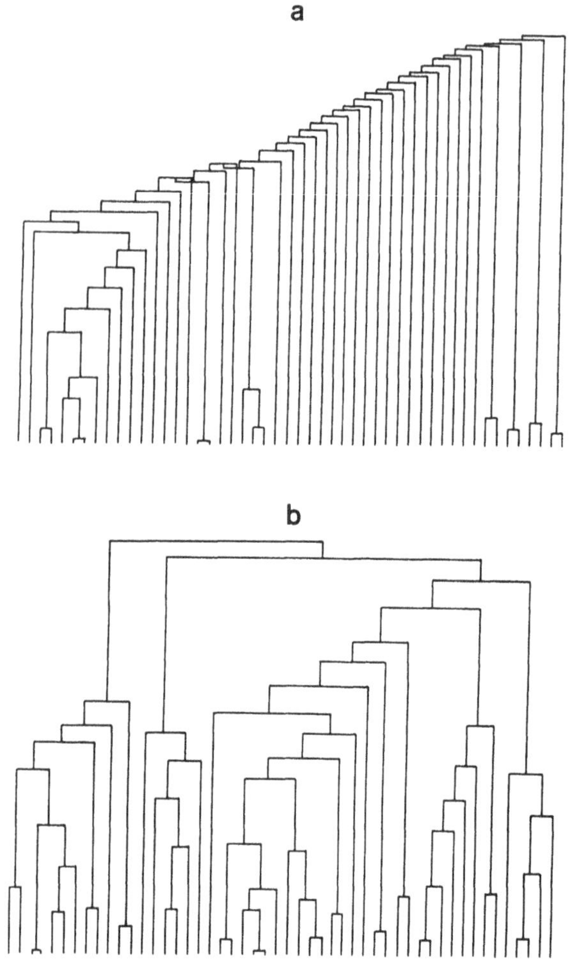

Fig. 1. Comparison of dendrograms produced by alternative algorithms of the MIVAR clustering strategy from a random data set. a) CP algorithm, b) RNN algorithm.

Although in this example only the CP algorithm produced reversals, in other analyses (not shown) both algorithms of MIVAR failed to satisfy monotonicity. This is in apparent contradiction with Diday's (1983) view that the reducibility condition holds for MIVAR and its results are always monotonic (see the table in his Appendix 2).

Analyses of random data (not illustrated) revealed that neither algorithm of WMIDIS has the property of producing reversal-free dendrograms. Also, the RNN algorithm of UMIDIS is also liable to failure of monotonicity. Whether the CP version of UMIDIS may also yield reversals, or it is always monotonic, is not known, however. I was unable to construct artificial or random data which led to reversals for this strategy; in the worst case complete chaining of objects with very small but monotonic increases of levels resulted. A proof is needed to substantiate the statement that UMIDIS-CP is always monotonic.

A flexible h-SAHN clustering strategy

Starting from the conditions (4a–d), which guarantee a monotonic fusion strategy, a new flexible method is defined by imposing the following constraints upon equation (2):

$$\alpha_i + \alpha_j + \beta + \gamma + \lambda_h + \lambda_i + \lambda_j = 1\,,$$
$$\lambda_h = \lambda_i = \lambda_j \leqq 0\,,$$
$$\alpha_i = \alpha_j = \beta, \quad \text{and}$$
$$\gamma = 0\,.$$

The change of parameters under these conditions provides an infinite number of results for the same set of objects. For $\lambda = 0$ (so that $\alpha = \beta = 1/3$), the firstly formed clusters will tend to attract single

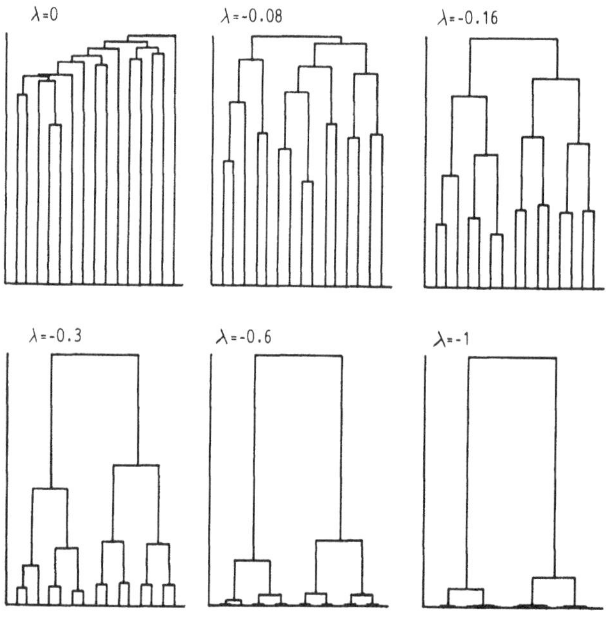

Fig. 2. The effect of λ on the results of a λ-flexible strategy (the CP algorithm is employed).

objects because cluster-to-object measures will generally be lower than object-to-object measures. The strategy seems to be space-contracting and the hierarchy has some degree of chaining so characteristic of single linkage dendrograms. Note that complete chaining observed for high values of β in β-flexible sorting (Lance & Williams 1967) does not occur for this maximum value of λ. As λ becomes large negative, the objects are more intensely grouped and the differences between the first and last fusion levels greatly increase. This is because cluster-to-cluster measures become usually much larger than object-to-object measures. This is demonstrated in Fig. 2 by a series of dendrograms obtained at 6 values of λ for a small data set (15 plots taken randomly from the whole set described below).

Application of combinatorial clustering methods to phytosociology

Data

80 sample plots, each of 4×4 m^2 size, were taken in the rock grassland communities of the Sashegy Nature Reserve, within the city limits of Budapest, Hungary, in 1976. The percentage cover of species was estimated in each plot, but in this paper only presence/absence scores will be used for classifications. The total number of species is 123. The 80 by 123 phytosociological table is not presented here; a copy is available from the author upon request.

Previous classifications

The grassland communities on the dolomite substrate of the study area have long been the subject of intensive phytosociological research. Based on the methods of the Zürich-Montpellier school, Zólyomi (1958) reported 4 community types (associations) from the area. These are: 1) *Festucetum pallentis hungaricum* mostly in southern exposition on rocks and steep slopes; 2) *Caricetum humilis balatonicum* and 3) *Festuco pallenti-Brometum pratensis*, both on hilltops and gentle slopes with some accumulated rendzina, and finally 4) *Seslerietum sadlerianae* on the north-eastern slopes with relatively cool microclimate. The objective of my surveys in the reserve was to examine whether this classification can be confirmed by cluster analysis and multidimensional scaling methods. Published results (Podani 1985, 1986, 1988a) seem to suggest that in the presence/absence situation there are 3 vegetational noda in the study area along a combined gradient of species richness and plant cover. The first nodum roughly corresponds to the open *Festucetum pallentis hungaricum*, but the distinction of the other types is less clear in the binary case.

In this paper the classification study mentioned above will be extended by applying all combinatorial methods to the Sashegy data. Euclidean distances between the 80 plots are used because this measure is compatible with every algorithm discussed here. d-SAHN clustering (SL, CL, UPGMA, WPGMA, UPGMC-CP, WPGMC-CP, and β-FLEX with $\beta = -.25$) was performed by program NCLAS2. h-SAHN clustering (MISSQ, MNSSQ, MIVAR-RNN and -CP, MNVAR, WMIDIS-RNN and -CP, UMIDIS-RNN and -CP, MNDIS-CP and λ-FLEX with $\lambda = 0, -.08, -.16, -.30, -.60$, and -1) was carried out by program HMCL2. The resulting dendrograms were compared in every pair based on three dendrogram descriptors (cluster membership divergence, subtree membership divergence, and cladistic difference, Podani & Dickinson 1984). Cophenetic difference and partition membership divergence were excluded from the comparison because of the presence of reversals in some dendrograms and the lack of commensurability in hierarchical levels. The distance matrix of dendrograms as prepared by program DENDAT was subjected to principal coordinates analysis (PCoA, program PRINCOOR) and complete linkage clustering (CL, program NCLAS2) to reveal structural relationships among dendrograms. All the programs used here are included in the SYN-TAX III package (Podani 1988b). The computations were performed on an IBM370 mainframe computer and an IBM AT compatible machine.

Clustering results

The 23 dendrograms (not shown) represent a wide range of classifications of the 80 sample plots. One extreme is the complete chaining of objects (MIVAR-CP, UMIDIS-CP, and WMIDIS-CP) without any groups indicated. This reflects an undesirable property of the CP algorithm of these strategies: the firstly formed cluster will tend to be fused with individual objects one by one in the subsequent clustering steps. Therefore, these extremely space-contracting strategies are not recommended for use in phytosociological classification and their results will be excluded from further comparisons in this paper. Other dendrograms (SL and WMIDIS-RNN) also exhibit a relatively high degree of chaining with the presence of nuclei for some interpretable clusters. UPGMC, WPGMC and λ-FLEX (with $\lambda = 0$) also produced chains, but a large cluster containing relevés from the species rich section of the study area is quite distinct in the hierarchy. The sample plots are subdivided into two large clusters by MIVAR-RNN, UMIDIS-RNN and λ-FLEX ($\lambda = -.08$); the first cluster interpretable as one of the three noda (either the richest or the poorest in species) and the other containing the remaining plots. 3 clusters, clearly identifiable as noda representing 3 levels of species richness (low, intermediate, and high) and total cover (open, transitional, and closed grassland), are depicted by MNVAR, MISSQ and β-FLEX ($\beta = -.25$). The 3 groups may be easily delineated in the map of the study area (see Podani 1985, 1986, 1988a) suggesting good phytosociological interpretability. The dendrograms obtained by λ-FLEX ($\lambda = -.16$, $-.3$, $-.6$, and -1), MNDIS, CL and MNSSQ imply a more refined group structure (four or five clusters) that can be derived by breaking the 3-cluster MISSQ or MNVAR classifications. Finally, the UPGMA and WPGMA dendrograms suggest the existence of even more small clusters.

Comparison of dendrograms

The quick evaluation of results revealed high similarities as well as considerable differences among the alternative classifications. However, a

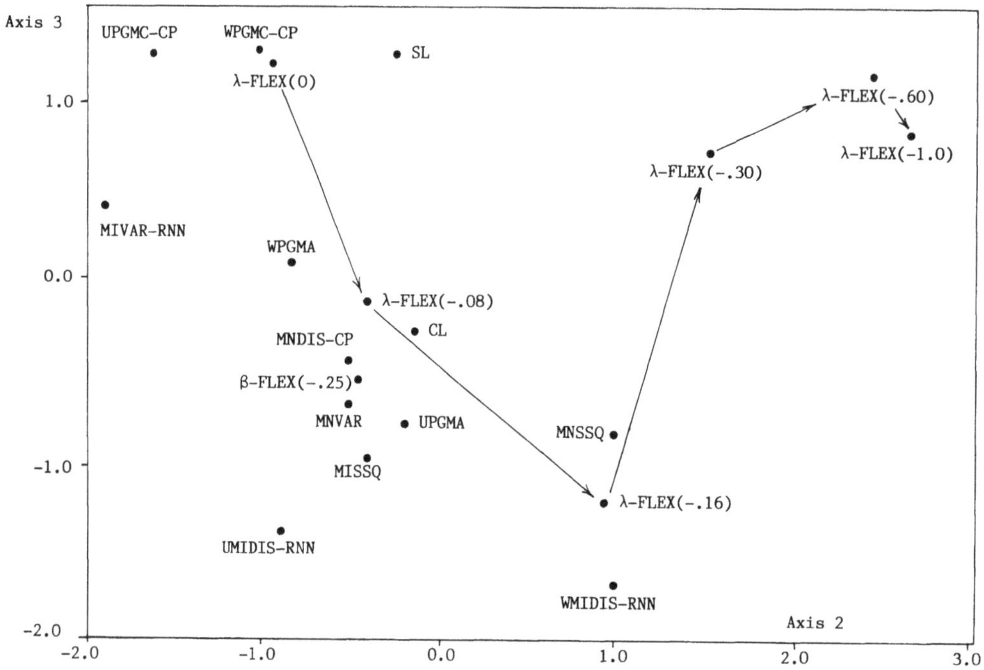

Fig. 3. Principal co-ordinates ordination of 20 dendrograms representing classifications of 80 sample plots from the rock grasslands of Sashegy Nature Reserve (see text for symbols).

more thoroughful analysis of the performance of methods calls for objective assessment. The comparison of partitions of relevés at a nearly constant number of clusters, as done by Gauch & Whittaker (1981) would be less useful in the present study because of the excessive differences in the topological structure of dendrograms. Instead, the multiple comparison strategy of Podani & Dickinson (1984) was adopted as it avoids problems of defining clusters in the hierarchy.

The PCoA ordination of dendrograms reveals that the most important underlying factor implies a tendency towards chaining. On the first component (15.5%) WMIDIS-RNN and SL have large positive scores (4.9 and 2.0, respectively). The other dendrograms are positioned around the centroid (scores ranging from − 1.0 to 0.7); therefore the first axis is not illustrated. The relationships among dendrograms are best explained by the next two axes (12.3% and 7.7%, respectively) shown in Fig. 3. SL, UPGMC, WPGMC and λ-FLEX (λ = 0) form a group of dendrograms already recognized by subjective scrutiny, but they considerably differ from WMIDIS (see also Fig. 4). Note the effect of changing the value of λ in λ-FLEX: for small negative values the dendrograms are close to the majority of results. However, further decrease of λ leads to a rather different classification. The phytosociologically most interpetable results form a cluster, from CL to MNDIS, in the dendrogram of Fig. 4.

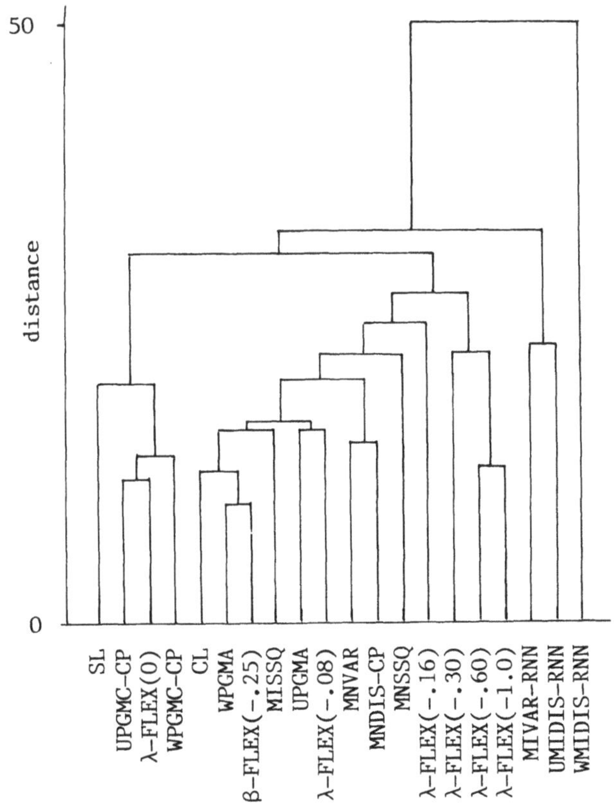

Fig. 4. Classification of 20 dendrograms of the Sashegy relevés by complete linkage clustering (see text for symbols).

Concluding remarks

The combinational SAHN clustering methods may be logically divided into two classes. The d-SAHN methods seek for minimal between-cluster distances and the h-SAHN techniques for maximal within-cluster homogeneity. The d-SAHN clustering methods are compatible with the well-known recurrence formula of Lance & Williams (1966) and make use of parameters α, β and γ. Podani (1978, 1979) used a separate formula for the h-SAHN methods using parameters α, β and λ, whereas Jambu (1978) suggested that all combinatorial methods should be included in the same general equation. The latter suggestion is elegant but a little unfortunate in the sense that the λ-s have no meaning for the d-SAHN methods and the γ parameter is never used by the h-SAHN strategies.

One aim of the present paper is to provide a comprehensive list of combinatorial clustering procedures, with relatively more emphasis placed on the h-SAHN methods. Within-cluster sum of squares, variance and average distance are used as measures of cluster homogeneity. The optimization of homogeneity may be achieved in two basically different ways: the fusion criterion is either the maximization of homogeneity of the new cluster created in a clustering step, or the minimization of the decrease of homogeneity. The combinations of homogeneity measure and fusion criterion define 6 clustering methods, one of them with weighted and unweighted alternatives. 5 of these clustering methods have been suggested earlier, only UMIDIS and WMIDIS appear to be new.

Based on the monotonicity constraints of MNSSQ, MNVAR and MNDIS, a new flexible strategy is defined. This method is apparently similar to the β-flexible method of Lance & Williams (1967) in producing a series of trees that represent transitions between space-contracting and space-dilating clusterings. To establish its potential utility in applied studies, this λ-flexible method deserves future research along the lines of DuBien and Warde's (1979) study on (β, γ)-flexible methods and Milligan's (1987) work evaluating the β-flexible method.

Special attention was paid to the algorithmic and monotonicity properties of combinatorial h-SAHN methods. The CP and RNN algorithm of the same method may produce radically different results, as examples based on random and actual data demonstrated.

The use of the RNN algorithm of MNDIS may destroy the ultrametric structure output by the CP algorithm of the same procedure. As far as I know, it has not been reported earlier that the ultrametric properties of a method depend on the algorithm employed. The explanation of this feature merits a detailed mathematical analysis of these algorithms.

The analysis of field data from the rock grassland communities of the Sashegy Nature Reserve served as a basis for the comparison of d- and h-SAHN procedures. The classifications were considerably different suggesting that there is no clear-cut group structure in the data. The phytosociological implication of the results is that the existence of the community types formerly described from the study area cannot be confirmed in the presence/absence case. Instead, there is a species richness gradient from the open grassland towards the completely closed communities of northern exposition. The congruence among many clustering results still suggests that three noda are worth distinguishing for descriptive purposes. The multiple comparison of dendrograms may help phytosociologists to select clustering procedures that reflect different aspects in the data. However, since the comparisons were based only on a single actual data set, a more extensive simulation study is needed to compare the performance characteristics of combinatorial clustering procedures.

Acknowledgements

I am grateful to A. D. Gordon (University of St Andrews) and P. Juhász-Nagy (L. Eötvös University) for their comments and suggestions. I would like to thank the anonymous referees for their thorough criticism of the manuscript.

References

Anderberg, M.R. 1973. Cluster analysis for applications. Wiley, New York.

Batagelj, V. 1981. Note on ultrametric clustering algorithms. Psychometrika 46: 351–352.

Bruynooghe, M. 1978. Classification ascendante hiérarchique des grands ensembles de données: une algorithme rapide fondé sur la construction des voisinages réductibles. Les Cahiers de l'analyse des Données 3: 7–33.

D'Andrade, R. 1978. U-statistic hierarchical clustering. Psychometrika 43: 59–68.

Day, W.H.E. & Edelsbrunner, H. 1984. Efficient algorithms for agglomerative hierarchical clustering. J. Classif. 1: 7–24.

Diday, E. 1983. Inversions en classification hiérarchique: application à la construction adaptive d'indices d'agrégation. Rev. Stat. Appl. 31: 45–62.

Diday, E., Lemaire, J., Pouget, J. & Testu, F. 1982. Eléments d'analyse de données. Dunod, Paris.

DuBien, J.L. & Warde, W.D. 1979. A mathematical comparison of an infinite family of agglomerative clustering algorithms. Can. J. Stat. 7: 29–38.

Gauch, H.G. & Whittaker, R.H. 1981. Hierarchical classification of community data. J. Ecol. 69: 537–557.

Greig-Smith, P. 1983. Quantitative plant ecology. 3rd ed. Blackwell, Oxford.

Gordon, A.D. 1987. A review of hierarchical classification. J. Roy. Stat. Soc., Ser. A. 150: 119–137.

Gower, J.C. 1967. A comparison of some methods of cluster analysis. Biometrics 23: 623–638.

Jambu, M. 1978. Classification automatique pour l'analyse des données. Tome 1. Dunod, Paris.

Jambu, M. & Lebeaux, M.-O. 1983. Cluster analysis and data analysis. North Holland Publ. Company, Amsterdam.

Johnson, S.C. 1967. Hierarchical clustering schemes. Psychometrika 32: 241–254.

Lance, G.N. & Williams, W.T. 1966. A generalized sorting strategy for computer classifications. Nature 212: 218.
Lance, G.N. & Williams, W.T. 1967. A general theory of classificatory sorting strategies. I. Hierarchical systems. Comput. J. 9: 373–380.
Morena-Casasola, P. & Espejel, I. 1986. Classification and ordination of coastal sand dune vegetation along the Gulf and Caribbean Sea of Mexico. Vegetatio 66: 147–182.
Mucina, L. 1982. Numerical classification and ordination of ruderal plant communities (Sisymbrietalia, Onopordetalia) in the western part of Slovakia. Vegetatio 48: 267–275.
Milligan, G.W. 1979. Ultrametric hierarchical clustering algorithms. Psychometrika 44: 343–346.
Milligan, G.W. 1987. A study of the beta-flexible clustering method. Working paper No WPS 87-61. College of Business, Ohio State University.
Murtagh, F. 1983. A survey of recent advances in hierarchical clustering. Comput. J. 26: 354–359.
Orlóci, L. 1967. An agglomerative method for classification of plant communities. J. Ecol. 55: 193–205.
Orlóci, L. 1978. Multivariate analysis in vegetation research, 2nd ed. Junk, The Hague.
Orlóci, L. & Stanek, W. 1979. Vegetation survey of the Alaska Highway, Yukon Territory: types and gradients. Vegetatio 41: 1–56.
Podani, J. 1978. Hierarchical clustering methods for the analysis of binary phytosociological data. Ph. D. thesis, L. Eötvös University, Budapest (manuscript in Hungarian).
Podani, J. 1979. Generalized strategy for homogeneity-optimizing hierarchical classificatory methods. In: Orlóci, L., Rao, C.R. & Stiteler, W.M. (eds), Multivariate methods in ecological work. pp. 203–209. International Co-operative Publishing House, Burtonsville, Maryland.
Podani, J. 1985. Syntaxonomical congruence in a small-scale vegetation survey. Abstr. Bot. 9: 99–128.
Podani, J. 1986. Comparison of partitions in vegetation studies. Abstr. Bot. 10: 235–290.
Podani, J. A method for generating consensus partitions and its application to community classification. Coenoses (in press).
Podani, J. 1988. SYN-TAX III. A package of programs for data analysis in ecology and systematics. Coenoses 3: 111–119.
Podani, J. & Dickinson, T.D. 1984. Comparison of dendrograms: a multivariate approach. Can. J. Bot. 62: 2765–2778.
Popma, J., Mucina, L., van Tongeren, O. & van der Maarel, E. 1983. On the determination of optimal levels in phytosociological classification. Vegetatio 52: 65–75.
Sneath, P.H.A. & Sokal, R.R. 1973. Numerical taxonomy. 2nd ed. Freeman, San Francisco.
Sokal, R.R. & Michener, C.D. 1958. A statistical method for evaluating systematic relationships. Univ. Kansas Sci. Bull. 38: 1409–1438.
van der Maarel, E. 1979. Multivariate methods in phytosociology, with reference to the Netherlands. In: Werger, M.J.A. (ed.), The study of vegetation. pp. 163–225. Junk, The Hague.
Wishart, D. 1969. An algorithm for hierarchical classifications. Biometrics 25: 165–170.
Zólyomi, B. 1958. The natural vegetation of Budapest and its surroundings. In: Pécsi, M. (ed.), Budapest természeti képe. pp. 508–642. Akadémiai, Budapest (in Hungarian).

Appendix

The derivation of updating parameters for h-SAHN clustering methods.

nh-SAHN strategies

As far as the derivation of parameters is concerned, this is the simpler situation. In general, a total of all inter-object distances is reproduced from the six w-s, and then this total is divided by some function of $n.$ ($n. = n_h + n_i + n_j$) to yield a homogeneity measure for cluster $C_h \cup C_i \cup C_j$.

MNSSQ

Since w_{ij} is obtained as the sum of all squared within-cluster distances divided by the number of objects in the cluster, we can write that

$$w_{ij} = \frac{1}{n_i + n_j} \sum_{p,q \in C_i \cup C_j} d_{pq}^2 \text{ and}$$

$$w_i = \frac{1}{n_i} \sum_{p,q \in C_i} d_{pq}^2 .$$

$n_i w_i$ is contained in both $(n_h + n_i)w_{hi}$ and $(n_i + n_j)w_{ij}$, and $n_h w_h$ and $n_j w_j$ are also present twice, therefore the sum of squared distances within $C_h \cup C_i \cup C_j$ is

$$\sum_{p,q \in C_h \cup C_i \cup C_j} d_{pq}^2 = (n_h + n_i)w_{hi} + (n_h + n_j)w_{hj} + (n_i + n_j)w_{ij} - n_h w_h - n_i w_i - n_j w_j .$$

This sum divided by $n.$ yields the sum of squares in cluster $C_h \cup C_i \cup C_j$.

MNVAR

The total of squared distances in cluster $C_i \cup C_j$ is reproduced as follows,

$$w_{ij} = \frac{1}{(n_i + n_j)^2} \sum_{p,q \in C_i \cup C_j} d_{pq}^2 \text{ and}$$

$$w_i = \frac{1}{n_i^2} \sum_{p,q \in C_i} d_{pq}^2 .$$

Then

$$\sum_{p,q \in C_h \cup C_i \cup C_j} d_{pq}^2 = (n_h + n_i)^2 w_{hi} + (n_h + n_j)^2 w_{hj} + $$
$$+ (n_i + n_j)^2 w_{ij} - n_h^2 w_h - n_i^2 w_i - n_j^2 w_j .$$

This sum divided by $n_.^2$ yields the variance in cluster $C_h \cup C_i \cup C_j$.

MNDIS

The derivation of the updating parameters, shown for distances, starts by reproducing the sum of distances within clusters:

$$w_{ij} = \frac{1}{b_{ij}} \sum_{p,q \in C_i \cup C_j} d_{pq} \text{ and}$$

$$w_i = \frac{1}{b_i} \sum_{p,q \in C_i} d_{pq} ,$$

where the binomial coefficient $b_i = \binom{n_i}{2}$ is the number of pairs within C_i, and $b_{ij} = \binom{n_i + n_j}{2}$. Then, by similar justification as in the above cases, the sum of all distances within cluster $C_h \cup C_i \cup C_j$ is

$$\sum_{p,q \in C_h \cup C_i \cup C_j} d_{pq} = b_{hi} w_{hi} + b_{hj} w_{hj} +$$
$$+ b_{ij} w_{ij} - b_h w_h - b_i w_i - b_j w_j .$$

This sum divided by $b_. = \binom{n_.}{2}$ yields the average within-cluster distance for $C_h \cup C_i \cup C_j$.

ch-SAHN strategies

The derivation of parameters starts by reproducing the total of pairwise distances for the new cluster $C_h \cup C_i \cup C_j$. This quantity is divided by a function of $n_.$ yielding the homogeneity measure for that cluster. Then, subtraction of unweighted or weighted homogeneity measures of clusters C_h and $C_i \cup C_j$ from this quantity gives the increment.

MISSQ

w_i and w_j are the sum of squares for clusters C_i and C_j, respectively, w_{ij} is the increase upon their fusion, therefore the sum of squares of cluster $C_i \cup C_j$ will be $w_{ij} + w_i + w_j$. Thus

$$\sum_{p,q \in C_i \cup C_j} d_{pq}^2 = (n_i + n_j)(w_{ij} + w_i + w_j)$$

and

$$\sum_{p,q \in C_h \cup C_i \cup C_j} d_{pq}^2 = (n_h + n_i)(w_{hi} + w_h + w_i) +$$
$$+ (n_h + n_j)(w_{hj} + w_h + w_j) +$$
$$+ (n_i + n_j)(w_{ij} + w_i + w_j) -$$
$$- n_h w_h - n_i w_i - n_j w_j .$$

This quantity divided by $n_.$ gives the sum of squares of cluster $C_h \cup C_i \cup C_j$. Then, the increase of sum of squares is obtained by subtracting the heterogeneity of clusters $C_i \cup C_j$ and C_h,

$$w_{h,ij} = \frac{1}{n_.} \sum_{p,q \in C_h \cup C_i \cup C_j} d_{pq}^2 - (w_{ij} + w_i + w_j + w_h) =$$
$$= \frac{1}{n_.}[(n_h + n_i)w_{hi} + (n_h + n_j)w_{hj} + ((n_i + n_j) - n_.)w_{ij} +$$
$$+ ((n_i + n_j) + (n_h + n_i) - n_i - n_.)w_i +$$
$$+ ((n_i + n_j) + (n_h + n_j) - n_j - n_.)w_j +$$
$$+ ((n_h + n_i) + (n_h + n_j) - n_h - n_.)w_h] .$$

The last three terms cancel, so the formula reduces to

$$w_{h,ij} = \frac{1}{n_.}[(n_h + n_i)w_{hi} + (n_h + n_j)w_{hj} - n_h w_{ij}] ,$$

therefore parameters λ_h, λ_i and λ_j are zero.

MNVAR

The variance of cluster $C_i \cup C_j$ may be determined according to the formula

$$VAR(C_i \cup C_j) = w_{ij} + \frac{n_i}{n_i + n_j} w_i + \frac{n_j}{n_i + n_j} w_j ,$$

and similar relations hold for $VAR(C_h \cup C_i)$ and $VAR(C_h \cup C_j)$. Then, the variance of $C_h \cup C_i \cup C_j$ will be

$$VAR(C_h \cup C_i \cup C_j) = \frac{1}{n_.^2}[(n_i + n_j)^2 \, VAR(C_i \cup C_j) +$$
$$+ (n_h + n_i)^2 \, VAR(C_h \cup C_i) + (n_h + n_j)^2 \, VAR(C_h \cup C_j) -$$
$$- n_h^2 w_h - n_i^2 w_i - n_j^2 w_j] . \tag{7}$$

The variance before the fusion of C_h with $C_i \cup C_j$ is

$$VAR(C_h; C_i \cup C_j) = \frac{n_h}{n_.} w_h + \frac{n_i + n_j}{n_.} VAR(C_i \cup C_j) . \tag{8}$$

Substitution of variances into (7) and (8), and subtraction of (8) from (7) yields the increment sought:

$$w_{h,ij} = \frac{(n_h + n_i)^2}{n_.^2} w_{hi} + \frac{(n_h + n_j)^2}{n_.^2} w_{hj} +$$
$$+ \left(\frac{(n_i + n_j)^2}{n_.^2} - \frac{n_i + n_j}{n_.} \right) w_{ij} +$$
$$+ \left(\frac{n_h^2 + n_j n_h + n_j n_i}{n_.^2} - \frac{n_j}{n_.} \right) w_h +$$

$$+\left(\frac{n_i^2+n_in_h+n_in_j}{n_.^2}-\frac{n_i}{n_.}\right)w_i +$$

$$+\left(\frac{n_j^2+n_jn_h+n_jn_i}{n_.^2}-\frac{n_j}{n_.}\right)w_j .$$

The last three terms cancel and the formula reduces to:

$$w_{h,ij} = \frac{(n_h+n_i)^2}{n_.^2}w_{hi} + \frac{(n_h+n_j)^2}{n_.^2}w_{hj} - \frac{n_h(n_i+n_j)}{n_.^2}w_{ij} .$$

WMIDIS

Cluster sizes are disregarded so that the average distance within cluster $C_i \cup C_j$ is

$$DIS(C_i \cup C_j) = w_{ij} + \frac{1}{2}w_i + \frac{1}{2}w_j , \quad (9)$$

and similar relations exist for $DIS(C_h \cup C_i)$ and $DIS(C_h \cup C_j)$. The average distance within $C_h \cup C_i \cup C_j$ is calculated as

$$DIS(C_h \cup C_i \cup C_j) = \frac{1}{b.}[b_{hi}DIS(C_h \cup C_i) + b_{hj}DIS(C_h \cup C_j) + b_{ij}DIS(C_i \cup C_j) - b_h w_h - b_i w_i - b_j w_j] . \quad (10)$$

The average distance before the fusion of C_h with $C_i \cup C_j$ is

$$DIS(C_h; C_i \cup C_j) = \frac{1}{2}w_h + \frac{1}{2}DIS(C_i \cup C_j) . \quad (11)$$

Substitution of average distances into (10) and (11), subtraction of (11) from (10) and subsequent rearrangement of the formula give the recurrence relation:

$$w_{h,ij} = \frac{1}{b.}[b_{hi}w_{hi} + b_{hj}w_{hj} + (b_{ij} - b./2)w_{ij} - ((b_h + n_in_j)/2)w_h + ((b. - 2b_i - 2n_hn_j)/4)w_i + ((b. - 2b_j - 2n_hn_i)/4)w_j] .$$

UMIDIS

Formula (9) is replaced by

$$DIS(C_i \cup C_j) = w_{ij} + \frac{b_i}{b_i + b_j}w_i + \frac{b_j}{b_i + b_j}w_j \quad (12)$$

and expression (11) is rewritten as

$$DIS(C_h; C_i \cup C_j) = \frac{b_h}{b_h + b_{ij}}w_h + \frac{b_{ij}}{b_h + b_{ij}}DIS(C_i \cup C_j) . \quad (13)$$

Substitution of average distances into (12) and (13) and subtraction of (13) from (12) leads to a very complicated formula which cannot be brought into a much simpler form because sums of binomial coefficients are present in the denominators. It is left to the reader to show that the recurrence relation for UMIDIS takes the form

$$w_{h,ij} = \frac{1}{b.}[b_{hi}w_{hi} + b_{hj}w_{hj} + (b_{ij} - b.b_{ij}/(b_h + b_{ij}))w_{ij} + (b_hb_{hi}/(b_h + b_i) + b_hb_{hj}/(b_h + b_j) - b_hb./(b_h + b_{ij}) - b_h)w_h + (b_ib_{ij}/(b_i + b_j) + b_ib_{hi}/(b_h + b_i) - b_ib.b_{ij}/((b_h + b_{ij})(b_i + b_j)) - b_i)w_i + (b_jb_{ij}/(b_i + b_j) + b_jb_{hj}/(b_h + b_j) - b_jb.b_{ij}/((b_h + b_{ij})(b_i + b_j)) - b_j)w_j] . \quad (14)$$

Vegetatio **81**: 79–94, 1989

On the use of grammars in vegetation analysis

M. B. Dale[1] & M. M. Barson[2]
[1]*C.S.I.R.O., Division of Tropical Crops and Pastures, Cunningham Laboratory, Carmody Rd., St Lucia, 4067, Australia;* [2] *Bureau of Rural Resources, Department of Primary Industries, Resources and Energy, P.O. Box 858, Canberra ACT 2601, Australia*

Accepted 10.9.1988

Keywords: Aquatic vegetation, Image processing, Inference, Lake, Rain forest, Spatial series, String, Succession

Abstract

Grammars provide a means of generating large amounts of variability from relatively small resources, and seem especially relevant when the variation to be modelled is discontinuous since they describe the organisation symbols from some alphabet. In order to use grammars, it is desirable that they be inferred from observational data, and we present three examples of such inference. Two involve temporal successions, while the third is based on spatial transects of aquatic vegetation in wetlands in Victoria, Australia. The temporal successions were adequately represented by simpler regular grammars, suggesting that the processes involved may be modelled in a relatively simple manner. The spatial data requires a context-free grammar, reflecting the greater complexity of the spatial series, the uniqueness of each site, and the difficulties of modelling parallel systems with serial models. Examining the grammars further leads to some suggestions concerning the features of them which are important for ecological interpretation. It further leads to a novel view of the nature of groups in classification.

Introduction

Since the early studies of Goodall (1953) and Williams & Lambert (1960) a large literature has accumulated on the methodologies and practice of numerical classification. In a separate and more ancient stream, classical phytosociologists have also proposed classifications and erected nomenclatural bastions to defend the results of their deliberations. (See Mueller-Dombois & Ellenberg 1974.) Combinatory approaches have been tried as well (Westhoff & van der Maarel 1973; van der Maarel *et al.* 1976). But when you have chosen your relevés, estimated your species performances, by Domin or density, selected your similarity, and indulged the arcane activities of tabular sorting, whether by man or by machine, what then?

The commonest continuation involves the characterisation of classes in terms of environmental information or possibly historical information. In contrast we shall here make use of relationships between the individual items classified, the quadrats or relevés, to examine further their patterns. In our particular examples the samples were arranged in linear series of transect. If each sample is assigned to a particular vegetation type, then the transect can be regarded as a sequence or string of symbols, each symbol representing a vegetation class of a sample. It is with the structures to be identified in these sequences that we shall be concerned. We shall concentrate on models of processes for generating these sequences in the form of grammars, rather than simply searching for convenient structure as, for example, do Aiello *et al.* (1973).

A grammar is simply a set of rules showing how certain, *nonterminal,* symbols may be replaced by other symbols which may be themselves *nonterminal,* and hence liable for further replacement, or *terminal.* Terminal symbols are never replaced and represent the elements of the sequence, in our case the vegetation types resulting from the initial classification. Starting from a single special symbol representing the entire transect we apply various rules in succession and finally derive the observed sequences. Thus the grammars are descriptive and the application of the rules is hierarchical. The rules may be variously restricted in their form and in their range of application giving various classes of grammars, which have been previously discussed by Dale (1981). It would be possible, at the expense of greater complexity and computing cost, to investigate arrays of symbols instead of linear sequences thereby studying the distribution of vegetation types in a landscape. However a detailed treatment of the various forms of grammars and their inter-relationships will not be given here.

Grammars provide, in fact, a very flexible means of generating great variety from limited resources. They are especially applicable when changes are primarily discontinuous since they are essentially sets of rules showing how symbol changes can be made, but Lindenmayer (1974) has indicated how to define continuous models in a grammatical context. In many cases it is possible to decide the form of the grammar a priori, because the nature of the sequences generated is already known. This is not true in the present study so we must consider if it is possible to infer grammars from vegetation data? If so, what do they tell you, and how? In this paper our primary purpose is to examine some examples of inferred grammars for vegetation data and to discuss how they might be interpreted.

Since there are many different kinds of grammars, before embarking on an inference procedure which serves to discover rules we shall obviously have to decide the limits of such inference, i.e. to determine which class or classes of grammar interest us. Angluin (1980) has shown that inference of grammars from examples of their productions is a computationally hard task, made more difficult because in general we have no examples of ungrammatical strings with which to limit the possibilities. Selection of an appropriate grammar can be based on the identification of functionality with specific rules or on the use of Occam's razor.

Since various kinds of grammar are related hierarchically we can determine if a *simpler* grammar is *relatively more adequate* as a model than a more complex one. This obviously requires us to measure both the complexity and adequacy of a grammar. Complexity reflects the number and nature of the rules, adequacy the extent to which the language generated by these rules is indeed that which we have observed. In this paper we have not sought such measures (see Cook & Rosenfeld 1976 for some possibilities), restricting our investigations to two classes of grammar, regular and context-free, which will be described later.

One major interest here, then, lies in the inference of a grammar, that is in deriving the grammar from observational data. However it should not be assumed that this is the only approach to using grammars, ecologically or biologically. For example, other uses for grammars have been proposed by Haeffner (1975, 1978, 1981) who uses a very powerful grammatical formalism to describe how an ecosystem may be constructed from component species. Grammars have also been used in studies of animal behaviour (Westman 1977), in generating images of plants in cinematic animation (Smith 1984), and most widely in studies of morphological development (Mayoh 1973; Frijters & Lindenmayer 1974; Hellendoorn & Lindenmayer 1974; Bell 1976; Frijters 1978; Niklas 1979). A good example of what we presently think to be a profitable approach to using grammars is the work of Hernandez & Arias (1984) who applied grammatical methods as part of a procedure for the automated identification of foetal stress during parturition. In most studies, however, the grammars are not inferred from the data automatically, but are derived, or constructed, by human observers.

In fact the difficulties of inference vary with the class of grammars being studied. In parallel grammars all instances of a specific symbol are changed simultaneously, which might be reasonable when studying spatially distributed processes. This simultaneity introduces severe constraints on the generating processes so that inference appears to be possible provided that the alphabet of terminal symbols is defined. Feliciangeli & Herman (1973, see also Doucet 1974) have examined the possibilities of inferring one class of such grammars, the totally parallel Lindenmayer systems, but no applications of such inference procedures have yet appeared in the literature. For serial grammars, where only a single symbol is changed at any one time, a wider variety of methods exist in the literature, but again most of these have not been applied to extensive data, and many are computationally very expensive. Westman (1977) and Dale (1979, 1981) provide a simple introduction to the literature and nomenclature.

Our analysis has four phases.

1. Initial classification of the observational units (quadrats) to obtain the letters of the alphabet. The alphabet is the set of basic symbols from which our vegetation is constructed. Here we have used numerical methods to establish the primary vegetation types, but intuitive methods could also be used. One point of interest is the effect of changing the alphabet, or more specifically changing the number of classes which we accept, on the resulting grammar. If possible we should wish to identify grammatical properties which were resistant to minor errors in the typification procedure.

2. Reorganisation of the observational units into sequences of alphabetic symbols. Such organisation may be based on temporal, spatial or any other kind of relationship between the quadrats. It is possible to have more complex relationships such as trees or arrays, but for the present we shall restrict ourselves to linear sequences.

3. Inference of the grammar for the sequences. Here we seek to determine a set of rules which will generate the observed sequences of symbols. In general without negative information, that is information on sequences which are impossible, the set of rules inferred will generate more sequences than were actually observed.

4. Interpretation of the rules so obtained. Having obtained a grammar we have to determine what information it holds which is of ecological or phytosociological interest, and how this information is represented in the rules.

Data sources

For our studies we shall use three sets of data. The first of these concerns secondary succession in rain forest after clearance. The data consist of records of the vegetation present on 10 plots at 12 times, although the temporal intervals between samplings are unequal. Fuller description is given in Williams *et al.* (1969). These data can also be regarded as five pairs of spatially adjacent plots, so that the number of vegetation types can be objectively determined using the test of Sandland & Young (1979). In this test each pair of plots is regarded as two replicate samples of the same vegetation type. The pair should therefore be found in the same final group of the numerical analysis. If pairs are allocated to different groups then the difference is due to 'between replicate' variation which is not regarded as significant. The Sandland-Young (1979) test for these data indicated groupings at 4 and 8 group levels which, as we shall see, fall within the acceptable limits of stability for these data.

The second set of data records primary succession on estuarine sandy flats with salt marsh, and has already been analysed by Hogeweg *et al.* (1985, see also van Noordwijk-Puyk *et al.* 1979). It actually is composed of two data sets at different spatial scales, with the number of time periods observed also different in the two sets. It is interesting that Hogeweg *et al.* actually present an analysis closely related to our present grammatical studies. They developed a stochastic automation as a model of succession. This is equivalent to obtaining a parser for a regular grammar, that is a program which determines if any particular sequence is indeed a product of a specific grammar.

The third set of data records the spatial distribution of vegetation in wetlands in Victoria, Australia. In each wetland area, two or three transects were laid and the presence of different species recorded in 1 m^2 quadrats placed at intervals along the transects. Detailed descriptions of locations and a classification of the wetlands are given in Barson (1984). We have used a subset of 36 wetlands, rejecting all the strongly saline sites. In most cases these latter had a uniformly species-poor vegetation, involving species which rarely occurred in other habitats. Their inclusion would simply have produced a subgrammar essentially discrete from that applicable to the other wetlands, and added extra complexity to the analysis. These basic floristic data were classified to form 12 groups using a monothetic divisive method (Lance & Williams 1968). The defining species are shown in Table 1 and the sequence of classes along the transects is given in Table 2. Thus the data for the grammar inference procedure are a list of vegetation types along each transect. Note that the transects are not all of equal length; in fact they were terminated in some cases because the water depth became too great.

Preliminary analysis

For the successional data, it is easy to see that each location can be represented as a temporal sequence of multivariate observations. There is an obvious directionality and the only problem is to reduce the multivariate observations on the several species to a single set of univariate symbols, the alphabet. In principle we could derive a grammar for each species independently but it is not obvious how these would be combined. We proceed instead by classifying, using the 'site-times' as the units to be classified. Each location falls into exactly one class at each time, so that the temporal changes at one location can be represented as a string of class symbols. (The possible use of fuzzy classes we reserve for a future study.) It is these symbol strings which form the input to the grammar inference procedure.

For the initial classification of the secondary successional data, we employed a polythetic agglomerative classification using the Jaccard similarity measure and flexible sorting with presence/absence data. Since it might be argued that the number of classes, and hence the number of symbols, could have considerable significance we have examined the results of varying this number. These preliminary studies showed that some properties of the inferred grammars were robust to changes in number of classes. There were, in fact, no major differences in the number of *alternative derivation sets* when the number of vegetation types was varied from 3 to 10. Following Fleck (1974) we define an alternative derivation sequence in three ways.

1. It may be a set of rules which together form a regular grammar, that is every rule in the sequence contains at most one nonterminal symbol in each alternative on its right-hand side. The sequence of rules

a:: = >**b:1** | **b:2**; **b**:: = >**c:1** | **c:3**; **c**:: = >**1:2** | **2:3** (1)

is of the required form. This is to be read as follows using the following conventions.

a. The nonterminal symbols are all alphabetic; thus **a, b, c.**

b. The terminal symbols are all numeric; thus **1, 2,**

Table 1. Victorian wetlands: monothetic division forming 12 groups.

1.	+ Potamogeton + Eleocharis
2.	.... – Eleocharis + Triglochin
3.	.... – Triglochin
4.	– Potamogeton + Azolla
5.	.... – Azolla + Triglochin + Myriophyllum
6.	.... – Myriophyllum
7.	.... – Triglochin + Baumea
8.	.... – Baumea + Myriophyllum
9.	.... – Myriophyllum + Sphagnum
10.	::: – Sphagnum + Lepidosperma
11.	::: – Lepidosperma + Hydrocotyle
12.	Residual group.

Table 2. Victorian wetlands data: transect data for 12 vegetation type analysis. Transect numbers precede the list of types.

Palparra Settlement Swamp	1/ 8 8 3 3 4 4 4 4 4 4 4 4 8;
	2/ 3 4 4 8 4 4 4 4 4
	3/ 3 3 3 3 3 2 3 4 4 4 4 4 4
Break No. 2 Swamp	4/ 2 6 6 5 6 6 6 6 6 6
	5/ 5 6 6 6 6 6 6 6 6
	6/ 6 2 6 6 6 6 12
Tremaines Swamp	7/ 7 8 8 7 8 8 8 8
	8/ 8 8 8 8 8 8 8 8
	9/ 12 12 12 6 8 5
Brown Reedy Swamp	10/ 5 8 6 12
	11/ 10 8 5 5 6 12 12 12 12 12
	12/ 3 5 5 5 12
Kanawinka Swamp	13/ 2 2 6 5 6 6 5 6
	14/ 2 6 5 2 5 6
	15/ 3 2 6 5 6 6
Floating Island Swamp	16/ 12 4 12 4 4
	17/ 12 4 4 4
	18/ 4 4 4 4
Cobrico Swamp	19/ 6 4
	20/ 6 12
	21/ 6 4 4
Lake Elusive	22/ 8
	23/ 12
Lake Bong Bong	24/ 7 6 7 7 7 7 6
	25/ 12 7 7 7 6
	26/ 7 7 7 7 7 7 6
Dock Inlet	27/ 5 12 6 6 6 6 6 6 6 6 5 6 6 6 6 6 6 6 6 6 6 6
	28/ 6 6 6 6 12 12
	29/ 7 6
Ewings Marsh	30/ 2 1 2 5 2 3 12 12 12 6
	31/ 12 12 12 12 12 12 12
	32/ 2 2 2 2 3 2 2 2 1 6 3 3 3 3 2 3 2 3 2 3 2 6
Swan Lake West	33/ 12 12 6 5 8 8 8
	34/ 8 12 5 5 2 12
	35/ 5 6 12 12 6 8 8 12
Lake Malseed	36/ 6 6 6 6 6 2
	37/ 2 2 12 6 12
	38/ 2 2 2 2 6 2 2 6 12
Gelliondale Swamp	39/ 6 6 4 4 4 4 4 4 4 4 4 4 4 6
	40/ 10 4 6 6 4 4 4 4 6 4 4
	41/ 10 8 5 4 4 4 4 4 4
The Long Swamp	42/ 3 3 3 3 3 3 12 12 12 3 12 3
	43/ 10 10 10 10 10 10
	44/ 10 10 10 10 10 10
Dereel Lagoon	45/ 5 8 12 3 3 3 7 12 7 7 7 7 7
	46/ 8 7 6 7 5 7 7 7 7 7 7 7
	47/ 10 7 7 7 7 7 7 7 7 7 7
Two Tree Swamp	48/ 8 8 8 8 8 8 8 8 8 8 8
	49/ 1 4 1 2 1 4 4 4 1 12 12
	50/ 4 4 4 4 4 4 4 4 4 4
Cockpit Lagoon	51/ 6 6 6 6 12 6 6 6 6 6 6 6 6 6 6 6
	52/ 6 12 12 6
	53/ 12 12 6 6 6 6 6 6 6 6 6 6 6 6 6 6
Bittern Lagoon	54/ 12 12 1 1 1 1 1 1 1 1 1 1 1 1 1 12 6
	55/ 3 1 1 1 1 1 1 1 1 1 3 3 3
	56/ 12 1 1 1 1 1 1 1 1 1 1 1 1 1
Victoria Lagoon	57/ 8 5 8 5 5 5 5 6 5 5 6 6 6 6 6 12 6
	58/ 5 5 12 6 5 5 5 5 5 5 5 5 5 5 5 6 6 6
	59/ 8 5 6 6 12 6 6 6 12
King River Meander	60/ 8 12 12 8 8 8 8 8 8
	61/ 8 8 8 12 12 12 12 12 12 12 12
Duck Holes	62/ 2 6 6 6 6 6 6
	63/ 2 2 2 2 2 2 6
	64/ 2 2 2 2 2 5 5
Loch Garry	65/ 12 4
	66/ 12 12 4 4 4 4
Middle Lake	67/ 8 4 4
	68/ 12 4 8 4 4 4
	69/ 12 8 12 12 12 12
The Stony	70/ 1 1 1 1 3 12 12 12
	71/ 1 1 1 1 1 1 12 3 3
Lake Kanagulk	72/ 4 4 4 4 4 12 12 12 12
	73/ 4 4 4 4 4 4 4 12 4 12 12
	74/ 12 4 4 4 4 4 4 4 4 4 12
Heywood Golf Club Swamp	75/ 8 8 12 12 12 12 12
	76/ 8 8 8 12 12 12 12 12 12 12
	77/ 8 8 12 12 12 12 12 8 12 12 8
Lake Bridgewater	78/ 12 7 7 6 7 7 7 7 7 7 7 12 12 12
	79/ 12 7 7 7 12 12
	80/ 6 7 7 7 7
Cotter's Lake	81/ 11 12 12 11 11 11 11 12 12 12 12
	82/ 7 11 11 11 11 11 11 7 7 7 7
	83/ 7 7 7 7 7 7 7 7 7 7 7
Lake Barracouta	84/ 6 6 6 6 6 6 7 7 7 7 7 7 7 7
	85/ 12 12 7 7 7
	86/ 12 12 12 6 12
Lake Beadle	87/ 6 6 6 6 6 6 6
	88/ 6 6 12 12 12 12
	89/ 6 6 12
Brisbane Ranges	90/ 8 5 3 5 12 12 12
	91/ 8 8 12 8 8 8 8 8 10
	92/ 8 5 8 3 5 5 5 5 5
Fernbank Swamp	93/ 3 3 3 3 3 3 3 3 3 3 3
	94/ 10 3 3 3 3 3 3 3 3 3 3
	95/ 10 3 3 3 3 3 3 3 3 3 3
Mackenzie Road Swamp	96/ 1 2 9 9 9 9 12 9 9 9 9 9
	97/ 9 9 9 9 9 9 9 9 9 9 9
	98/ 9 9 9 9 9 9 9 9 9 9 9
Kangaroo Swamp	99/ 7 8 8 10 10 12 12 7 7 7 7 7 7 7
	100/ 10 10 10 10 12 10 7 7 7 7 7
	101/ 3 3 3 3 3 3 7 10 12 7 7
Banyule Billabong	102/ 6 6 6
	103/ 4 6 6 6 6 6
	104/ 6 6 6 6 6 6

3. We shall refer to the rule for symbol '**c**' as a *terminating rule* since the right-hand side contains no nonterminals. Such rules terminate the derivation set; they do not necessarily terminate a specific string. Indeed the inference procedure used here is such that terminating rules most often provide symbols which initiate the strings, though this is not universally true. Rules of the form

a:: = >1:b | b:1 (2)

are indeed found occasionally.
c. The metasymbol '::= >' is to be read as 'can be replaced by'.
d. The metasymbol ' |' is read as 'or' and separates alternative substitutions.
e. The metasymbol ';' separates rules in a single derivation set when these occur on a single text line. Continuation onto subsequent lines of text is indicated by indentation.
f. The metasymbol ':' is a separator placed between symbols.

Thus our example reads:
a. The nonterminal '**a**' can be replaced by either '**b:1**' or '**b:2**'. This could of course be regarded as two separate rules (also known technically as productions) but it is convenient to combine rules which have a common left-hand side.
b. The nonterminal '**b**' can be replaced by either '**c:1**' or '**c:3**'.
c. The nonterminal '**c**' can be replaced by either '**1:2**' or '**2:3**'.
If we therefore start with a symbol '**a**' we can expand it first to, say, '**b:1**'. We can then further expand the symbol '**b**' to obtain '**c:3:1**' and finally expand the '**c**' to obtain the final sequence '**2:3:3:1**'. This is obviously but one of the possible sequences derivable from these rules. The sequence of rules need not be entered from the symbol '**a**'; that is other rules may involve '**a**', '**b**' or '**c**'. The basic notion here is that the application of the series of rules finally results in a sequence of symbols and we may have observed only some portion of that sequence in a particular example.

2. It may be a single rule which does not fit into the previous category because it contains somewhere on its right-hand side a production involving two non-terminal symbols.

z:: = >a:b | a:1 | 2:b (3)

is thus another type of derivation sequence.

3. Rules where symbols from the left-hand side occur on the right-hand side, giving *recursive* definitions, are sometimes found although our inference procedure is designed only to identify a few kinds of possible recursion. These are also regarded as derivation sets. For example

a:: = >1:2 | 1:a (4)

The unique rule which initiates the grammar and involves the symbol '**S**' representing an entire sentence is always regarded as a separate derivation sequence.

For the primary successions in salt marsh, classes were already available from the previous studies. Thus for these data sets both the establishment of the alphabet of symbols and the generation of the symbol strings were relatively easy. The interest here lies with the differences between grammars at different scales of sampling, and the contrast of primary succession with the previous secondary succession in the rain forest.

With the spatial transect data, there are more difficult problems. First, there is no obvious directionality to the transects. We have chosen to determine this by proceeding from the shoreline into the wetland but there is no necessity about such a choice. In fact some transects had been terminated because the water depth was too great for data collection. However, in some shallow wetlands the transects traversed the water body, from shore to shore, which would lead to the possibility of palindromic structures in the patterns of symbols. Thus if there is a strong zonation of types we might expect a string of the form:

A B C B A (5)

In fact this is a symptom of the more general problem that we could expect the spatial pattern to be better represented by a parallel grammar. Inference of parallel grammars has not proceeded far. The difficulty is not that inference is impossible; it is certainly possible for many cases. However, few effective algorithms have yet been developed and implemented. Jürgensen & Lindenmayer (1986) have very recently developed a working method for one class of L-systems, but this implies that all changes are synchronised which is clearly unacceptable.

If a serial grammar is used to simulate a parallel grammar we can expect extra complexity, for even if an underlying parallel grammar is essentially context-free, (see below) a serial grammar may need to be context-sensitive in order to simulate the synchronisation implicit in the parallel changes. The simplest form of grammar with which we shall be concerned is a *regular* grammar, where every alternative of the right-hand side of a rule contains at most 1 non-terminal symbol. Next we distinguish *context-free* grammars, where the left-hand side of any rule contains a single non-terminal symbol, from context-sensitive grammars where the left-hand side of a rule can contain several symbols some of which define the context in which the rule may be applied. Thus the role

a: : = >**b**:**c** (5)

is context-free whereas the rule

z a x: : = >**z b**:**c x** (6)

is *context-sensitive*, for it is taken to mean that the symbol **a** can be replaced by **b:c** only if it is preceded by symbol **z** and followed by symbol **x**. Context-free grammars incorporate hierarchical structure while context-sensitive grammars, as their name implies, additionally import context into the processes of change. It has also been shown for context-sensitive grammars (Klein *et al.* 1984) that neither directionality of context nor extension to random contexts modify the generating power of a grammar, so that our arbitrary determination of directionality may not be overly important. In general though, the class of context-sensitive grammars is not inferable, or rather there are an infinite number of alternatives, and little is known concerning which subclasses can be uniquely inferred, and under what conditions. More complex grammars with rules permitting erasure of symbols will also be ignored, although it is worth noting in passing that there is always a context-free grammar with added erasing rules which can simulate any specified context sensitive grammar.

We might equally consider using two-dimensional arrays made up of the several transects from each wetland. There are proposals for the inference of so-called *tree-grammars* for pictures, which are regarded as two-dimensional symbol patterns, but such grammars would seem to require more extensive and regular sampling within each wetland than is available in the present dataset. The work of Jensén & van der Maarel (1980) would provide such two-dimensional data for each wetland, though, and would thus permit the use of such tree grammars.

One further distinction which requires comment is that of *programmed* grammars (Lu & Fu 1984). In many grammars the choice of which rule to apply is deliberately made nondeterministic, that is essentially left to choice or chance. In programmed grammars each rule specifies the changes permitted and additionally specifies which other rules may be used immediately after it. These grammars in context-free form are capable of simulating certain aspects of context sensitivity, while remaining inferable. The notion that the previous change can control what following changes are allowed seems ecologically reasonable and deserves further attention. Here we have used an implicit rule for determining the ordering, which is simply take the nearest following rule which is applicable.

To provide added difficulties, the initial classification of the wetlands quadrats proved extremely tenuous because the sample quadrats are largely species-poor, often monospecific, and the required number of classes would seem to be very large. Attempts to use polythetic methods gave, if anything, worse grouping than monothetic methods and the use of alternative descriptors, such as dominance or life form did not alleviate the poverty of entries. We have, for the purposes of experiment, chosen to work with 12 classes, but this should probably be increased. In hindsight it would also have been desirable to have a greater density of samples in each transect so that the sequence of vegetation is better defined.

Our difficulties are not yet over, for the grammar inference procedure will be sensitive to the order of presentation of the sequences. We are trying to learn the grammar, and the occurrence of certain structures in the sequences first presented may lead to the inference of rules which are inappropriate later. Inappropriate here means that they block the formation of other, simpler, rules and lead to the development of overly complex grammars. We have experimented with varying the order of presentation, with the secondary successional data, from the rain forest, although with a fixed alphabet. The results are presented later.

One further possibility should be mentioned. We could seek simply to classify the transects, which is possible provided only that a suitable similarity measure can be found. We could use also the several transects in each wetland as replicates and avail ourselves of the Sandland & Young (1979) test to determine the number of groups. If we should attempt to infer parallel grammars, then some such approach is probably necessary in order to systematise the order of presentation of the transects. Of course with temporal data the order of presentation is fixed.

Grammar inference procedures

Depending on the class of grammar which is thought appropriate, various inference procedures are available in the literature. Most of these have been applied only to toy problems, while some require considerable human intervention. In order to simplify our first attempts, we have chosen to seek a serial grammar, specifically a context-free grammar. The alternatives of parallel grammars have been noted in the previous paragraph. In choosing to seek a context-free grammar, we are allowing the possibility that we shall find only a regular grammar, which is a less complex generating system. The more general class of context-sensitive grammars is, in the general case, not uniquely inferable. This is not to say that specific kinds of context-sensitive grammars cannot be inferred, for context-free grammars are themselves one such subset. However, it is not at all obvious what the desired form of context-sensitivity would be, so that we can reject context-sensitive grammars in this preliminary study. As we noted earlier, a context-sensitive serial grammar may be necessary even if an equivalent parallel grammar is context-free. We have chosen to initially emphasise deterministic grammars. *Probabilistic* grammars, where each alternative in a rule is assigned some probability, might have simplified the description somewhat, in that improbable alternatives could have been neglected.

There still remains the choice of inference methods. Various suggestions have been made in the literature, for example Cook & Rosenfeld (1976) and Liou & Dubes (1977). We chose to use the Chirathamjaree & Ackoryd's (1980) method with some few modifications. The original paper has a large number of minor typographical errors and the proposed algorithm can be made much more space-efficient than seems apparent at first glance. Thus the method can be applied to large sets of sequences, whereas most other methods described in the literature are restricted to small problems. In any case the procedure would form a good starting point for an optimisation of the grammar. We do not claim that the grammars we infer are in fact optimal. Improvements can certainly be made which would probably result in smaller and thus more efficient grammars. The biggest merit of the method used is speed; this is probably the largest grammar inference exercise ever undertaken. For the stochastic regular grammar, derived only for the wetlands data, we used the method of Richetin & Venadata (1984). Both inference procedures are incorporated into the TAXON classification package (Ross *et al.* 1983).

As we have just noted, the manner in which a grammar will present interesting information is not obvious. Since changing the number of classes, and hence symbols, must change the grammar somewhat, simple rules will probably not be very interesting. We have concentrated on three other features as being potentially more useful. Two of these are concerned with the specific form of the individual rules, and in our later examples occurrences of these features are marked by '!!!'. The first feature is the presence of recursively defined rules, that is of rules defined in part using self reference. This can provide considerable reduction in the number of rules necessary to describe a particular body of data. Where the recursion simply implies repetition of some terminal symbol, as in the example

a::=>1 | 1:a (7)

it is not overly interesting. If the rule involves more complex components, as when the symbol **'a'** represents some more elaborate construct, then recursion does become interesting. In this context it should be noted that the inference procedure used is not designed to identify complex recursive properties, so our results may underestimate the potential recursiveness of the data. Thus rules of the form

$$\mathbf{a}::=>\mathbf{b}{:}1 \mid \mathbf{b}::=>\mathbf{a}{:}2 \mid 2 \tag{8}$$

with mutual recursion, are unlikely to be found using the inference procedure adopted.

The second feature is the occurrence of rules where the right-hand side contains only *nonterminal* symbols. We shall call these context-free rules since they can only occur in grammars of at least that complexity. Such nonterminal symbols represent combinations of classes and permit hierarchical decomposition of structure. Recursion represents a repeating pattern in the vegetation classes, but context-free rules indicate concatenation of patterns.

The third feature of interest is the number of derivation sets, defined earlier. If we start with a long sequence, then it is possible that several other observed sequences will be substrings of it, so that the same sequence of rules, entered at different initial positions will generate several sequences. For example, the several rules:

$$\mathbf{a}::=>\mathbf{b}{:}4;\ \mathbf{b}::=>\mathbf{c}{:}5;\ \mathbf{c}::=>3{:}2 \tag{9}$$

form a single sequence, which might be entered at **a**, **b**, or **c**, but follow the same continuation path once entered. Such structure could be the result of different rates of development, with our observations recording different stages along the same path. What is critical is less the precise nature of the path than the number of such paths which occur. In presenting the results, we have grouped rules into derivation sequences, with the individual rules separated by semicolons (;). Where the rule has some special significance we have marked it with !!!. In all cases terminal symbols are represented by *digits*, nonterminals are *alphabetic symbols*.

Results

The first, and most important distinction to be made, is between the temporal sequences and the spatial sequences. In all cases the temporal sequences were adequately described by a regular grammar, *without context-free rules*. This suggests that they can be adequately modelled by stochastic finite state automata, as Hogeweg *et al.* (1985) attempted. This conclusion, if substantiated by further studies, is of considerable interest for these are mathematically tractable models about which a considerable theory is available.

The spatial patterns, in contrast, are much less well structured and seem to require context-free rules. This may of course simply reflect the underlying need for a parallel grammar in order to cope with spatial processes, or alternatively could indicate the existence of complex structures such as palindromic sequences. These latter involve patterns of the form ABCBA, and might occur if the sequence of types follows some repetitive sequence. Transects within an individual wetland usually have some common structure, and where they differ there is often convincing evidence of environmental variation; for example influence of tidal flow and associated salinity changes, or influx of moving fresh water. But each wetland is largely independent, and very little structure can be gleaned which is general to a class of wetlands. Structure common to a class of wetlands was remarkable by its absence at least on preliminary examination. The interpretation of this result we shall leave until later.

Turning to the specific results, those for the secondary succession in the rain forest are particularly clear. Table 3 shows the results using alphabets of between 2 and 15 symbols, and also some examples showing the effects of changing the order of presentation of the sequences, with a fixed, 6 group, alphabet. In almost all cases there are, apart from the initial sentential rule, only two derivation paths, which can both be recognised from the previous study. One leads to rain forest, the other to the impeded stage dominated by *Lantana camara*. If we take the 6-group permutation 1 result as an example (Fig. 1) we can see that the sequences may be entered at 7 places, 4 in one sequence and 3 in the other. Presumably this variability in the initial vegetation types reflects in part the seed bank and regrowth from remnant trees. There is no recursion of importance and significantly no context-free rules are needed. Thus we can interpret the result as showing the existence of two distinct seral trends, which do not interconnect, at least during the period over which our data were collected. In effect there are two distinct grammars, which do not interact.

There are some few exceptions. One 6-group result has 3 derivation paths presumably because the infer-

Table 3. Mount Glorious rain forest data: inferred grammars for various numbers of vegetation types. In all grammars terminal symbols are represented by numbers, nonterminal symbols by letters. The symbol ':: = >' represents 'may be written as'. The symbol ' |' separates alternatives; **a** :: = > | **2** is read as '**a**' may be replaced by **1** or **2**. The symbol ';' indicates the end of a rule. The symbols '( )' are used when a terminal symbol is defined recursively. Thus **(1)** represents a string of **1**'s whose length is indefinite. The symbol '**S**' represents the starting symbol. This symbol is always on its own line. Rules on the same line, or indented represent a single derivation sequence. Thus the number of lines is indicative of the complexity of the rule structure. In most of the examples in this table only 3 lines are present.

Terminal alphabet : **2 group symbols**

```
S::=> a | b
a::=> c : 2 ; c::=> 2 : 1
b::=> 1 : 2
```

Terminal alphabet : **3 group symbols**

```
S::=> a | b | c | d
a::=> b : 2 ; b::=> e : 3 ; e::=> 3 : 1
c::=> d : 2 ; d::=> 1 : 3
```

Terminal alphabet : **4 group symbols**

```
S::=> a | b | c | d
a::=> b : 3 ; b::=> c : 4 | 3 : c ; c ::=> e : 3 | e : 2 ; e::=> g : 4 ;
      g::=> 4 : 1
d::=> f : 2 | f : 3 ; f::=> 1 : 4
```

Terminal alphabet : **5 group symbols**

```
S::=> a | b | c | d | e
a::=> b : 4 ; b::=> c : 5 | 4 : c ; c::=> f : 4 | f : 3 ; f::=> g : 5 ;
      g::=> 5 : 2 | 5 : 1
e::=> 5 : d ; d::=> h : 4 ; h::=> i : 3 | i : 5 ; i::=> 1 : 5 | 1 : 2
```

Terminal alphabet : **6 group symbols, permutation 1**

```
S::=> a | b | c | d | e | f | g
a::=> b : 4 ; b::=> c : 5 ; c::=> g : 4 | 5 : g ; g::=> h : 3 | h : 6 ;
      h::=> 1 : 6 | 1 : 2
d::=> e : 4 ; e::=> f : 6 | f : 3 | 4 : f ; f::=> i : 3 | i : 4 | i : 6 ;
      i::=> j : 6 | j : 2 ; j::=> 5 : 2 | 5 : 1 | 5 : 6
```

Terminal alphabet : **6 group symbols, permutation 2**

```
S::=> a | b | c | d | e
a::=> b : 4 ; b::=> c : 3 | c : 6 | 4 : c ; c::=> f : 3 | f : 4 | f : 6;
     f::=> h : 2 | h : 6 ; h::=> 5 : 6 | 5 : 2 | 5 : 1
d::=> e : 4 ; e::=> g : 3 | g : 6 ; g::=> 1 : 6 | 1 : 2
```

Terminal alphabet : **6 group symbols, permutation 3**

```
S::=> a | b | c | d | e
a::=> f : 4 ; f::=> b : 5 ; b::=> g : 4 | 5 : g;
     g::=> i : 3 | i : 6 ; i::=> 1 : 6 | 1 : 2
c::=> d : 4 ; d::=> e : 3 | e : 6 | 4 : e ; e::=> h : 3 | h : 4 | h : 6 ;
      h::=> j : 6 | j : 2 ; j::=> 5 : 2 | 5 : 6
```


Terminal alphabet : **6 group symbols, permutation 4**

```
S::=> a | b | c | d | e
a::=> b : 4 ; b::=> c : 3 | c : 5 | c : 6 | 4 : c ; c::=> f : 4 | f : 6;
      f::=> h : 2 | h : 6 ; h::=> 5 : 1 | 5 : 2 | 5 : 6
d::=> e : 4 ; e::=> g : 3 | g : 6 ; g::=> 1 : 2 | 1 : 6
```

Terminal alphabet : **6 group symbols, permutation 5**

```
S::=> a | b | c | d | f | h
a::=> b : 4 ; b::=> c : 3 | c : 6 | 4 : c ; c::=> g : 3 | g : 4 | g : 6 ;
      g::=> i : 2 | i : 6; i::=> 5 : 1 | 5 : 2 | 5 : 6
d::=> e : 4 ; e::=> f : 5 ; f::=> h : 4 | 5 : h ; h::=> j : 3 | j : 6 ;
      j: => 1 : 2 | 1 : 6
```

Terminal alphabet : **6 group symbols permutation 6**

```
S::=> a | b | c | d | e | f
a::=> b : 4 ; b::=> c : 3 | c : 5 | c : 6 ; c::=> g : 3 | g : 4 | g : 6 ;
     g::=> j : 2 | j : 6 ; j: => 5 : 1 | 5 : 2 | 5 : 6
d::=> h : 4 ; h::=> k : 6 ; k::=> m : 2 ; m::=> 4 : 5
e::=> f : 4 ; f::=> i : 3 | i : 6 ; i::=> 1 : 2 | 1 : 6
```

Terminal alphabet : **10 group symbols**

```
S::=> a | b | c | d | e
a::=> f : 6 ; f::=> b : 8 ; b::=> c : 4 | 8 : c ; c::=> g : 3 | g : 10 ;
      g::=> 1 : 10 | 1 : 2
d::=> h : 3 | h : 5 | h : 6 ; h::=> e : 7 | e : 9 | e : 10;
      e::=> i : 3 | i : 5 | i : 7 | i : 9 | i : 10 : 7 : i ;
      i::=> j : 2 | j : 9 | j : 10 ; j::=> 8 : 1 | 8 : 2 | 8 : 4 | 8 : 10
```

Terminal alphabet : **15 group symbols**

```
S::=> a | b | c | d | e | f | g
a::=> h : 7 ; h::=> b : 12 ; b::=> c : 13 | c : 15 ;
    c::=> i : 4 | i : 15 | 12 : i; i::=> 1 : 3 | 1 : 14
d::=> j : 7 ; j::=> k : 9 ; k::=> m : 14 ; m::=> n : 3 ; n::=> 8 : 12
e::=> v : 4 ; v::=> f : 13 | f : 15 ; f::=> p : 4 | p : 8 | p : 13 | p : 15;
    p::=> q : 3 | q : 13 | q : 14 | q : 15 ; q::=> 11 : 1 | 11 : 3 | 11 : 15
g::=> r : 6 ; r::=> s : 15 ; s::=> t : 6 ; t::=> u : 14 ; u::=> 10 : 2
```

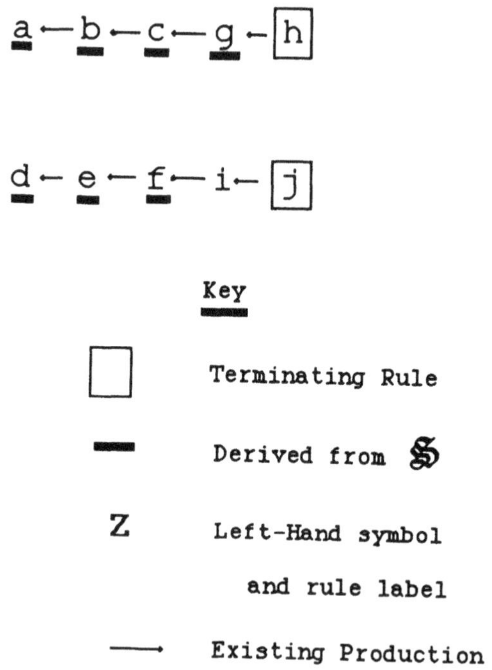

Fig. 1. Derivation set for rain forest data 6-group permutation 1. Key included.

ence procedure is inadequate and cannot 'unlearn' a rule structure. The 15-group result has 4 derivation sequences. Since the temporal sequences were only 10 units long, with 15 groups it is impossible for all groups to occur in every sequence, and the extra derivation paths are necessary to incorporate the resulting complexity.

For the primary successions, the results are presented in Table 4, where again no context-free rules are required. In both analyses symbol 6 appears to be some sort of terminal state, as far as our data go. However, the individual plots obviously proceed at very different rates towards this 'goal', with some plots never changing at all! Excluding these invariant stands, both datasets identically suggest the same *two* derivation sequences (Fig. 2). The longer 'abcd' set usually starts with vegetation type 2, although type 1 can occur initially, the shorter 'efg' set starts with vegetation type 1 only. Since both sequences can start with the same terminal symbol, the differentiation must be made in terms of earlier stages in the process of inference. The impression is of two routes converging to the same (climax?) type. However the derivation sets have captured the simplicity of the succession rather elegantly.

The wetland data are much more complicated. In Table 2 we showed the actual sequences obtained at the 12 group level, recognising that the 12th group is still rather heterogeneous. Table 5 contains the rather massive context-free grammar which was inferred, while Table 6 shows a stochastic regular grammar for the same data. In Fig. 3 we show the derivation sequences for the context free grammar.

Table 4. Dutch islands primary succession data inferred context-free grammars. Format as Table 3.

```
Data set 1.
  S::=> (1) | a | c | d | e
  a::=> b : 6 ; b::=> c : 3 ; c::=> d : 2 | d : 4 | d : 5 | d : 6 | d : 8 | 1 : d;
      d::=> 2 : 1 | 2 : 3 | 2 : 7
  e::=> f : 6 ; f::=> g : 3 ; g::=> 1 : 2
Data set 2.
  S::=> (1) | a | c | e | f
  a::=> b : 6 ; b::=> c : 4 ; c::=> d : 2 | d : 5 | d : 6 | d : 7 ;
      d::=> 2 : 1 | 2 : 3 | 2 : 4 | 2 : 5
  e::=> f : 6 ; f::=> g : 4 | g : 6 ; g::=> 1 : 2
```

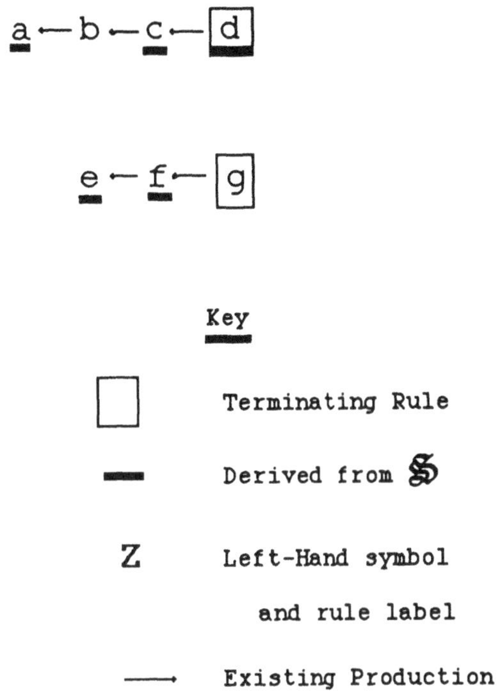

Fig. 2. Derivation set for primary succession. Key included.

Looking first at the grammar, there are a few context-free rules and a few significant recursions, but the result is dominated by the large number of rules needed. In fact there are slightly more rules than wetlands! Closer inspection reveals that each wetland tends to have its own set of rules, which apply to transects from that lake only. However, there are a few aberrant transects, due to such causes as the influx of fresh water or saline or tidal influences so that a few additional rules are needed. There are some few wetlands which are somewhat alike and thus share some rules, but the overall impression is certainly one of independent vegetation processes in each wetland. This might be expected for in effect lakes are inverted islands, so that all the dynamism of the theories of colonisation and extinction applicable to islands would be equally applicable to lakes. Indeed the situation is probably more dynamic since the lakes themselves may be ephemeral, or show widely fluctuating water levels so that the environment is in continuous flux.

There are also cases where parts of transects are repeated in several transects. Rule **V**, for example,

Table 5. Victorian wetlands data: inferred context-free grammar. Format as Table 3. Note that the sizes of symbols are used to differentiate certain nonterminals.

```
S::=> (3) | (4) | (6) | (7) | (8) | (9) | (10) | (12) |
    a | b | c | f | g | j | n | p | q | r | t | v | x | y | z |
    A | B | D | E | G | H | I | J | K | L | M | N | P | Q | R | S | T | U | V | W | Y |
    z | A | B | D | E | F

a::=> T : R   !!!

b::=> F : B | B : R   !!!

c::=> d : 12 ; d::=> e : 11 ; e::=> 11 : 12
f::=> 4 : 6 | 4 : 12
g::=> h : x   !!!

h::=> i : 5 ; i::=> 8 : V ; V ::=> 10 | 10 : V   !!!

j::=> k : 12 ; k::=> l : 1 ; l::=> m : 4 ; m::=> n : Y   !!!
n::=> o : 1 | o : 3 ; o::=> 1 : 4 | 1 : 12
p::=> q : 7 ; q::=> S : 5 | S : 12 ; S::=> T : 6 | T : 7 | T : 8 | 8 : T ;
T::=> U : 7 | U : 10 | 12 : U ; U::=> 7 : 6 | 7 : 8 | 7 : 11

r::=> s : 7 ; s::=> z : 12 ; z::=> D : 5 | D : 7 | D : 12 | 8 : D
t::=> u : 4 ; u::=> M : 5
v::=> w : 4 ; w::=> M : x   !!!
x::=> H : 6 ; H::=> 6 : 2 | 6 : 4 | 6 : 7 | 6 : 12
y::=> G : 12 | 8 : G ; G::=> Q : V | V : Q   !!!
I::=> W : 2 | W : 12 | 3 : W
A::=> R : V | R : R   !!!

B::=> C : 6 ; C::=> D : Z ; Z::=> 6 | 6 : Z   !!!

D::=> A : 3 | A : 6

E::=> F : 6 ; F::=> I : B   !!!
J::=> W : V ; K::=> L : 12 ; L::=> M : V | M : M   !!!
M::=> 10 : 3 | 10 : 4 | 10 : 7 | 10 : 8 | 10 : 12
N::=> O : 12 ; O::=> V : 6
P::=> Q : 4 | Q : 5 | 2 : Q | 4 : Q | Q : W   !!!
Q::=> R : 8 | R : 12 | 5 : R ; R::=> 12 : 1 | 12 : 4 | 12 : 6 | 12 : 7 | 12 : 9
V::=> 5 : 6 | 5 : 8   !!!

            What comes after the classification
W::=> X : 2 | X : 5 | X : 6 | X : 12 ; X::=> Y : 2 | Y : 5 | 6 : Y ;
    Y::=> 2 : 1 | 2 : 5 | 2 : 6

Z::=> C : C   !!! ; C::=> 3 : 1 | 3 : 2 | 3 : 4 | 3 : 5 | 3 : 7 | 3 : 12

A ::=> B : 1 | B : 4 | B : 12 ;

    B::=> C : 3 | C : 8 | C : 10 | C : 12 | 1 : C | 2 : C;


D::=> E : 1 | E: 3 | E : 8 | E : 10

    E::=> F: 4 | F : 5 | F : 8 | 5 : F | 12 : F ;

    F::=> 8 : 3 | 8 : 4 | 8 : 5 | 8 : 12
```

Table 6. Victorian wetlands data: inferred stochastic regular grammar. Probabilities of transitions between the types are indicated in parenthesis.

```
S::=> a (.03) | b (.1) | c (.08) | d (.05) | e (.06) | f (.15) | g (0.07) | h (.16) | i (.02) |
    j (.09) | k (.01) | l (.19)
a::=> 1 (0.2) | 1 : a (.77) | 1 : b (.04) | 1 : c (.04) | 1 : d (.04) | 1 : f (.02) | 1 : l (06)
b::=> 2 (.02) | 2 : a (.07) | 2 : b (.43) | 2 : c (.13) | 2 : e (.07) | 2 : f (.22) | 2 : l (.06)
c::=> 3 (.08) | 3 : a (.01) | 3 : b (.10) | 3 : c (.66) | 3 : d (.04) | 3 : e (.04) | 3 : g (.03) |
    3 : l (.04)
d::=> 4 (.13) | 4 : a (.02) | 4 : d (.73) | 4 : f (.04) | 4 : h (.03) | 4 : l (.05)
e::=> 5 (.05) | 5 : b (.05) | 5 : c (.02) | 5 : d (.02) | 5 : e (.44) | 5 : f (.24) | 5 : g (.02) |
    5 : h (.09) | 5 : l (.07)
f::=> 6 (.14) | 6 : b (.02) | 6 : c (.01) | 6 : d (.03) | 6 : e (.05) | 6 : f (.62) | 6 : g (.03) |
    6 : h (.01) | 6 : l (.09)
g::=> g (.11) | 7 : e (.01) | 7 : f (.07) | 7 : g (.73) | 7 : h (.03) | 7 : j (.01) | 7 : k (.01) |
    7 : l (.03)
h::=> 8 (.10) | 8 : c (.03) | 8 : d (.04) | 8 : e (.10) | 8 : f (.01) | 8 : g (.03) | 8 : h (.53) |
    8 : j (.03) | 8 : l (.14)
i::=> 9 (.10) | 9 : i (.87) | 9 : l (.03)
j::=> 10 (.11) | 10 : c (.07) | 10 : d (.04) | 10 : g (.07) | 10 : h (.07) | 10 : j (.51) |
    10 : l (.11)
k::=> 11 : g (.09 | 11 : k (.73) | 11 : l (.18)
l::=> 12 (.20) | 12 : a (.01) | 12 : c (.02) | 12 : d (.05) | 12 : f (.10) | 12 : g (.05) |
    12 : h (.03) | 12 : i (.01) | 12 : j (.01) | 12 : k (.01) | 12 : l (.49)
```

terminates several alternative pathways, starting variously at **y**, **A**, **J**, **K** and **N** as well as at **V** itself. There are also some recursive rules involving nonterminals, such as **V**, indicating the existence of repetitive structures. There seem to be no palindromic structures; these might be expected if a regular zonation pattern existed and the transect crossed the entire wetland area from side to side. Thus the recursive or context-free rules do not, of themselves provide much interesting information. It is therefore necessary to examine the derivation sets.

Given this superfluity of rules, it is perhaps surprising that the derivation sets can be reasonably neatly arranged. In the figure the terminating rules have been placed so that the lines indicating the derivations do not cross. Remember that these will usually represent the initial parts of a sequence, and hence the vegetation nearest the shore. Of the derivation sets 8 represent transects with no variation in vegetation type. (These are the numeric terminating rules at the bottom of the diagram.) Type 9 vegetation, essentially *Sphagnum*-defined, is found only in one of these invariant sets. There are two derivation sets disconnected from the majority, with terminating rules **e** and **f**; the former is associated with the presence of *Hydrocotyle,* the latter with *Azolla.* Both are relevant to very short transects only.

This leaves 11 terminating rules, which have been arranged so that, in general, context-free rules which

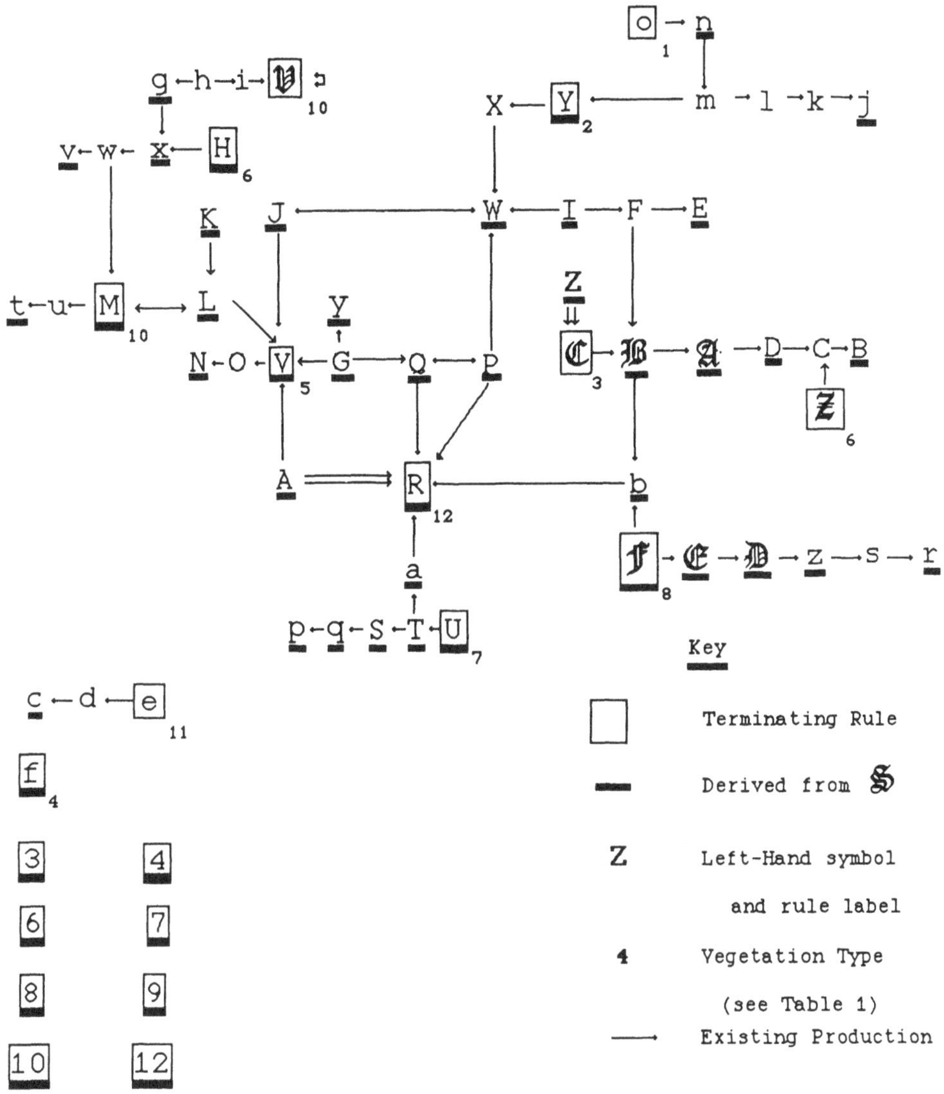

Fig. 3. Derivation sets wetland data. Key included. Nonterminals represented by larger fonts in Table 7 are here represented by Gothic script.

cross-connect the derivation sets join adjacent sets. Only five rules break this arrangement; **P**, **J**, **F**, **C**, and **b**. The residual group associated with rule **R** seems to be largely involved in these complexities, an indication perhaps that our initial classification was insufficiently detailed. However rule **V** also shows much cross-connection, for which at present we have no adequate explanation. The general arrangement of connections suggests some sort of interplay between *Potamogeton* (**C**) and *Myriophyllum* (**F**) types, alone or in combination with *Triglochin* types (**V** & **Y**). Indeed the diagram seems to distinguish emergent vegetation types (largely placed peripherally) from floating-leaved and submerged types. It is as if the wetlands exhibit a tri-polar structure with these three types at the apices and various combinations between, although this is disturbed somewhat by the residual type.

Of the rules pertaining to the remaining types, *Eleocharis* (**o**) is connected to the *Potamogeton/Triglochin* type, *Baumea* (**U**) to the residual types and *Lepidosperma* (**M** & **V**) with *Triglochin* or

Triglochin/Myriophyllum types. Types **H** and **Z**, representing 'pure' *Triglochin* types, appear in widely separated positions suggesting that this type may have at least two different roles, one as a peripheral zone in wetlands which have sufficiently deep water to contain *Potamogeton* or *Myriophyllum*, the other in situations which are marsh-like rather than lake-like associated with various *Lepidosperma* types. The reason for the duplication of this last is not apparent, although **V** is *Lepidosperma* unmixed with any other type, whereas **M** represents more complex mixtures.

Besides the context-free grammar, we also experimented with a stochastic regular grammar, with the results shown in Table 6. In fact the 'typical' transect seems to be lacking in variability; in many cases a single type is predominant throughout the transect. This means that the probability of changing between vegetation types seems to be rather small. This is a somewhat surprising result since it suggests either that each wetland in fact represents predominantly a single vegetation type or that extended areas are of a single type with occasional transitions. This does *not* mean that the wetlands cannot be classified. Provided that a suitable similarity measure can be found, classification can be made numerically and can be usefully employed in relating vegetational variation to environmental differences. For the temporal data we might use the suggestions of Yassouridis & Hansert (1986). As an example, we have here classified the spatial transects using a Levenshtein measure of dissimilarity (Levenshtein 1966; Little & Ross 1985; Dale 1988) with the results shown in Table 7. For each group we also indicate the major type(s) associated with the transects included in the group, as far as these could be determined visually. It is important to note that in many cases, in fact 17 of 36, all transects from a single wetland site are found in the same group, and of the remaining sites, many pairs of transects are similarly allocated. So classification, or ordination, can be accomplished if this is desired.

The results obtained here from the grammar inference procedures strongly suggest that each wetland has its own history and pattern of vegetation which is not very closely related to other wetlands. This is perhaps not surprising given their 'inverted island' nature. We have not accounted for seasonal variation, which for some of the wetlands results in highly variable water depths. In at least one case a 'wetland' had vanished during the period between selection and the time when sampling was attempted. Thus some of the differentiation may reflect recent environmental fluctuation. However the derivation sets do seem to provide a means of organising the mass of rules into coherent and sensible patterns. We are certainly not satisfied with the level of interpretation we have so far obtained. However it would seem that work on the interactions between the three dominant species, *Potamogeton, Myriophyllum* and *Triglochin,* would serve to clarify the relationships between many of the wetlands.

Table 7. Victorian wetlands data: classification using Levenshtein distances showing typical vegetation types.

Predominantly quadrat group 4: *Azolla*
1; 2; 3; 16; 17; 18; 39; 40; 41; 50; 66; 68; 72; 73; 74;
Predominantly quadrat group 6: *Triglochin* without *Myriophyllum*
4; 5; 6; 27; 28; 36; 51; 53; 59; 62; 87; 102; 103; 104
Quadrat groups 2 and 5: *Potamogeton* + *Triglochin, Myriophyllum* + *Triglochin*
12; 13; 14; 15; 30; 34; 37; 38; 57; 58; 63; 64; 90; 92
Quadrat groups 8 and 12: *Myriophyllum,* Residual
9; 10; 11; 19; 20; 21; 22; 23; 29; 31; 33; 35; 52; 61; 65; 67; 69; 75; 76; 77; 81; 86; 88; 89
Quadrat groups 1 and 3: *Potamogeton* + *Eleocharis, Potamogeton*
32; 42; 49; 54; 55; 56; 70; 71; 93; 94; 95; 101
Quadrat group 10: *Lepidosperma*
43; 44
Predominantly quadrat group 9: *Sphagnum*
96; 97; 98
Predominantly quadrat group 8: *Myriophyllum*
7; 8; 48; 60; 91
Quadrat groups 7 and 11: *Baumea, Hydrocotyle*
24; 25; 26; 45; 46; 47; 78; 79; 80; 82; 83; 84; 85; 99; 100

Conclusions

Our initial studies of grammatical inference procedures would suggest that successional data do not require an elaborate generating mechanism. In all three cases examined, once allowance has been made for possible misclassification in the alphabet determination, a regular grammar is sufficient. Certainly this should be probabilistic, but it does not seem that complex models are appropriate. Indeed the derivation sets provide an elegant method of determining the paths of succession.

In contrast, the spatial data provide many difficulties. It seems likely that some parallel grammar is desirable and might provide some simplification, but it seems that each wetland is essentially a unique processing device; each site acts as an island in a 'sea' of dry land. Presumably this reflects both chance and historical factors, together with the very important principle of 'what is mine I hold'. Add to this the expected variability due to seasonal fluctuations and the various rates of recovery of elements of the vegetation after such changes and the situation becomes one of great complexity. For all that the derivation sets do seem to provide an interesting means of organising this profusion and suggest some possibilities for further study. It is certainly possible to classify the wetlands and this could provide useful comparative information. But it does not lead to any great specificity in identifying the processes which operate within any single wetland. As a first exercise in the inference of grammars for vegetation data, the results seem to be interesting. There does not seem to be any great problem with the initial classification to form the alphabet. We have identified some properties of the grammars as important, and these seem robust to any misclassification, though the aberrant 6-group result may reflect some interference. Order of presentation of sequences does seem to be of some importance, but several different orders could be tried and the simplest grammar accepted. Furthermore the elegance of the derivation sets for the temporal data suggests some interesting possibilities for classification.

We shall assume that a group is not defined by some notion of a typical member, but by a grammar of some form. Thus members of a group are simply those sequences which can be legitimately derived using the grammar. Both the rain-forest data and the primary succession data here show two groups each associated with its own derivation set. It seems likely that if the samples from highly saline wetlands had been included they too would have produced a very simple, but disjoint, derivation set and hence would be regarded as providing a group defined by its own grammar. However there is no specific limit on the complexity, nor is there any assumption that differences between members of the same group are due to random causes.

To assign new members we would first attempt to parse them to determine which, if any, of the grammars could have generated them. If such a grammar exists then the new item can be assigned to its associated group. If no such grammar exists, then we can either introduce error-correcting rules and seek the grammar which is closest to being acceptable (closest in the sense of requiring the least number of errors to be corrected), or if we have stochastic grammars then we can attempt to define some sort of probability associated with possible new rules and hence determine the probability that a slightly modified grammar could have produced this sequence. If the new item is too different then we should start a search for a new grammar for it alone.

Such a view of group structure and classification is rather different to the usual notions of clustering with their emphasis on the typical item and measuring distances from it. It is therefore interesting that the temporal data sets used here suggest that such a procedure may well be computationally feasible. We are now actively investigating this possibility.

References

Aiello, A., Burattini, E., Caianiello, E. R. & Masarotti, A. 1973. An algorithm for feature classification and structural memorization. Kybernetik 12: 145–153.

Angluin, D. 1980. Finding patterns common to a set of strings. J. Comput. Syst. Stud. 21: 46–62.

Barson, M. M. 1984. Numerical analysis of macrophyte vegetation in Victorian wetlands. Ph.D. Thesis, University of Melbourne.

Bell, A. 1976. Computerized vegetative mobility in rhizomatous plants. In: Lindenmayer, A. & Rozenberg, G. (eds), Automata, development, languages, pp. 3–14. North Holland, Amsterdam.

Chirathamjaree, C. & Ackroyd, M. H. 1980. A method for the inference of non-recursive context-free grammars. Internl. J. Man-Machine Stud. 12: 379 – 387.

Cook, C. M. & Rosenfeld, A. 1976. Some experiments in grammatical inference. In: Simon, J. C. (ed.), Computer oriented learning processes. pp. 157 – 174.

Dale, M. B. 1979. On linguistic approaches to ecosystems and their classification. In: Orlóci, L., Rao, G. R. & Stiteler, W. M. (eds.), Statistical ecology 7: 11 – 20. Internatl. Coop. Publishing House, Maryland.

Dale, M. B. 1981. A grammatical approach to vegetation classification. In: Gillison, A. N. & Anderson, D. J. (eds), Vegetation classification in Australia, pp. 141 – 149. CSIRO & ANU Press, Canberra.

Dale, M. B. 1989. Similarity measures for structured data: a general framework and some applications to vegetation data. Vegetatio 81: 41–60.

Doucet, P. G. 1974. The syntactic inference problem for DOL sequences. In: Goos, G. & Hartmannis, J. (eds), L-systems. Lecture notes in computer science No. 15, pp. 146 – 161. Springer-Verlag, New York.

Feliciangeli, H. & Herman, G. T. 1973. Algorithms for producing grammars from sample derivations: a common problem of formal language theory and developmental biology. J. Comput. Syst. Sci. 7: 97 – 118.

Fleck, A. C. 1974. An analysis of grammars by their derivation sets. Inform. Control 24: 389 – 398.

Frijters, D. 1978. Principles of simulation of inflorescence development. Ann. Bot. 42: 549 – 560.

Frijters, D. & Lindenmayer, A. 1974. A model for the growth and flowering of *Aster novae-angliae* of the basis of table <1,0> L-systems. In: Goos, G. & Hartmanis, J. (eds), L-systems: Lecture notes in computer science 15, pp. 24 – 52. Springer-Verlag, Berlin.

Goodall, D. W. 1953. Objective methods for the classification of vegetation I. The use of positive interspecific correlation. Austr. J. Bot. 1: 39 – 63.

Haefner, J. W. 1975. Generative grammars that simulate ecosystems. Simulation Council Proc. 5: 189 – 211.

Haefner, J. W. 1978. Ecosystem assembly grammars: generative capacity and empirical adequacy. J. Theor. Biol. 73: 293 – 318.

Haefner, J. W. 1981. Avian community assemblage rules: the foliage gleaning guild. Oecologia 50: 131 – 142.

Hellendoorn, P. H. & Lindenmayer, A. 1974. Phyllotaxis in *Bryophyllum tubiflorum*: morphogenetic studies and computer simulation. Acta Bot. Neerl. 23: 473 – 493.

Hernandez, C. & Arias, J. E. 1984. Syntactic pattern recognition of foetal stress. J. Biomed. Eng. 6: 97 – 101.

Hogeweg, P., Hesper, B., van Schaik, C. P. & Beeftink, W. G. 1985. Patterns in vegetation succession. An ecomorphological study. In: White, J. (ed.), The population structure of vegetation, pp. 673 – 666. W. Junk, Dordrecht.

Jensén, S. & van der Maarel, E. 1980. Numerical approaches to wetland classification with special reference to macrophyte communities. Vegetatio 42: 117 – 128.

Jürgensen, H. & Lindenmayer, A. 1986. Modelling development by OL systems: modelling algorithms for developmental systems using cell lineages. Univ. Western Ontario, Dept. Computer Sci. Rep. 121. 53 pp.

Klein, H. C. M., Penttonen, M., Rozenberg, G. & Salommaa, K. 1984. Direction independent context-sensitive grammars. Inform. & Control 63: 113 – 117.

Lance, G. N. & Williams, W. T. 1968. Note on a new information statistic classification program. Comput. J. 11: 195.

Levenshtein, A. 1966. Binary codes capable of correcting deletions, insertions and reversals. Sov. Phys. Dokl. 10: 707 – 710.

Lindenmayer, A. 1974. Adding continuous components to L-systems. In: Goos, G. & Hartmanis, J. (eds), L-systems: Lecture Notes in Computer Science 15, pp. 52 – 68. Springer-Verlag, Berlin.

Liou, J-I. & Dubes, R. 1977. A constructive method for grammatical inference based on clustering. Research Report, Dept. Computer Science, Michigan State University.

Little, I. P. & Ross, D. R. 1985. The Levenshtein metric, a new means for soil classification tested by data from a sand-podzol chronosequence and evaluated by discriminant analysis. Aust. J. Soil Res. 23: 115 – 130.

Lu, H. R. & Fu, K-S. 1984. Inferability of context-free programmed grammars. Inter. J. Comput. Inform. Sci. 13: 33 – 58.

Mayoh, B. 1973. Mathematical models for cellular organisms. University of Aarhus, Computer Science Dept., Techn. Rep. 12, 38 pp.

Mueller-Dombois, D. & Ellenberg, H. 1974. Aims and methods of vegetation ecology. Wiley, New York.

Niklas, K. J. 1979. Simulation of apical developmental sequences in Bryophytes. Ann. Bot. 44: 339 – 352.

Richetin, M. & Vernadata, F. 1984. Efficient regular grammatical inference for pattern recognition. Pattern Recognition 17: 245 – 250.

Ross, D., Dale, M. B., Shields, K. & Hulett, C. 1983. TAXON Users manual edition P4 Ref. Man. 6. CSIRO Division Computing Res., Canberra.

Sandland R. L. & Young, P. C. 1979. Probabilistic tests and stopping rules associated with hierarchical classification. Austr. J. Ecol. 4: 399 – 406.

Smith, A. M. 1984. Plants, fractals and formal languages. Computer Graphics 18: 1 – 7.

Tai, J. W. & Fu, K-S. 1982. Inference of a class of CFPG by means of semantic rules. Internatl. J. Comput. Inform. Sci. 11: 1 – 23.

van der Maarel, E., Orlóci, L. & Pignatti, S. 1976. Data-processing in phytosociology, retrospect and anticipation. Vegetatio 32: 65 – 72.

van Noordwijk-Puyk, K., Beeftink, W. G. & Hogeweg, P. 1979. Vegetation development on salt-marsh flats after disappearance of the tidal factor. Vegetatio 39: 1 – 13.

Westhoff, V. & van der Maarel, E. 1973. The Braun-Blanquet approach. In: Whittaker, R. H. (ed.), Ordination and classification of communities, pp. 617 – 726. Junk, The Hague.

Westman, R. S. 1977. Environmental languages and the functional bases of animal behaviour. In: Hazlitt, B. A. (ed.), Quantitative methods in the study of animal behaviour, pp. 145 – 201. Academic Press, London.

Williams, W. T. & Lambert, J. M. 1960. Multivariate methods in plant ecology II. The use of an electronic digital computer for Association analysis. J. Ecol. 48: 689 – 710.

Williams, W. T., Lance, G. N., Webb, L. J., Tracey, J. G. & Dale, M. B. 1969. Studies in the numerical analysis of complex rain forest communities III. The analysis of successional data. J. Ecol. 57: 515 – 535.

Yassouridis, A. & Hansert, E. 1986. Equidirection: a measure of similarity among time series. Biom. J. 6: 747 – 758.

Vegetatio 81: 95–106, 1989

A new numerical solution to traditional phytosociological tabular classification

Otto Wildi
Swiss Federal Institute of Forestry Research, 8903 Birmensdorf, Switzerland

Accepted 14.7.1988

Keywords: Discriminant analysis, Forest vegetation, Gradient, Indicator value, Multivariate analysis, Numerical syntaxonomy, Outlier, Switzerland, Tabular sorting

Abstract

Often, manually and numerically derived phytosociological classifications yield different results. Hitherto, a twostep procedure has been suggested in which numerical analysis of the data is followed by the revision of the resulting table (c.f. van der Maarel 1982). In this paper a new methodology is presented which makes manual refinements superfluous. Objectives are derived from phytosociological paradigms and conclusions drawn for the analytical process. The problems to be solved are: data transformation, detection of outliers, selection of clustering methods, checking within-group diversity, analysis of the resulting group structure, rearrangement of relevés and species within the groups, and finally the selection of differential species. The method has been derived using the well known example of Ellenberg (Mueller-Dombois & Ellenberg 1974). The results almost perfectly reproduce the intuitively widely accepted manual refinements in structure and presentation. A test with plant sociological data from Swiss forests (Ellenberg & Klötzli 1972) proves that the method can also classify complex gradient- and group systems and that the numerical result matches Landolt's (1977) system of indicator values. Since the solutions can be exactly reproduced, it is no longer necessary to combine numerical analysis with additional editing.

Nomenclature: Hess, H. E., Landolt, E. & Hirzel, R. 1976–1980. Flora der Schweiz und angrenzender Gebiete. (2nd ed.). Vol. I–III. Birkhäuser, Basel.

Introduction

Large-scale studies in plant community ecology require the collection of large amounts of field data to describe specific aspects of the vegetation cover under investigation. In the beginning phytosociological tables often just documented observations. As soon as results had to be compared, however, data processing became necessary. Ellenberg's method for ordering vegetation tables from 1956 (Mueller-Dombois & Ellenberg 1974) is one of the first synoptical methodological outlines aimed at standardising the view of experts. Only a few years later the first numerical classification programs came into use. These were quite promising in principle since they offered a strictly formal approach to the classification of vegetation. They could be accepted by those who agree with Braun-Blanquet's (1964) concept of associations as well as by those who object to it. Results can be reproduced at any time and by any investigator and they fit the less restrictive concept of Poore's (1955) noda. Unfortunately, various severe drawbacks have been encountered, but rarely ever published. One of the few exceptions is van der Maarel's (1982) review on the manipulating and editing of phytosociological tables. One of his conclusions is: 'In the presentation of results the structured table with ordered blocks of relevés and species is a central issue. Still, many tables resulting from numerical classification are far from optimal in this respect. Thus, this element of data editing deserves further attention.' This statement clearly indicates the existence of conventions or even paradigms in phytosociological methodology, which are not taken into account by most statistical methods. This point becomes evident, when comparing the numerous attempts to numerically rearrange the grassland data set of Ellenberg (c.f. van der Maarel *et al.* 1978; Gauch 1982; Kuhn 1983, Popma *et al.* 1983). Most authors found that their solution is 'satisfactory' (Gauch 1982, p. 25), but the appearance of all these tables is far from what Ellenberg originally suggested. An important point is the interpretability of the results which largely depends on their presentation, hence the quality of the 'synthesis table' (Mueller-Dombois & Ellenberg 1974). Even if the findings are roughly the same in numerical- and traditional approaches (Grabherr 1985), they are extremely difficult to compare, when displayed according to different conventions. Editing may therefore be needed, even though this is undesirable, since it represents a mixture of manual and formal analysis. The problem might be solved by combining several methods, within one program package. Examples are TABORD (van der Maarel *et al.* 1978), CLUSLA (Louppen & van der Maarel 1979), TWINSPAN (Hill 1979a) and also DECORANA (Hill 1979b). In the latter, correspondence analysis is 'refined' by detrending (Hill & Gauch 1980). In the package of Wildi & Orlóci (1983) classification followed by analysis of concentration (AOC) is an example of another two-step analysis (Feoli & Orlóci 1979). The success of AOC indicates that the convergence of traditional and statistical approaches should be sought in the coupling of methods. The objective of this is of course not only to confirm old results, but much more to offer solutions which can be reproduced.

The aim of this paper is to define the criteria for a 'perfectly structured' vegetation table. For each criterion, a method has to be found by which this goal can be achieved. Then, a step by step procedure has to be designed which combines all these methods. The results have finally to be compared with previous findings throughout the entire analytical process. Weaknesses may be detected on either side: Manual sorting is never a unique process, but it can be done with more or less skill and sophistication. The same is true (yet often overlooked) for numerical methods, where only properly evaluated methods and options yield satisfactory results. There is no unambiguous means available for measuring the success of the analysis in ecological terms, but the example of Ellenberg will at least serve as reference system for structural comparisons.

Sorting conventions

While some of the conventions observed in the process of refining classifications have been explicitly formulated (c.f. Mueller-Dombois & Ellenberg 1974), others can be detected by inspecting the results. From the left hand side presentation in Table 1, the final 'synthesis table' of Ellenberg, the following can be concluded:

1. An outlier (relevé no. 19) is removed from the set. Regarding most multivariate methods, this is a useful manipulation, since it improves their efficiency in resolving group or gradient structures.
2. The relevés are classified into three groups. The average group size is eight and all the groups are of a similar size.
3. The set of species is divided into a smaller portion (24) of those having differentiating power and a majority (70) without.
4. The set of differentiating species is classified into four groups, but these exhibit considerable internal variation. More and less abundant species may occur in the same group. Unlike in many numerical solutions, very rare species are not classified.
5. The resulting blocks of relevé- and species groups are ordered in a way that an underlying floristic gradient emerges from the data.
6. The same gradient is further accentuated by the

Table 1. Original classification of Ellenberg's grassland data (left hand table). Centre: A numerical solution not conforming with phytosociological conventions (relevés clustered by minimum variance analysis, species by complete linkage based on covariance). Right hand side: The solution proposed in this paper.

RELEVE GROUP NO.			11111111	222222	3333333333
RELEVE NO.			1 1 2	1112 2	2111 1 221
			40195342	682052	3937618154 7
4	BROM EREC	1	9999999		
42	SCAB COLU	1	111 1 2		
46	THYM SEPY	1	121 1 1		
58	SALV PRAT	1	5 2144		
8	KOEL PYRA	1	2333		
5	FEST OVIN	1	221 1		
50	CAMP GLOM	2	1 11 11	1 1	
60	VIOL HIRT	2	3 2111	1	
7	BRIZ MEDI	2	112 1	2	
43	LINU CATH	2	1 11 1	1	
70	GEUM RIVA	3		1111 1	11123511 1
13	HOLC LANA	3		1	1112122292 1
71	MELA DIUR	3		1 1 1	1112 1411
12	ALOP PRAT	3		2 29	4 86991
73	LYSI NUMM	3		11	111111
77	LYCH FLOS	3		1	1 1 1111
67	GLEC HEDE	3		1 11	11 1
65	CIRS OLER	4		111	1112939239 9
14	DESC CAES	4			2 9 925 19
72	ANGE SILV	4			1 1 1 112
24	CARE ACUT	4			2 4 12 9
69	FILI ULMA	4			1 3 2 1
75	PIMP MAGN	4			11 1 1
76	POLY BIST	4			1 4 1
1	ARRE ELAT	5	28 59 91	999949	9999494999 4
2	DACT GLOM	5	56526985	999598	8199999999 4
40	GALI MOLL	5	19323732	566533	9293954129 6
6	POA PRAT	5	58499999	999649	9125264195 3
38	PLAN LANC	5	11111111	211122	8141114112 4
10	FEST PRAT	5	2 56325	599299	2929398932 2
41	CHRY LEUC	5	15121361	1 3311	11114111 22
45	RANU ACER	5	3111111	1 1112	1121211111 2
51	VERO CHAM	5	11111 1	1 1111	211111111 1
35	ACHI MILL	5	18622331	591624	1 41 1191
36	DAUC CARO	5	111111 1	111111	2 211 51 1
44	RUME ACET	5	1 1 1	1 1111	1111223112 1
49	TARA OFFI	5	11 1 1	11131	4 211 1 113
28	TRIF PRAT	5	1 1 1121	11 141	2 1 1 11 1
30	MEDI LUPU	5	111111	111 21	1 1111 1
3	HELI PUBE	5	1 4 11	941992	448 1 3
57	AJUG REPT	5	1 1	11111	111131111
11	TRIS FLAV	5	34522	95 889	4565 8
55	CREP BIEN	5	11	11121	6 11181 1
29	TRIF REPE	5	111	11 12	1 1 1 1111
39	HERA SPHO	5	1 1	11 131	1 911 1 4
48	CENT JACE	5	1621	2 12	1 22 1 43
52	PLAN MEDI	5	1 1 1111	111 1	1 1 1 1
37	CAMP ROTU	5	1111111	1 1	1 1 11
9	FEST RUBR	5	24936 2	221	1112 21
47	CERA CAES	5	1 111 1	11	11 1 1 1
74	PRUN VULG	5	1 1 1	11 1	1 11 21 1
61	BELL PERE	5	1 1 1	111 1	111 1
33	LATH PRAT	5	1	11	11 1111 1
31	VICI SEPI	5	11 11	1 1	1
66	TRAG PRAT	5	1	1 1 2	1 1 1
23	CARE FLAC	5	3 2 1	3	2 2
78	SENE JACO	5	11	11	1 1
68	ANTH SILV	5	1	2	1 1 1
34	VICI CRAC	5	1 1 1	1	1
32	LOTU CORN	5	11		1 1
54	LEON HISP	5	1 1	1	4
62	DIAN SUPE	5	1 1	1 1	
64	GALI BORE	5	11		1 1
53	SILE INFL	5	1 2		1
79	POTE REPT	5	1	1	1
83	PAST SATI	5		3	1 1
87	SILA PRAT	5		1 1	1
25	CARE HIRT	5			2 6
27	CARE GRAC	5			4 1
56	MYOS ARVE	5	1		1
63	PIMP SAXI	5	1 1		
80	CARD PRTE	5		1	1
81	MYOS PALU	5		1	1
86	GALI VERU	5	1 1		
89	EUPH ODON	5		1	1
91	RUME CRIS	5			1 1
15	POA TRIV	5			2
16	PHLE PRAT	5		1	
17	FEST ARUN	5			1
18	LOLI PERE	5		2	
19	GLYC FLUI	5			9
20	PHAL ARUN	5			9
21	PHRA COMM	5			1
22	BRAC PINN	5	5		
26	CARE PANI	5		2	
82	GERA PRAT	5		1	
84	GALI ULIG	5			1
85	SANG OFFI	5			1
88	RANU REPE	5			1
90	LAMI ALBU	5			1
92	POLY CONV	5			1
93	CHEN ALBU	5			1
94	ALCH VULG	5	1		

RELEVE GROUP NO.			1111111	222222222	33333333
RELEVE NO.			1 1 2	21 212 11	112 1 2
			4109534	322062583	71186475
69	FILI ULMA	1		1	231
65	CIRS OLER	1		1 1 11 1	23299993
44	RUME ACET	1	11 1	11 1111 1	12132211
73	LYSI NUMM	1		1 1	11111 1
70	GEUM RIVA	1		1 111 11	13152 11
33	LATH PRAT	1	1	1 1 1	1111 1
57	AJUG REPT	1	1	111 1111	11131111
12	ALOP PRAT	1		9224	6998 1
88	RANU REPE	2			1
87	SILA PRAT	2		11	1
25	CARE HIRT	2			26
14	DESC CAES	2			9952 19
77	LYCH FLOS	2		1	1111 1
10	FEST PRAT	2	25632	295259992	99983223
91	RUME CRIS	3			1
27	CARE GRAC	3			1
76	POLY BIST	3			14
13	HOLC LANA	3		1 1 1	22221219
71	MELA DIUR	3		1 11 1	1 1 2114
54	LEON HISP	3	1	1 1	4
15	POA TRIV	3			2
56	MYOS ARVE	3		1	1
24	CARE ACUT	3			41 92
72	ANGE SILV	3		1 1	1121
75	PIMP MAGN	4		1	1 1 1
39	HERA SPHO	4	1 1	1 111319	1 1 41
11	TRIS FLAV	4	3452	4 2899856	5 8
52	PLAN MEDI	4	11 111	111 1 111	1 1
28	TRIF PRAT	4	11 112	2 1111411	111 1
74	PRUN VULG	4	11 1	1 11 11	1 12 1
36	DAUC CARO	4	111111	211111112	1 151 1
55	CREP BIEN	4	1	6111 121	111811
66	TRAG PRAT	4	1	1 1 2 1	1 1
3	HELI PUBE	4	14 1	11992944	4 1 8 3
61	BELL PERE	4	11 1	1 1 1111	1 11
30	MEDI LUPU	4	111111	11 11211	11 11
29	TRIF REPE	4	1 11	1 11 211	1111 1
45	RANU ACER	4	131111	1111121 2	11112121
40	GALI MOLL	4	1392373	962553369	35149962
68	ANTH SILV	4	1	1 2 1	1
49	TARA OFFI	4	1 1 1 1	41 1 1312	1 11131
38	PLAN LANC	4	1111111	811122214	11141241
51	VERO CHAM	4	11111 1	21 1111 1	11111 11
83	PAST SATI	5		31	1
18	LOLI PERE	5		2	
89	EUPH ODON	5		1 1	
59	KNAU ARVE	5	1	1	
31	VICI SEPI	5	1 1 11	11	1
82	GERA PRAT	6		1	
16	PHLE PRAT	6		1	
67	GLEC HEDE	6		1 11	1 1 1
81	MYOS PALU	6		1	1
21	PHRA COMM	6			1
78	SENE JACO	6	11	1 1	11
2	DACT GLOM	6	5562698	895598999	99999949
46	THYM SEPY	7	112 1	1	
42	SCAB COLU	7	111 1	2	
8	KOEL PYRA	7	2333		
5	FEST OVIN	7	221	1	
7	BRIZ MEDI	7	112	1 2	
50	CAMP GLOM	7	11 1 11	1 1	
43	LINU CATH	7	11 1 1	1	
63	PIMP SAXI	7	1 1		
4	BROM EREC	7	9999999		
60	VIOL HIRT	7	32111	1	
58	SALV PRAT	7	52144		
37	CAMP ROTU	7	1111111	11 1	1 1 1
47	CERA CAES	8	11 11	11 1 1	11 1
9	FEST RUBR	8	29436 2	12 2 1 1	22 1
53	SILE INFL	8	1	2	1
6	POA PRAT	8	5489999	999699492	56142539
80	CARD PRTE	8		1	1
26	CARE PANI	8		2	
62	DIAN SUPE	8	1 1	11	
23	CARE FLAC	8	32 1	3	2 2
85	SANG OFFI	8			1
84	GALI ULIG	8			1
17	FEST ARUN	8			1
79	POTE REPT	8	1	1	1
64	GALI BORE	8	1 1		1 1
35	ACHI MILL	8	1682233	111654294	1 1 911
94	ALCH VULG	8	1		
22	BRAC PINN	8	5		
86	GALI VERU	8	1 1		
34	VICI CRAC	8	1 1 1	1	1
48	CENT JACE	8	162	11122	2 1 243
32	LOTU CORN	8	11		1 1
41	CHRY LEUC	8	1152136	1313111 1	1111422

RELEVE GROUP NO.			1111111	222222222	33333333
RELEVE NO.			1 2 1	1122 211	1 1 212
			4013495	268305223	48677115
63	PIMP SAXI	7	1 1		
8	KOEL PYRA	7	233 3		
4	BROM EREC	7	9999999		
46	THYM SEPY	7	1211	1	
42	SCAB COLU	7	111 1	2	
5	FEST OVIN	7	22 1	1	
43	LINU CATH	7	1 11 1	1	
50	CAMP GLOM	7	1 1111	1 1	
37	CAMP ROTU	7	1111111	1 1 1	11 1
58	SALV PRAT	8	5 4421		
86	GALI VERU	8	1 1		
60	VIOL HIRT	8	3 1121	1	
52	PLAN MEDI	6	1 111 1	1111 1 11	1 1
11	TRIS FLAV	6	5234	2954889 6	5 8
35	ACHI MILL	6	1863322	159162414	9 111 1
2	DACT GLOM	6	5659826	599859899	99994999
10	FEST PRAT	6	2 3256	559229992	28392993
55	CREP BIEN	6	1	1 161211	1811 11
44	RUME ACET	6	1 1 1	1 111111	23211121
72	ANGE SILV	1		1 1	1 1 2 1
57	AJUG REPT	5	1	1 1 11111	13111111
70	GEUM RIVA	5		11 1 111	5211131
71	MELA DIUR	5		1 111	1 2111 4
65	CIRS OLER	5		1111 1	99929233
12	ALOP PRAT	5		2 29 4	98 961
13	HOLC LANA	5		1 1 1	22121229
77	LYCH FLOS	5		1	1 1 111
14	DESC CAES	5			12 9959
73	LYSI NUMM	5		1 1	111 111
24	CARE ACUT	5			9142
6	POA PRAT		5849999999996499254253169		
38	PLAN LANC		1111111121812214241141 11		
40	GALI MOLL		193732325695336994936152		
41	CHRY LEUC		151362111 1311312141211		
45	RANU ACER		31111111 111212112121 11		
51	VERO CHAM		111 111 1 211111 1111111		
36	DAUC CARO		1111 11111211112 51111		
49	TARA OFFI		11 1 1 141311211113 1		
28	TRIF PRAT		1 112 11112141 1 11 11		
30	MEDI LUPU		1111 11 111 21111 11 1		
3	HELI PUBE		1 14 194 99214 8431		
9	FEST RUBR		249 236 121 21 1 2 2		
29	TRIF REPE		1 11 11112 1111 1 1		
39	HERA SPHO		1 1 111131 9 114 1		
48	CENT JACE		62 112 12 1 4 2231		
47	CERA CAES		1 1 111 11 11 11		
74	PRUN VULG		1 11 111 1 1 2 111		
61	BELL PERE		1 1 1 11 1 11 111		
33	LATH PRAT		1 1 11 11 111		
31	VICI SEPI		11 1 1 1 1 1		
66	TRAG PRAT		1 1 2 11 1 1		
23	CARE FLAC		3 2 1 3 2 2		
67	GLEC HEDE		111 1 1 1		
78	SENE JACO		11 1 1 11		
7	BRIZ MEDI		11 2 12		
34	VICI CRAC		1 1 1 1 1		
68	ANTH SILV		1 1 2 1 1		
32	LOTU CORN		11 11		
54	LEON HISP		1 11 4		
62	DIAN SUPE		1 1 1 1		
64	GALI BORE		11 11		
69	FILI ULMA		1123		
75	PIMP MAGN		1 111		
53	SILE INFL		12 1		
59	KNAU ARVE		1 1		
76	POLY BIST		1 4		
79	POTE REPT		1 1 1		
83	PAST SATI		3 11		
87	SILA PRAT		1 1 1		
25	CARE HIRT		2 6		
56	MYOS ARVE		1 1		
80	CARD PRTE		1 1		
81	MYOS PALU		1 1		
89	EUPH ODON		11		
15	POA TRIV		2		
16	PHLE PRAT		1		
17	FEST ARUN		1		
18	LOLI PERE		2		
21	PHRA COMM		1		
22	BRAC PINN		5		
26	CARE PANI		2		
27	CARE GRAC		1		
82	GERA PRAT		1		
84	GALI ULIG		1		
85	SANG OFFI		1		
88	RANU REPE		1		
91	RUME CRIS		1		
94	ALCH VULG		1		

order of the relevés and species within the groups. (A normal agglomerative cluster analysis would of course never do this.)

There are suggestions for the order in which all these features should be achieved (Mueller-Dombois & Ellenberg 1974). This, however, is part of the method. The numerical strategy differs from it. While the selection of differentiating species is an early step in manual sorting, numerical classification should of course be based on the entire data set (Wildi 1986). The operational sequence influences the results in any case.

Methods

All the methods shown in this paper are part of the newly revised program package of Wildi & Orlóci (1983). For the description of many algorithms we refer to Orlóci (1978). Correspondence analysis is the same as Hill's (1974), even though the computation method differs. For normal ordinations of relevés, the D-algorithm of principal component analysis with stand norm transformation is preferred (Orlóci 1978). The method for selecting differentiating species is that of Jancey (1979). Analysis of concentration (AOC) is described by Feoli & Orlóci (1979).

A special problem is posed by the detection of outliers. The method developed is related to single linkage clustering. For each observation (relevé or species), the similarity to the nearest neighbour is determined. As a result, single (but not groups of) outliers can be detected and eventually eliminated.

For printing the vegetation table in condensed form, the scores are rounded to figures from 1 through 9. All scores greater than 9 will therefore appear as 9 (see Table 1).

The chaining of programs is an important element in this concept, and it is supported by the program package. The usage follows the scheme in Table 2. Any analysis starts at the level of initialization (INIT) and may include a large number of program runs. Chains are programmed at the level of the operating system. The entire method shown in Table 5 is an example. It is available from the author as a small test-version for IBM AT-compatibles at no cost.

Table 2. Structure of the program package used. Operations proceed from top to bottom. The programs represent modules within the analytical process.

<table>
<tr><td colspan="3">Preparation of the data and data simulation
CREO (1) SPAG (2)</td></tr>
<tr><td colspan="3">Initializing the data system and description of the data
INIT (3)</td></tr>
<tr><td rowspan="3">Data-manipulation
EDGR (4)
EDDA (5)
EDCO (6)</td><td>Resemblance structure
RESE (7)</td><td rowspan="3">Presentation of data
TABS (17)
GRAP (18)
ORDB (19)</td></tr>
<tr><td>Primary analysis
RANK (8) CLTR (9)
PCAB (10) MREG (11)</td></tr>
<tr><td>Secondary analysis
GRID (12) AOCL (13)
DIAN (14) IDEN (15)
REAL (16)</td></tr>
</table>

The analysis of Ellenberg's grassland data

Detection of outliers

Outliers reduce the efficiency of most multivariate methods in unfolding gradient- and group structures. In Ellenberg's example relevé 19 is such a case. A method should be devised for the numerical analysis, which is sufficiently sensitive for outliers, even when little is known about the data. A first attempt has been made with cluster analysis. In Gauch (1982, p. 26) a dendrogram is presented from a TWINSPAN analysis (Hill 1979a). There, rel. 19 appears at the right hand side of the dendrogram, but it is closely related to rel. 25. With minimum variance clustering (not shown here), rel. 25 is separated from all the others. We conclude, that cluster analysis is not a safe method to detect outliers.

In a second attempt, ordination is used, with presence-absence transformation tolerated in this case. Correspondence analysis (Fig. 1) indeed separates rel. 19 clearly from all the others within

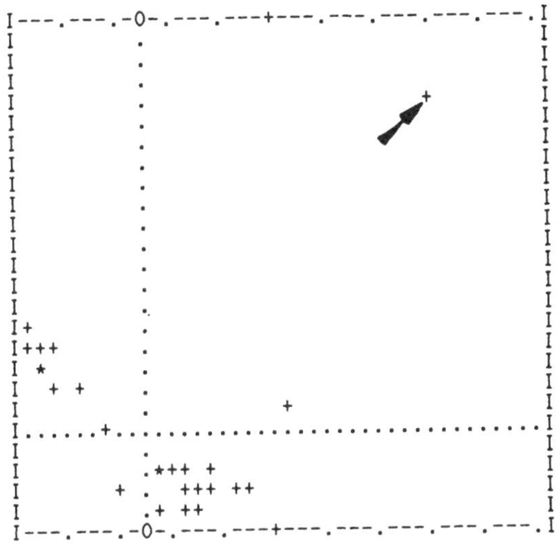

Fig. 1. Ordination of relevés in Ellenberg's grassland data set. Correspondence analysis of presence-absence scores is used. The arrow points to rel. 19.

two dimensions. Principal component analysis (not shown here), which usually better resolves detailed structural features, does the same, but only on the third axis. Even though distortions are uncontrolled, ordination appears here as a more useful tool than cluster analysis.

An even simpler method which does not require visual inspection is the analysis of nearest neighbours. Low similarity values indicate high isolation of the individual. We find that rel. 19 is by far the most aberrant with a similarity ratio of 0.402 to its nearest neighbour, while all other nearest neighbour values are between 0.56 and 0.82. According to general experience within-group values usually remain above 0.50 (see also Gauch 1982, p. 102). As a consequence rel. 19 is now removed from the data.

Classification of relevés

Most investigators of Ellenberg's data found that in the classification of relevés, manual and numerical approaches often hardly differ. What should be chosen is a clustering method which forms groups of low internal variation. Prior to that, data transformation may be needed (van der Maarel 1979) and a resemblance measure evaluated (c.f. Legendre & Legendre 1979). To get a summary of the results of the chosen variations, a series of systematic tests has been made and compared with Ellenberg's solution within a contingency table. The latter has been analysed using analysis of concentration which measures the fit by the mean square contingency coefficient.

The result of this is given in Table 3. Two clustering methods (complete linkage and minimum variance), two resemblance measures (covariance, similarity ratio) and eight different scalar transformations are compared. Less than half of the 32 analyses gave solutions which are close to Ellenberg's. All of these are based on strong transformations. No transformation at all as well as extreme solutions like log(x+16) or presence-absence, clearly result in a loss of information. Complete linkage analysis is usually less sensitive and hence 'safer' than minimum variance analysis. The same is true for the similarity ratio. These divergences disappear almost, if the relevé vectors are normalized previous to the analysis (not shown here). We therefore conclude that in the present (not yet really complex) case, data transformation is the important point, while any of the clustering methods tested can yield the desired results.

The comparison of the most stable classification with Ellenberg's shows, that the results differ slight-

Table 3. Clustering relevés with minimum variance analysis (MVA) or complete linkage analysis (CLA). Resemblance measures are covariance (cov) and the similarity ratio (sir). Eight transformations are tested. The resulting classifications are compared with Ellenberg's using the mean square contingency coefficient as a measure of fit. (*) – identical (stable) solutions with three differently classified relevés. (**) – solution with four differences. (***) – best solution, but also three differences.

Transformation	MVA, cov	CLA, cov	MVA, sir	CLA, sir
no	0.366	0.670	0.615	0.457
square root	0.704	0.755*	0.755*	0.755*
log(x+1)	0.755*	0.755*	0.755*	0.755*
log(x+2)	0.755*	0.670	0.755*	0.755*
log(x+4)	0.718	0.755**	0.755*	0.755*
log(x+8)	0.568	0.788***	0.568	0.654
log(x+16)	0.448	0.524	0.708	0.708
presence-absence	0.420	0.524	0.448	0.708

ly. Rel. 2 from Ellenberg's group 1 (group 3 in the numerical solution) falls into group 2. This is in fact the only one where no *Bromus erectus* is found. All the relevés from Ellenberg's group 2 are also in the numerical group 2. The difference lies in the boundary between the two last groups.

Classification of species

Classification of species is crucial, as there is more variation in the frequency of species vectors than in the relevé vectors. According to the conventions mentioned, groups should consist of species with more or less the same frequency. Therefore, 1. a non-centering resemblance measure should be used, 2. the number of groups must be high in order to avoid internal heterogeneity, and 3. the scalar transformation ought to have a tendency towards a more qualitative view. These rules can be confirmed through a series of tests. In all cases, species vectors are normalized. From the phytosociological point of view, the most undesirable result is achieved when using the (non centering) covariance as the resemblance measure (Table 1, centre). The block structure of the table is then gradually improved by applying chord distance and replacing $\log(x+1)$ transformation by $\log(x+64)$ or even presence-absence.

The number of species groups, eight in Table 1, centre, is twice the one of Ellenberg's solution. This is still a very small number, since all the groups are somewhat heterogeneous. We accept this as a reasonable solution though it is still far from what we aim at in plant sociology.

Species selection

A set of species can now be determined which is highly discriminating. Jancey's (1979) ranking is supposed to accomplish this. As a first step, the method should be tested and compared with Ellenberg's findings. The transformation used is $\log(x+64)$, which is the same as for the classification of the species. Only the 10 highest ranked species are shown (Table 4). Of the 24 first, 22 correspond with Ellenberg's differential species. This indicates that Jancey's method really reflects the plant ecologist's view well. As there is no justification to limit the list to 24 ranks, 30 are used for the final example (Table 1, right hand side).

Table 4. Ranking species with Jancey's (1979) method, based on $\log(x+64)$ transformation of the scores. The underlying classification is Ellenberg's (1956). The first 10 ranks are shown.

Rank	Species	*F*-value
1	*Holcus lanatus*	64.575
2	*Bromus erectus*	49.000
3	*Cirsium oleraceum*	31.208
4	*Geum rivale*	15.211
5	*Salvia pratensis*	11.667
6	*Thymus serpyllifolium*	11.667
7	*Scabiosa columbaria*	11.667
8	*Angelica silvestris*	9.1875
9	*Deschampsia caespitosa*	9.1875
10	*Koeleria pyramidata*	7.0000

Secondary analysis

Further refinements are achieved by secondary analyses of the previous results. First, the dense blocks within the structured vegetation table should be grouped along the diagonal. This way, a main gradient may emerge if it exists in the data. Analysis of concentration will accomplish this. A 3×8 contingency table is constructed (3 relevé groups, 8 species groups). Its elements are occurrences of species within the table's blocks. The analysis of concentration adjusts for different block sizes and yields a variety of information through an eigenvalue procedure, e.g. ordination coordinates for the relevé and species groups. We now order the blocks along the first axis and find that dense blocks appear in the diagonal of the vegetation table.

It appears that Ellenberg's table (Table 1, left hand side) strongly reveals the gradient structure in the data. The relevés and species are ordered within the groups as well. An analysis of subsets, namely the row- and column groups of the table would be desirable. This, however may become rather time consuming. It is easier to look for the main trend within the entire data set and to order the relevés and species within the groups according to this. Correspondence analysis is used as it is an appropriate method of ordination in which the main gradient appears right on the first axis and structural details are suppressed.

As a final step in the entire analysis (Table 5) the species with the highest discriminating power are

Table 5. Standard strategy for the analysis of plant sociological data. The suggestions hold for ordinal data with a scale of around 10 steps and for frequency counts. Typical matrices are supposed to have 70% through 90% of zero entries.

Objective	Step no.	Computer program	Decisions	Suggestions
Classification of the relevés	1	INIT (initialize)	scalar transformation vector transformation	square root or log(x+1) normalize
	2	RESE (resemblance matrix)	resemblance mesure for relevés	covariance or similarity ratio
	3	CLTR (cluster analysis)	clustering algorithm number of groups	minimum variance (typical sample) or complete linkage analysis (group centroids) according to objective (e.g. around $\sqrt{n}$)
Ordination of species and relevés (correspondence analysis)	4	INIT (initialize)	scalar transformation vector transformation	square root, presence–absence special option for correspondence analysis (opt. 3)
	5	RESE (resemblance matrix)	resemblance measure, for rows or columns (which ever is smaller)	scalar product without centering (mandatory)
	6	PCAB (eigen analysis)	method	normal correspondence analysis
Classification of species	7	INIT (initialize)	scalar transformation vector transformation	log(x+64) for percentages, log(x+10) for ranks normalize
	8	RESE (resemblance matrix)	resemblance measure for species	distance measure without centering
	9	CLTR (cluster analysis)	clustering algorithm number of groups	complete linkage analysis or minimum variance clustering 2–5 times as many as there are relevé groups
Ordering dense blocks along the diagonal	10	AOCL (concentration analysis)	–	–
Rearrange relevés and species within groups	11	EDGR (rearrange rows and columns)	order of groups order of relevés/species within the groups	according to AOCL (step 10) according to 1st axis in correspondence analysis
Printing of the entire vegetation table	12	TABS (printing of the vegetation table)	ordering criteria	according to EDGR (step 11)
Reducing the species set	13	DIAN (Jancey's ranking)	number of species selected classification for ranking	20–50% of all species cluster analysis (step 9)
Printing the reduced vegetation table	14	TABS (printing of the vegetation table)	ordering criteria	according to Jancey's ranking (step 13)

moved into the top portion of the table. To further follow the conventions, the remaining are added in the order of decreasing frequency.

The final (but not sole) solution (Table 1, right hand side) can now be compared with Ellenberg's (Table 1, left hand side). We find that the way of presenting the result is identical, while the classification of the relevés as well as the selection of differentiating species are very similar. Some details differ. At least initially, more than the number of species groups suggested by Ellenberg were needed to resolve the complex structure. In Jancey's ranking, some very frequent species are included as well, such as *Dactylis glomerata* and *Festuca pratensis,* which really display differentiating power, yet only if their scores are considered quantitatively.

An application to forest sociology

A recent plant sociological study of Swiss forests has been published by Ellenberg & Klötzli (1972) who describe 71 associations. Of these, a summary table is added in their appendix, with 71 columns (vegetation types) and 416 rows (species). The associations are further classified to form 11 higher-level types. The latter are aimed at improving the overview. They more or less agree with terminological conventions but are not the result of a skilled analysis (Ellenberg & Klötzli 1972, p. 602). A true analysis would yield a well-structured table, and with it, a set of species that would help explain the broad variation of the woodland vegetation in Switzerland. To achieve this is the objective of the new method presented here. Contrary to Ellenberg's grassland example the traditional way of classification of this data is more critical for the following reasons:

1. There are more than five times as many descriptors (species) than individuals (associations).
2. Since the individuals are already the result of a clustering process, no dense clusters (noda) can be expected.
3. In Switzerland, the great variation in bedrock, climate and topography produces complex gradients and vegetation types, making the task of classification quite tedious.
4. Non-linearity is almost certain to occur due to the broad vegetation spectrum.

The analysis follows exactly the scheme used for Ellenberg's example. As it is presumed that the resemblance space of the associations is more evenly populated than a normal relevé space, and within group variation should be about the same, complete linkage analysis is used instead of minimum variance clustering. The number of groups is 11, the same as published. Species groups are arbitrarily set to as many as 60, taking account of the high variation which may be expected. The number of discriminating species is limited to 30 in order to yield a classification which is easy to interpret.

Before inspection of the final result, an ordination of the 71 associations is examined (Fig. 2). Obviously, there are no dense clusters which would facilitate the classification. Also, several gradients occur. These are only resolved in three dimensions, as the overlapping of groups indicates. Data of this kind are in fact extremely difficult to handle in a vegetation table. However, after reducing the species number by more than a factor of ten, a clear structure emerges (Table 6). The main types of Swiss forests can be identified:

Group number 9 (the first in Table 6) is a cluster of wet woods with deciduous trees, dominated by *Alnus*.

Group 10 includes typical *Fagus silvatica* stands

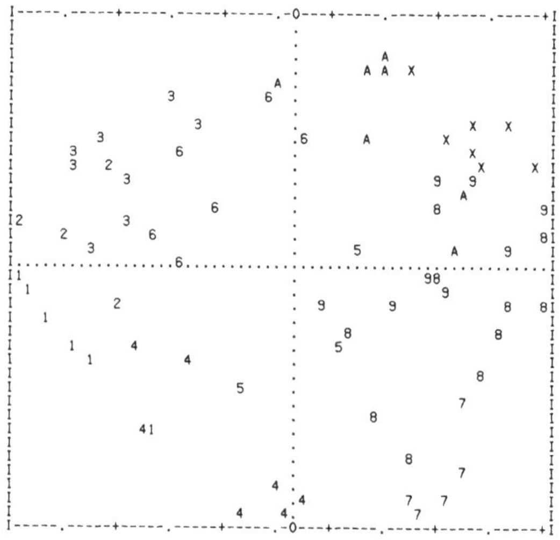

Fig. 2. Ordination of 71 associations from Swiss forests according to Ellenberg & Klötzli (1972). The groups result from cluster analysis. Symbol X is used for group 10, A for group 11.

Table 6. Classification of Swiss forest associations. The numbers match with those of Ellenberg & Klötzli (1972). The species set is reduced to 50. Codes '%' and '*' are used for planted species (shrubs, trees).

RELEVE GROUP NO.	99999999	1111111 0000000	77777	1111111 1111111	8888888888	555	666666	33333333	4444444	2222	111111
RELEVE NO.	43322324 31278094	2121 1 6123728	33334 68970	22221 4 4310859	1 21131461 0955457126	334 342	1 691432	55564554 10208457	6666566 4135386	4455 6567	656577 789901
249 CARE ACUT41	+3 1+1 4										
54 HUMU LUPU41	+3+ ++ +										
245 FILI ULMA36	++ 1++++	+		+							
14 ALNU INCA36	++41+++			1					+		
15 PRUN PADU36	++3+331+	+									
74 SAMB NIGR37	++ ++++	+++++++		+	+	+					
210 FEST GIGA37	3+333++	++ +++		+							
290 SATU CALA23			+3+++		+						
65 VIBU LANT58	++ + +	++ +	+3333		3+++3++343				34++3		
36 CRAT OXYA58	+	1 +	+ +		+++3+1+3++						
146 HEDE HELI58	1++3	+4+44+1	43+1+	++	+35+34+31+	11	++ +				
60 LIGU VULG58	++ +++	+1 +	34333		31+3+++33+				+ +		
35 CORY AVEL58	3+4+4	11+3+ +	44344	14 +++	++4+31+++3	34+		+	+		
39 ROSA ARVE58	+ +	++ + +	++ +		3+33+++31+						
37 LONI XYLO58	++++3 3	+++3+++	+33+3	+3	3+333+3313			+			
51 CORN SANG58	+3+ 4 3	++ + +	333+3		++3+1+ +4+				+		
66 VIBU OPUL58	++++++++	++ + +	+		++3++++ ++						
123 PHYT SPIC51	3	+3+3+++	++	3+3+++3	+++3+3+++		33	+4 +			
6 ACER PSEU51	++ 3	33++++4		+34++	+3++++3 33	+					
271 CARE MONT22			+ +1+		+++514155	3			+4 3		
272 MELI MELI22			3333		3++33 +33+						
4 PIRU MALU22		+	++3++		+++ ++ +						
55 PRUN SPIN22	+++		++3++		++++++ ++				+		
147 LATH VERN57		++3 +	3+3++	3+	+3+3+++3 +						
294 SILE NUTA24			+++3			+					
281 POLY OFFI21			+3333		+ + +++	+++			3 +		
265 BETO OFFI21			++3++		+ + 3+	+++					
1016 QUER PUBE21			+5131		+ +	11					
297 VINC OFFI21			+33+3		+++++ +	+++					
1029 CAST SATI31			%			***					
29 CAST SATI31					%	+44					
270 SERR TINC31			+			+++					
93 PREN PURP52		+ +++3+	+	+14443+	+++33 3333	++	++++++	+5+1++34	3	++	
84 SOLI VIRG52	+	+++ +++	3++ 3	+3+++33	++3+3+33+3	+33	+3++3+	+333++3+	33		+
1010 FAGU SILV52		15+5555	3	+55551	55 55+5 5		555555	+			
82 CYTI SCOP30						++3	++				
258 LATH MONT30					3	+3+	+ + ++				
361 PLAG NEGL 6							+ + ++				
360 BRAC VELU 6							+ 3+++				
383 MNIU SPIN 7								+++ +1++		+	
352 DICR SCOP 9				3	3 3		+ 3++3	333333+3	+ ++	+ 34	+33333
350 HYLO SPLE 9			1	+ + 5	+ 3		+33+++	54515554	11+3 4	5455	+5455+
387 SPHA DIVS40	1									1153	13
312 VACC VITI 3								+34+	31	+33	353343
372 PLEU SCHR 3					+			1++14341	+ 14	4444	+44454
307 MELA SILV 3								114 +441		++	3+3++
388 BAZZ TRIL 1										1 ++	
375 CETR ISLA 5									1	+	++3+41
376 CLAD ARBU 5									1	+	++1+++
377 CLAD RANG 5								+	1	+ +	++4++4

with only few discriminating species but a clear relationship to the wet and rich *Alnus* stands (Group 9).

Group 7 includes deciduous woodlands with *Quercus pubescens,* growing on warm sites as species group 21 indicates.

Group 11 consists of mixed stands of *Fagus silvatica,* and (not apparent in the reduced list) *Abies alba.* They occur at higher altitudes and are characterized in Table 6 by the lack of species group 58.

In group number 8 we find again *Fagus* communities on nutrient-rich soils, often exposed to warmer conditions.

Castanea sativa woods of the south are in group 5.

Group number 6 largely matches the *Luzulo-Fagion* of Ellenberg & Klötzli (1972), with *Brachythecium velutinum* as a typical species.

Abies and *Picea* woods are comprised in group 3, with *Mnium spinosum* as a typical species.

Pinus woods are found in group 4. Table 6 indicates the almost entire lack of characteristic species and Ellenberg and Klötzli in fact assign these associations to different higher sociological categories.

In group 2 there are *Picea* stands on acid sites.

Group 1, finally, includes *Pinus montana* and *Pinus cembra* stands growing at high altitude.

As described, the numerically elaborated classification can be interpreted based on the discriminating species found. It is difficult to discern if the traditional order is more successful. One indication can be derived from ecological indicator values as presented by Landolt (1977). After multiplication of the association × species matrix (presence-absence scores) by the matrix of species × indicator values, a new data set is available characterizing the associations based on indicator values. After adjustments, the latter is analysed with discriminant analysis. This is an attempt to resolve the group structures with the help of the new descriptors. The result is shown in Fig. 3. Obviously, the numerical solution is resolved more successfully (lower graph, Wilk's lambda = 0.548) by the indicator values than in the traditional way (upper graph, Wilk's lambda = 0.411) and hence can be preferred in applications where ecological factors are of interest.

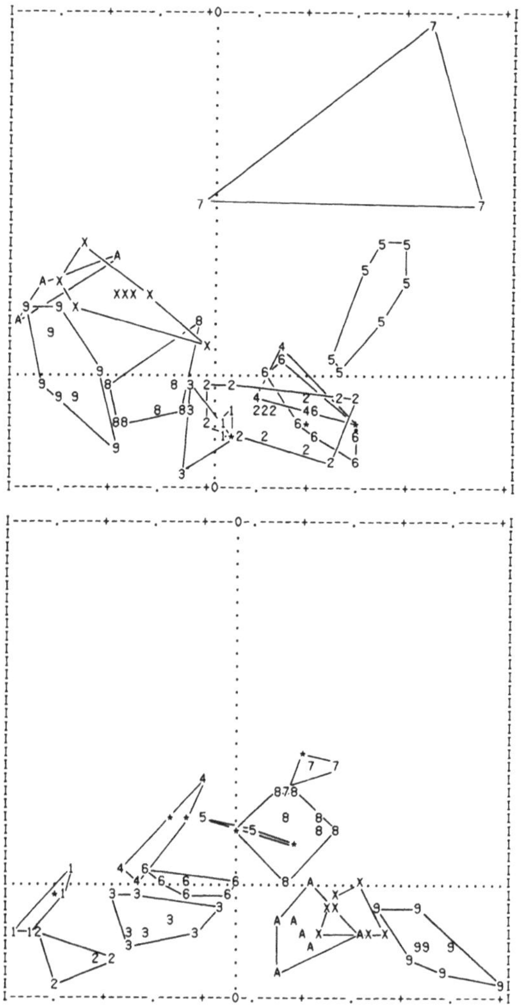

Fig. 3. Discriminant analyses of grouped Swiss forest associations based on average indicator values of Landolt (1977). Sociological criteria as shown in Ellenberg & Klötzli (1972) are used in the upper graph, the numerical solution in the lower graph.

Discussion

Whenever methods are developed and tested, the question arises whether real or artificial data should be used. The latter have the advantage to be known in advance (Lagonegro 1984; Kenkel & Orlóci 1986) and the success can be measured more objectively. On the other hand, many problems arise with real data. There, groups and gradients tend to occur together in great diversity. More or less isolated individuals reduce the efficiency of most statistical models in use. Finally an ecological interpretation is requested. A method only qualifies as being of general relevance if it passes the test with real data a considerable number of times. This is true for the approach presented in this paper, although for practical reasons only two examples are presented.

The method developed here (Table 5) gives convincing results with the Ellenberg data (right hand side in Table 1). Yet this example is quite simple. As the ordination in Fig. 1 illustrates, there is an almost linear gradient and one outlier. Many numerical methods can detect this. The second example, in contrast, is extremely complex with several gradients and also some fairly isolated individuals. The question arises how to put the data points of the configuration shown in Fig. 2 into a logical order.

In general terms, there are three principles which are all applied (Fig. 4). First, the multidimensional space is broken up into subspaces representing groups (ellipses in Fig. 4). Normal cluster analyses

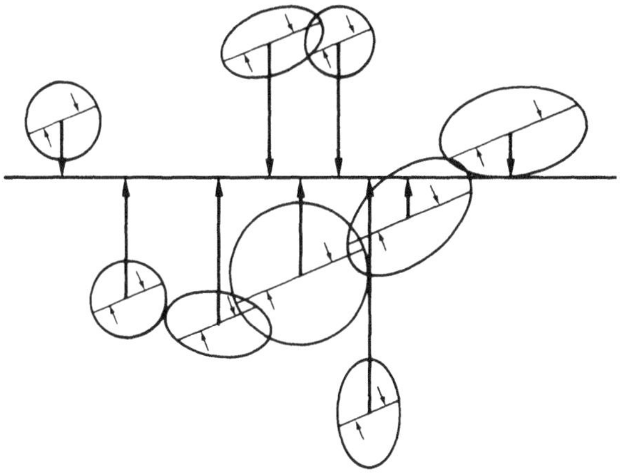

Fig. 4. The underlying principle for breaking up the *n*-dimensional space to bring the relevés and species into a meaningful order.

achieve this. There are many different solutions, with variable numbers of groups. Many will produce a well-structured table. Second, the groups - and not the data points - are squeezed into a unidimensional order (arrows in Fig. 4). This means partitioning the gradients and treating them like natural clusters. Since the number of groups is always much smaller than the numbers of relevés and species, it should not be too difficult to recognize the elements belonging to the same gradient. Third, the order of the rows and columns within the groups also enhances the gradient structure. This emphasizes internal variation. Figure 4 shows schematically the idea of ordering according to a correspondence analysis. The direction of all the projections (small arrows) is given by just one axis. This may be close to an optimum for some, but not efficient for others. The solution suggested represents an approximation which can be further optimized.

A further approximation to an ideal solution is the use of Jancey's (1979) ranking. The variance computed there is based on all groups. As already mentioned by Jancey, species may have high resolving power for some relevé groups, but low power for others. Using the sum is just a compromise. Optimization would of course once more increase the already big computational effort.

Why this effort? The answer is that our findings force us to operate in many different resemblance spaces. Considering correspondence analysis, as an example, there is just one - for relevés and species. Here, a first is used for clustering relevés, a second for clustering species. This requires different transformations as well as different resemblance measures. A third is used for groups in the analysis of concentration. It has a much lower dimensionality. Finally, a fourth is used for ordering the relevés and species within the groups. It is the space defined by correspondence analysis and it concentrates much variation on the first axis.

Selecting discriminating species is part of the traditional phytosociological approach. In the present case it also has a general meaning. As it is based on the classification of the relevés, it suppresses variation between the groups. Noise is thus removed so that the result becomes easier to interpret.

References

Braun-Blanquet, J. 1964. Pflanzensoziologie. Grundzüge der Vegetationskunde. (3rd ed.). Springer, Wien.

Ellenberg, H. & Klötzli, F. 1972. Waldgesellschaften und Waldstandorte der Schweiz. Mitt. Schweiz. Anst. Forstl. Vers. w. 48: 589-930.

Feoli, E. & Orlóci, L. 1979. Analysis of concentration and detection of underlying factors in structured tables. Vegetatio 40: 49-54.

Gauch, H. G. 1982. Multivariate analysis in community ecology. Cambridge University. Press, Cambridge.

Grabherr, G. 1985. Numerische Klassifikation und Ordination in der Alpinen Vegetationsökologie als Beitrag zur Verknüpfung moderner 'Computermethoden' mit der pflanzensoziologischen Tradition. Tuexenia 5: 181-190.

Hill, M. O. 1974. Correspondence analysis: A neglected multivariate method. Appl. Statist. 23: 340-354.

Hill, M. O. 1979a. TWINSPAN - a FORTRAN program for arranging multivariate data in an ordered two way table by classification of individuals and attributes. Cornell University, Ithaca.

Hill, M. O. 1979b. DECORANA - a FORTRAN program for detrended correspondence analysis and reciprocal averaging. Cornell University, Ithaca.

Hill, M. O. & Gauch, H. G. 1980. Detrended correspondence analysis, an improved ordination technique, Vegetatio 42: 47-58.

Jancey, R. C. 1979. Species ordering on a variance criterion. Vegetatio 39: 59-63.

Kenkel, N. & Orlóci, L. 1986. Applying metric and nonmetric multidimensional scaling to ecological studies: some new results. Ecology 67: 919-928.

Kuhn, N. 1983. VEGTAB, ein Computer-Programm als Hilfe zur

tabellarischen Vegetationsgliederung. Tuexenia 3: 499-522.

Lagonegro, M. 1984. SPAGHET: A coenocline simulator useful to calibrate software detectors. Studia Geobot 4: 63-99.

Landolt, E. 1977. Oekologische Zeigerwerte zur Schweizer Flora. Veröff. Geobot. Inst. ETH 64.

Legendre, L. & Legendre, P. 1979. Ecologie numérique. Masson, Paris.

Louppen, J. M. W. & van der Maarel, E. 1979. CLUSLA: A computer program for the clustering of large phytosociological data sets. Vegetatio 40: 107-114.

Minchin, P. R. 1987. An evaluation of the relative robustness of techniques for ecological ordination. Vegetatio 69: 89-107.

Mueller-Dombois, D. & Ellenberg, E. 1974. Aims and methods of vegetation ecology. John Wiley Sons, New York.

Orlóci, L. 1978. Multivariate analysis in vegetation research. (2nd ed.). Junk, The Hague.

Poore, M. E. D. 1955. The use of phytosociological methods in ecological investigations. I-III. J. Ecol. 43: 226-244, 245-269, 606-651.

Popma, J., Mucina, L., van Tongeren, O. & van der Maarel, E. 1983. On the determination of optimal levels in phytosociological classification. Vegetatio 52: 65–75.

van der Maarel, E. 1979. Transformation of cover-abundance values in phytosociology and its effects on community similarity. Vegetatio 39: 97-114.

van der Maarel, E. 1982. On the manipulation and editing of phytosociological and ecological data. Vegetatio 50: 71-76.

van der Maarel, E., Janssen, J. G. M. & Louppen, J. M. W. 1978. TABORD, a program for structuring phytosociological tables. Vegetatio 38: 143-156.

Wildi, O. 1986. Analyse vegetationskundlicher Daten. Theorie und Einsatz statistischer Methoden. Veröff. Geobot. Inst. ETH 90.

Wildi, O. Orlóci, L. 1983. Management and multivariate analysis of vegetation data. (2nd rev. ed.). Eidg. Anst. forstl. Versuchswes. Ber. 215.

Vegetatio **81**: 107–115, 1989.

Syntaxonomy of the *Onopordum acanthium* communities in temperate and continental Europe

Ladislav Mucina
Department of Vegetation Ecology & Nature Conservancy, Institute of Plant Physiology, University of Vienna, Althanstraße 14, A-1091 Vienna, Austria

Accepted 12.12.1988

Keywords: Association, Numerical classification, *Onopordetalia,* Ordination, Phytogeography, Prodromus

Abstract

Ruderal communities dominated by *Onopordum acanthium* are encountered in those regions of Europe with a dry and continental climate. Variability in the climate and local phytogeography within these regions is reflected in the floristic variability of the communities studied. A floristic cline within the communities was summarised by means of multivariate classification and ordination methods. Two associations distributed in temperate Europe and belonging to the *Onopordetalia,* the *Onopordetum acanthii* s.str. and *Carduo acanthoidis-Onopordetum acanthii,* were distinguished; The *Carduo acanthoidis-Onopordetum* is varied and occurs in several geographic races and sub-races. The *Onopordetum acanthii* s.str. is confined to deep inner-montane valleys with a continental climate. The other *Onopordum acanthium* communities described from Spain belong to the *Scolymo hispanici-Onopordetalia nervosi.*

Abbreviations: CA = correspondence analysis; CLC = complete linkage clustering

Nomenclature: Moore, D.M. 1982. Flora Europaea. Check list and chromosome index. Cambridge Univ. Press, Cambridge; for *Agropyron, Atriplex, Cannabis, Matricaria* and *Tripleurospermum:* Ehrendorfer, F. (ed.) 1973. Liste der Gefäßpflanzen Mitteleuropas. 2. Aufl. Gustav Fischer Verlag, Stuttgart. For mosses: Frahm, J.-P. & Frey, W. 1983. Moosflora. Eugen Ulmer Verlag, Stuttgart. The names of syntaxa other than the associations with *Onopordum acanthium* follow the nomenclature of Mucina, L. & Maglocký, Š. (eds) 1985. A list of vegetation units of Slovakia. Doc. Phytosociol., Camerino, 9: 175–200.

Source reference: The paper was presented at the Symposium no 6–45 'Distribution of plants and plant communities: recent anthropogenic changes' during the XIVth International Botanical Congress in Berlin (West).

Introduction

Onopordum acanthium L. is a tall, biennial monocarpic thistle forming a leaf rosette in the first year of its life cycle and setting flowers and seeds in the second. The species is both anemochorous and epichorous (Müller-Schneider 1986). Like other thistles such as *Cirsium vulgare, C. eriophorum,*

Carduus nutans, C. acanthoides (Asteraceae) and *Eryngium campestre* (Apiaceae), it tends to colonise disturbed pastures. *Onopordium acanthium* is a weak competitor; availability of the regeneration gaps is imperative for the development and maintenance of *Onopordum acanthium*-dominated stands. It retreats quickly when the disturbance by grazing, poaching and browsing of animals or the disturbance of the sward by heavy machines ceases. Otherwise it is able to maintain a dominance community for several years (Brandes 1977a).

Onopordum acanthium is well established especially in continental areas characterised by a summer-dry climate such as the Pannonian Basin, inner-Alpic basins and valleys (Braun-Blanquet *et al.* 1936; Braun-Blanquet 1961) and the south-Ucraine steppe zone (Solomakha *et al.* 1986; Kostil'ov 1987). *Onopordum acanthium, O. tauricum* and *O. acaule* are the only representatives of the genus in temperate Europe; most of the *Onopordum* species occur in the submediterranean and mediterranean regions (do Amaral Franco 1976).

In most of its natural distribution area *Onopordum acanthium* is considered a character-species of the *Onopordion acanthii* (for diagnosis see Brandes 1977a, 1977b; Mucina 1981). Under influence of dry continental climates the species shows a broader coenologic amplitude, while in subhumid and humid temperate climates it occurs secondarily also in *Artemisietalia* communities (Brandes 1977b). Several *Onopordum* communities were described (see below). They show a great floristic variability and extensive distribution over Europe (see Seybold & Müller 1972).

The aim of the paper is to numerically describe and interpret the variability within the complex of *Onopordum acanthium* communities in temperate and continental Europe in relation to geography and climate.

Materials and methods

The study was based on 380 relevés and another 9 constancy tables (the latter based on ca 140 relevés) of *Onopordum acanthium*-dominated communities and communities assigned to the *Onopordetum acanthii* by diverse authors. The following countries are covered: Austria, Bulgaria, Czechoslovakia, Federal Republic of Germany, France, German Democratic Republic, Hungary, Italy, the Netherlands, Poland, Rumania, Spain, Switzerland, USSR (Ucraine, Byelorussia and Bashkiria) and Yugoslavia.

First the relevés from the same geographically well-defined areas (sharing a similar climate and phytogeographic history) were combined into local tables (sensu Westhoff & van der Maarel 1978) which served for the construction of an initial synoptic table. Heterogenous and incomplete relevés (made on very large or very small sample plots) were discarded, as well as local tables composed of fewer than 5 relevés were not considered. The synoptic table was completed by adding nine constancy tables of the *Onopordum acanthium* communities from literature sources and represented only by synoptic values (columns). All species in the final synoptic table were expressed in the constancy-class scale of Braun-Blanquet (1964).

Numerical classification and ordination were performed on the final synoptic table with the constancy-class values 1–5 as scores. The clustering strategy was a complete linkage method (CLC) based on the Similarity Ratio. The computer program used was NCLAS of Podani (1984).

Ordination on the local tables collected in the synoptic table was performed by correspondence analysis CA (Hill 1974, 1979). No downweighting of rare species was carried out.

Climatic data on average monthly precipitation, minimal and maximal mean monthly temperature were compiled from Müller (1982) for climatic stations located in the areas of origin of the local tables. Two climatic indices were computed. Gorczynski's K' for thermic continentality (e.g. Daget 1977b, p. 10) and the Emberger's pluviometric Q combining humidity and temperature character of climate (e.g. Daget 1977a, p. 88, equation 3) were computed. Both K' and Q values on the ordination plane of CA (axes 1 and 2) were fitted using a Kriging algorithm (Ripley 1981; Jongman *et al.* 1987) and plotted by program package SURFER (Golden Software Inc.).

Results and discussion

Mediterranean Group

A great number of communities dominated by various *Onopordum* species is found in the Mediterranean which is considered to be the developmental centre of the genus (do Amaral Franco 1976). Some of them, occurring in the 'continental' meseta of central Spain (Fig. 1) contain *Onopordum acanthium* as the dominant species.

The Mediterranean group includes the *Verbasco pulverulenti-Onopordetum* from Extremadura (Rivas Goday 1964) and Avila (Rivas-Martínez 1975), the *Carthamo lanati-Onopordetum* from Salamanca and Zamora (Ladero Alvarez *et al.* 1983; Navarro Andrés & Valle Gutiérrez 1984), the *Carduo carpetani-Onopordetum* from the provinces of León, Ávila, Segovia and Madrid (Llamos Garcia 1984; Rivas-Martínez *et al.* 1986), and perhaps also the *'Onopordetum acanthii'* of Bellot Rodriguez (1966) from the mediterranean regions of Galicia. The *Onopordum acanthium*-rich subunit within the *Onopordetum illyrici* described from the surroundings of

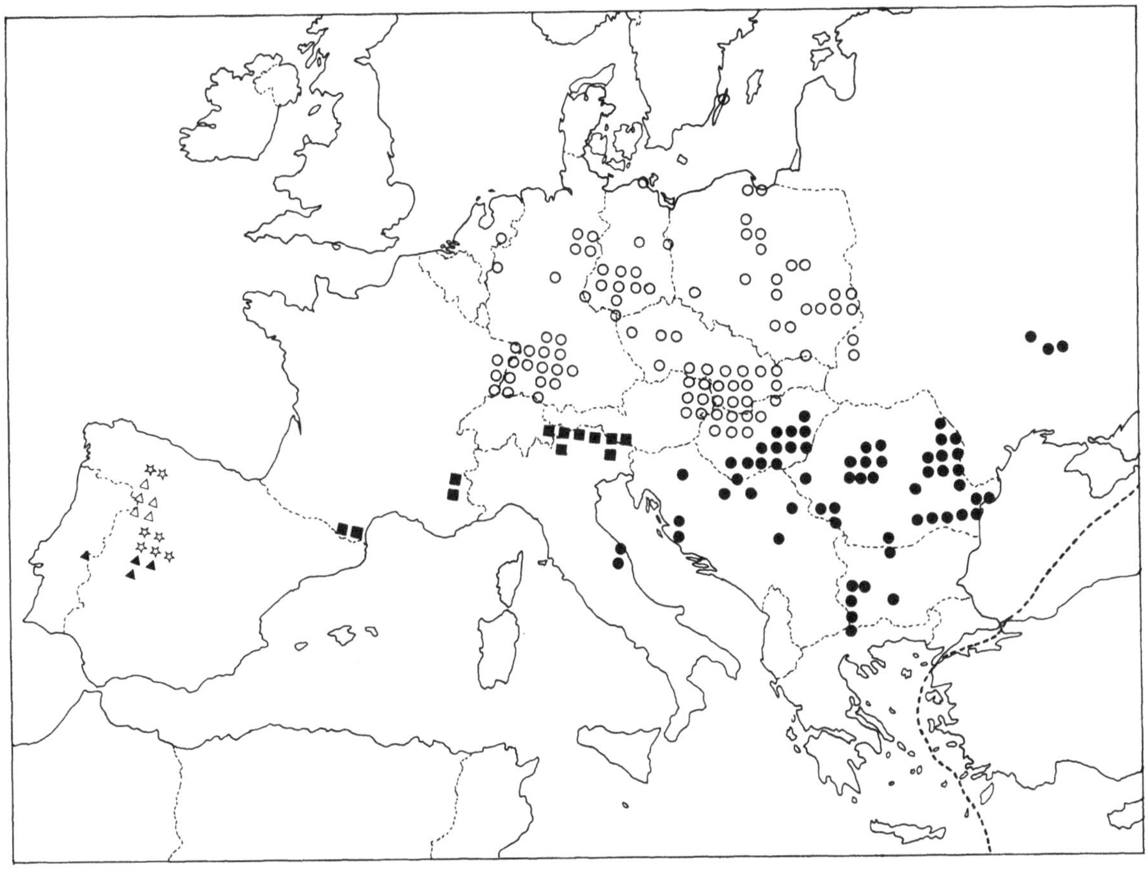

Fig. 1. Distribution of *Onopordum acanthium* communities in Europe superimposed on the UTM grid used for mapping in Flora Europaea. Empty circles – western race of the *Carduo acanthoidis-Onopordetum*, full circles – eastern race of the latter association, full squares – *Onopordetum acanthii s.str.*, stars – *Carduo carpetani-Onopordetum*, full triangles – *Verbasco pulverulenti-Onopordetum*, empty triangles – *Carthamo lanati-Onopordetum*.

Thessaloniké in northern Greece (Oberdorfer 1953/1954) also shows some similarity with the group discussed. These communities, assigned to the *Carduo carpetani-Cirsion* (the *Scolymo hispanici-Onopordetalia nervosi;* Ladero Alvarez *et al.* 1983; see also *Carthametalia lanati* by Brullo & Marcenó 1985) differ markedly in species composition from the *Onopordum acanthium* communities of temperate Europe, which belong to the *Onopordion acanthii* (the *Onopordetalia acanthii*). The latter two can be considered to be vicariant orders (Mucina in press). Due to the great floristic differences between the Mediterranean and temperate groups of the communities, the former is disregarded subsequently, as it requires a separate study.

Temperate and Continental Group

The original description of the only recognised community of this group to date the *Onopordetum acanthii,* was based on material from deep valleys of the Pyrenées (Braun-Blanquet *et al.* 1936). These locations, like the valleys of the Alps, are characterised by frequent temperature inversions and an overall continentality of the climate (Braun-Blanquet 1961; Izard *et al.* 1985). The Alpic localities of the *Onopordetum acanthii* range from the Dauphiné Alps in France to the eastern Alps of Switzerland (Engadin) and south and east Tyrolia. Braun-Blanquet *et al.* (1936) also indicated the occurrence of this community in the continental regions of eastern Europe.

The locations of the *Onopordum acanthium* communities are characterised by a dry (semi-arid to sub-humid) summer climate with high summer temperatures and cold winters. In the inner-Alpic basins the climate ranges from submediterranean (western Alps) to temperate subcontinental. As a rule, the well-developed *Onopordetum acanthii* occurs even in precipitation-rich western Europe in basins or broad valleys which lie in the rain shadows of adjacent mountain ranges running from north to south. These ranges form natural barriers for the precipitation-rich air masses as they move inland. Dry 'islands' are thus formed in the basins and valleys, which house xeric vegetation complexes showing an exclave distribution in western Europe. This is true in particular for the Rhine Valley from Baden to Westphalia lying in the shadow of the Vosges, Nordpfalz and Eifel; the so called 'mitteldeutsche Trockengebiet' (e.g. Schubert & Mahn 1959; Gutte 1972; Brandes 1977a), sheltered by the Harz and Thüringen Wald from the west; the dry north-Bohemian basins at the foot of the Krušné Hory; e.g. Kolbek 1975); the extensive system of the inner-Alpic basins (Braun-Blanquet 1961) and the Pannonian Basin surrounded by the Alps from the west, the Carpathians from the north and east, and the Dinarides from the south. In these regions, the *Onopordetum* and other *Onopordion acanthii* communities occur in a vegetation complex of a submediterranean-subcontinental character which includes the *Quercetalia pubescenti-petraeae, Festucetalia valesiacae,* and *Onopordion acanthii.* In other, more subhumid, regions the *Geranion sanguinei* is also involved (van Gils & Keysers 1977).

As shown by the synoptic table (full table and details on the sources available upon request from the author), the east-European type should be considered a separate unit, called *Carduo acanthoidis-Onopordetum acanthii* (Soó 1947) as it differs by a large number of differential-species. Probably *Centaurea solstitialis, C. calcitrapa, Anchusa ochroleuca, A. gmelinii, Echium italicum* and *Artemisia austriaca* could also be considered as regional character-species of the *Carduo acanthoidis-Onopordetum* in the Balkan part of its distribution area (Slavnić 1951; Oberdorfer 1953/1954; Dihoru & Doniţă 1970; Coste 1975). Raabe & Brandes (1988) consider this type an eastern race with *Carduus acanthoides* (of the *Onopordetum acanthii* sensu lato). Also the *Onopordetum acanthii* s.str. (the name was preserved for the vegetation type of continental mountain valleys of the Pyrenées and the Alps) is differentiated by an own group of species.

The *Onopordetum acanthii* s.str. is floristically most similar to the east-European race of the *Carduo acanthoidis-Onopordetum* (see below) as is

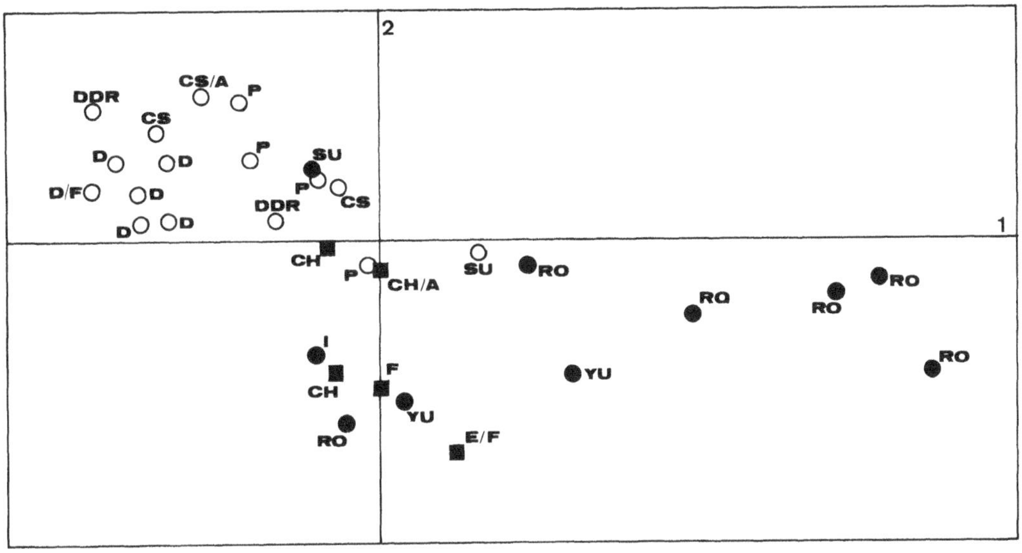

Fig. 2. CA ordination (axes 1 and 2) of the local tables of the communities studied. For symbols see Fig. 1. The abbreviations refer to the European countries involved.

also shown by the ordination diagram of axes 1 and 2 (Fig. 2). It is separated from the *Carduo-Onopordetum* clearly along axis 3 of the CA ordination (not reproduced).

Onopordum acanthium can be considered a regional character-species of the *Onopordetum acanthii* s.str. in the inner-mountain valley localities, but it also has high diagnostic value in the *Carduo acanthoidis-Onopordetum* in the continental lowland regions of eastern Europe (perhaps as a weak character-species). Its fidelity to the *Carduo acanthoidis-Onopordetum* is more pronounced at the edges of its distribution, e.g. in northwestern Europe (Brandes 1977a). The other character-taxa of the *Onopordetum acanthii* s.str. are *Carduus nutans* ssp. *platylepis,* limited to the Alpic area of the association (here having the rank of a local character-species, sensu Westhoff & van der Maarel 1978), and *Onopordum acaule* (limited to the Pyrenean area of the *Onopordetum acanthii* s.str.; cf. Braun-Blanquet *et al.* 1936; Braun-Blanquet 1948, p. 144). Accordingly, two geographic races can be recognised within the *Onopordetum acanthii* s.str., the Alpic (*Carduus platylepis* race) and the Pyrenean *(Onopordum acaule)* race (Fig. 1).

The within-association variability of the *Carduo acanthoidis-Onopordetum* is more complicated, as the unit covers a wide distribution (Fig. 1), the largest of the *Onopordion acanthii* communities. Two races with great floristic differences were revealed by a cluster analysis (Fig. 3).

The *Arrhenatherum* race is limited to the northern and western regions of central Europe and its distribution area lies between the Rhine River and the Polesie region in Byelorussia and Ucraine. This unit houses many species with a broad coenologic amplitude (the *Artemisietea* and *Galio-Urticetea* species), mesic species recruited

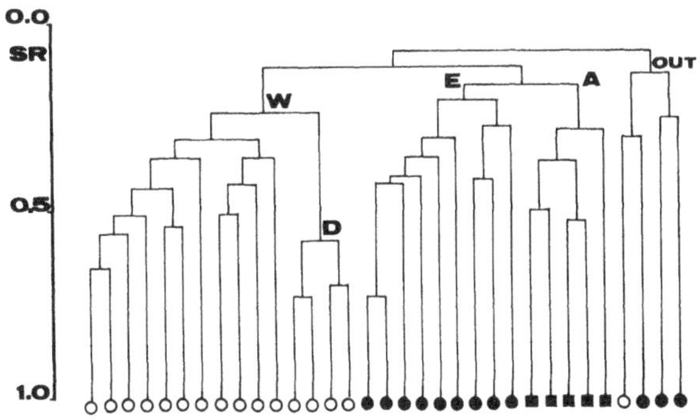

Fig. 3. CLC dendrogram of the local synoptic tables of the *Onopordum acanthium* communities in Europe. W – western race of the *Carduo acanthoidis-Onopordetum*, E – eastern race of the latter association, D – sub-race with *Malva alcea* from south-west Germany, A – the *Onopordetum acanthii* s.str., OUT – outlying tables (of transitional or fragmentary character). For symbols see Fig. 1.

from the elements of the *Molinio-Arrhenatheretea*, elements of nitrophilous fringes *(Glechometalia hederaceae)* and even some semi-hygrophilous species such as those of the *Convolvuletalia sepium*, *Lolio-Potentillion* and *Phragmiti-Magnocaricetea*, all of which penetrate the *Onopordum acanthium* communities from adjacent stands. The diagnostic species of the *Onopordion acanthii* are replaced by *Dauco-Melilotion* species. A shift in the phytosociological behaviour of several species of the *Onopordetalia acanthii* (considered as such in southeast Europe) can be observed within the *Carduo acanthoidis-Onopordetum* in west Europe. The species that possess broader amplitude in south-east Europe (*Onopordetalia* species) became more limited to this community (and the *Onopordion acanthii* in milder climate areas). The group of these species includes *Anchusa officinalis*, *Verbascum phlomoides*, *Reseda luteola* and *Echinops sphaerocephalus* (see also Brandes 1977a). In general, the community varies in mesic habitats, and impoverished forms of the community with some *Arction lappae* species are the rule (Brandes 1977b; Mucina 1981). A general retreat as a consequence of changing management practices is also observed in many regions of central Europe (for a review see Mucina in press).

Two subunits (sub-races) according to the presence of xerophilous elements can be distinguished within the western race of the community: the *Malva alcea* sub-race and the central race. The *Malva* sub-race is a local phenomenon of the upper Rhine Valley and close surroundings in south-western Germany (Baden and Schwaben). Submediterranean influence is demonstrated in this unit by presence of species such as *Ballota alba*, *Malva alcea*, *Isatis tinctoria*, *Rhynchosinapis cheiranthos* and *Geranium dissectum*. The central sub-race is again very varied, and covers a large area ranging from east Germany to northern Hungary and having more species able to survive harsh subcontinental winters. It is well-developed in dry regions such as the mitteldeutsche Trockengebiet, and the Pannonian Basin of south Slovakia and north Hungary, thus in the rain shadows of mountain ranges such as the Harz and Thüringer Wald, the Alps and the Carpathians.

In contrast, the east-European race (with *Xanthium* spp.) is more xeric and is characterised by summer annuals of (sub)continental distribution (such as *Amaranthus albus*, *A. blitoides*, *Euclidium syriacum*, *Atriplici-Sisymbrion* species) or by some C4-assimilation grasses *(Cynodon dactylon, Setaria viridis, Digitaria sanguinalis)*. This unit is the most species-rich within the group discussed. It is found in the south Pannonian Basin, in Transylvania, from Moldavia to the steppe regions of south Ucraine, in the Balkans and northern Italy (Hruška 1985; Brandes 1987). The extreme eastern area of its distribution in Bashkiria is considered to support a separate subrace differentiated by continental species such as *Axyris amaranthoides*, *Bassia sedoides*, *Cirsium setosum*, *Dracocephalum thymiflorum* and *Leonurus* spp. (Mirkin *et al.* 1986).

The separation of two races within the *Carduo acanthoidis-Onopordetum* is also supported by an ordination of the tables belonging to this association, which shows the two races to occupy two separate geographic regions (Fig. 4). Only one table from the Ucraine geographically belonging rather to the east-European race has joined the west-European unit (Fig. 3). A characteristically low number of taxa were found in that table (Solomakha *et al.* 1986).

The occurrence of *Onopordum acanthium*-dominated communities or their fragments north of the central-European mountain systems, particularly in the north-west European lowlands (Sissingh

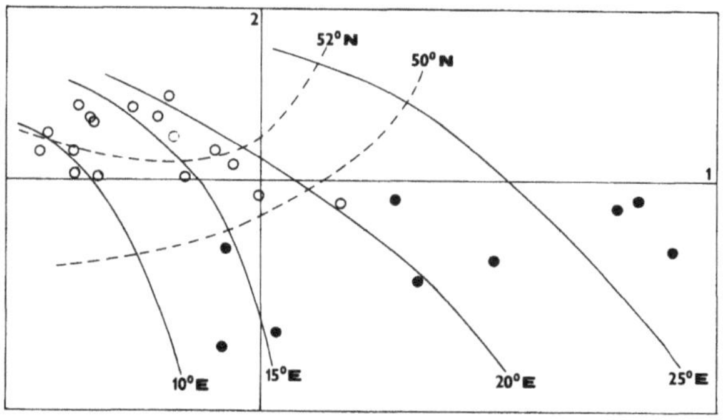

Fig. 4. CA ordination (axes 1 and 2) of the tables belonging to the *Carduo acanthoidis-Onopordetum* only. The geographic longitude (full lines) and latitude (dashed lines) are plotted. For symbols see Fig. 2.

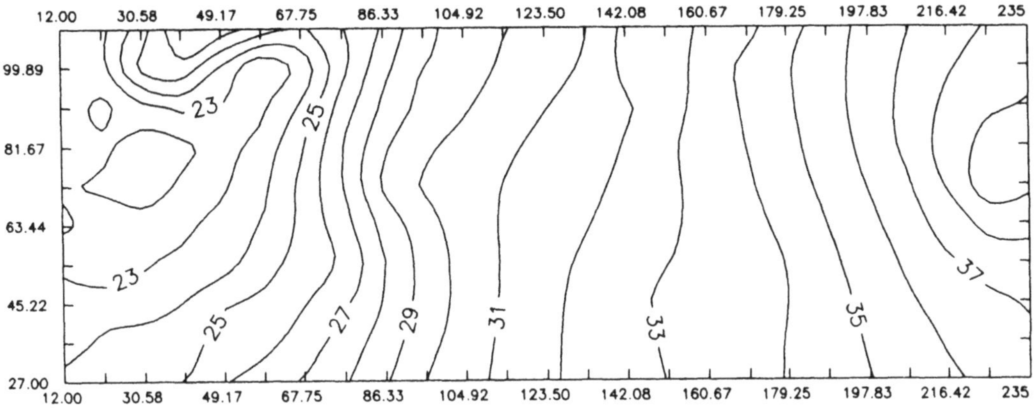

Fig. 5. Pattern of thermic continentality according to Gorczynski (K') on the ordination plane of axes 1 and 2 of the Correspondence Analysis. The interpolation method used was kriging and the plotting was carried out with program package SURFER (Golden Software Inc.).

1950; Westhoff & den Held 1969; Géhu *et al.* 1985; is supposedly of recent neophytic character (Fig. 1). Towards the east of the regions discussed, the climate has a subcontinental to continental character as is also shown by rising values of thermic continentality (Fig. 5) which favours the occurrence of a floristically rich *Carduo-Onopordetum acanthii.* The pluviothermic coefficient of Emberger generally decreases along axis 1 (Fig. 4) towards the climatically extreme of continental Europe (Fig. 6). The highest value of Q was scored for Ancona which has a mild submediterranean climate.

A similar pattern of two basic races with an opposite distribution pattern following the climatic differences between atlantic west Europe and continental east Europe was also observed with other ruderal communities such as the *Berteroetum incanae* (Mucina & Brandes 1985), *Malvetum neglectae* (Brandes 1981), *Sambucetum ebuli* (Brandes 1982), *Torilidetum japonicae* (Mucina in press) and can be considered a rather general phytogeographic phenomenon.

Acknowledgements

Data treatment was carried out and the first drafts of this paper prepared during the author's stay at the Dept of Geobotany, University of Nijmegen, the Netherlands in 1980–1981 (Fellowship in Biology) and later at the Dept of Geobotany of the

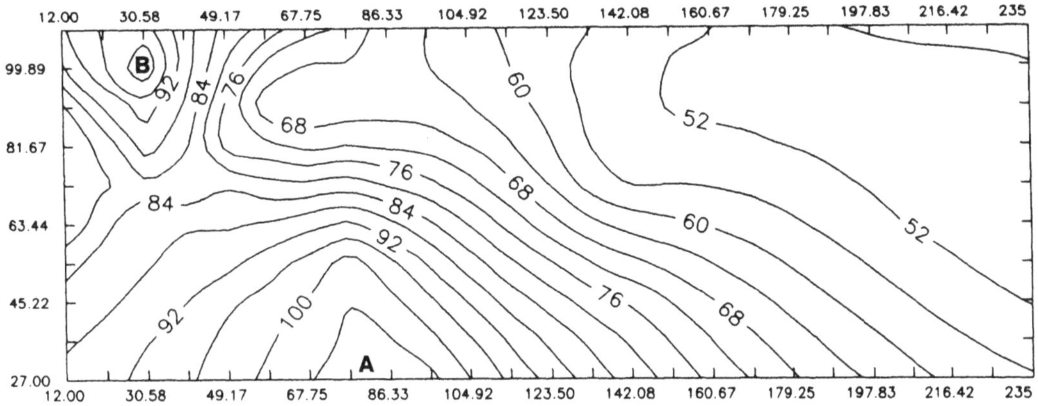

Fig. 6. Pattern of Emberger's pluviometric coefficient (Q) on the ordination plane of axes 1 and 2 of correspondence analysis A – Ancona, B – Berlin. For further details see Fig. 5.

Institute of Experimental Biology & Ecology, Slovak Academy of Sciences, Bratislava, Czechoslovakia.

Unpublished relevé material as well as data from the literature were generously provided by D. Brandes, G. Coldea, J. Czaplewska, W. Forstner, P. Gutte, K. Hruška, I. Jarolímek, P. Julve, S. Klotz, K. Kopecký, J. Kornaś, T. Krippelová, W. Kunick, W. Lohmeyer, L. Marković, A. Pyšek, A. Szabó, A. Terpó, V. Tlusták, I. Ullmann, V. Westhoff and M. Zaliberová. G. Schlosser kindly assisted with some computations. A. Szabó and E. van der Maarel contributed helpful comments on the paper and Jo Hughes corrected the English.

The financial support of the Deutsche Akademische Austauschdienst (DAAD) which made the presentation at the XIVth International Botanical Congress possible is much appreciated.

References

Bellot Rodriguez, T. 1966. La vegetación de Galicia. Anal. Inst. Bot. Cavanilles 24: 1–306.

Brandes, D. 1977a. Die *Onopordion*-Gesellschaften der Umgebung Braunschweigs. Mitt. Florist.-Soziol. Arbeitsgem. N.F. 19/20: 103–113.

Brandes, D. 1977b. Über *Onopordum acanthium*-Gesellschaften in Mitteleuropa. Doc. Phytosociol. N.S. 23–31.

Brandes, D. 1981. Über einige Ruderaigesellschaften von Verkehrsanlagen im Kölner Raum. Decheniana 134: 49–60.

Brandes, D. 1982. Das *Sambucetum ebuli* Felf. 1942 im südlichen Mitteleuropa und seine geographische Gliederung. Tuexenia 2: 47–60.

Brandes, D. 1987. Zur Kenntnis der Ruderalvegetation des Alpensüdrandes. Tuexenia 7: 121–138.

Braun-Blanquet, J. 1948. La végétation alpine des Pyrénées Orientales. Consejo Superior de Investigaciones Scientificas, Barcelona.

Braun-Blanquet, J. 1961. Die inneralpine Trockenvegetation. Eugen Ulmer Verlag, Stuttgart.

Braun-Blanquet, J. 1964. Pflanzensoziologie. Grundzüge der Vegetationskunde. 3. Aufl. Springer-Verlag, Wien, New York.

Braun-Blanquet, J., Gajewski, W., Wraber, M. & Walas, J. 1936. Prodromus des groupements végétaux. Fasc. 3. Classe des Rudereto-Secalinetales. Comm. S.I.G.M.A., Montpellier: 1–37.

Brullo, S. & Marcenó, C. 1985. Contributo alla conoscenza della vegetazione nitrofila della Sicilia. Coll. Phytosociol., 12: 23–148.

Coste, I. 1975. Contribution à l'étude de la végétation dans les Monts de Locva (Sud-Ouest de la Roumanie). – I. Doc. Phytosociol. 9–14: 63–71.

Daget, P. 1977a. Le bioclimat méditerranéen: Analyse des formes climatiques par le système d'Emberger. Vegetatio 34: 87–103.

Daget, P. 1977b. Le bioclimat méditerranéen: Caracter généraux, modes de caractérisation. Vegetatio 34: 1–20.

Dihoru, G. & Doniţă, N. 1970. Flora şi vegetaţia podişului Babădăg. Edit. Acad. Sci. R.P.R., Bucureşti.

do Amaral Franco, J. 1976. *Onopordum* L. in: Tutin, T.G. *et al.* (eds), Flora Europaea. Vol. 4. Plantaginaceae to Compositae (and Rubiaceae). pp. 244–248. Cambridge Univ. Press, Cambridge.

Géhu, J.-M., Géhu-Franck, J. & Scoppola, A. 1985. Schéma syntaxonomique des végétations nitrophiles et subnitrophiles de la Région Nord/Pas-de-Calais. Coll. Phytosociol. 12: 567–575.

Gutte, P. 1972. Ruderalpflanzengesellschaften West- und Mittelsachsens. Feddes Repert. 83: 11–122.

Hill, M.O. 1974. Correspondence analysis: a neglected multivariate method. J. Roy. Stat. Soc., Ser. C 23: 340–354.

Hill, M.O. 1979. DECORANA. A Fortran program for detrended correspondence analysis and reciprocal averaging. Ecology and Systematics, Cornell University, Ithaca.

Hruška, K. 1985. Ruderal xerothermic vegetation in the Marche (central Italy). Coll. Phytosociol. 12: 149–154.

Izard, M., Casanova, H., Devau, B. & Pautou, G. 1985. Continentalité et notion de zone interne dans les Pyrénées. Acta Ecol., Ecol. Gener. 6: 317–344.

Jongman, R.H.G., ter Braak, C.J.F. & van Tongeren, O.F.R. 1987. Data analysis in community and landscape ecology. Pudoc, Wageningen.

Kolbek, J. 1975. Die Festucetalia valesiacae-Gesellschaften im Ostteil des Gebirges České středohoří (Böhmisches Mittelgebirge). 1. Die Pflanzengesellschaften. Fol. Geobot. Phytotax. 10: 1–57.

Kostil'ov O.V. 1987. Rostinnist' skhiliv Kayal'nits'kogo limanu. Ukr. Bot. Zh., Kiiv, 44: 81–84.

Ladero Alvarez, M., Navarro Andrés, F. & Valle Gutiérrez, C.J. 1983. Comunidades nitrófilas Salamantinas. Stvd. Bot. Salamanca 2: 7–67.

Llamos García, F. 1984. Flora y vegetación de la Maragateria (León). Institution Fray Bernardino de Sahagura de la Excma. Diputación Provincial de León, León.

Mirkin, B.M., Abramova, L.M., Ishbirdin, A.R., Rudakov, K.M., Sakhapov, M.T. & Solomeshch, A.I., 1986. Sintaksonomiya ruderal'noi rastitel'nosti Bashkirii. V. Klass Artemisietea vulgaris. Min. Vyssh. Sred. Spetsial. Obrazov. SSSR, Moskva.

Mucina, L. 1981. Die Ruderalvegetation des nördlichen Teils der Donau-Tiefebene. 1. *Onopordion acanthii*-Verband. Fol. Geobot. Phytotax. 16: 225–263.

Mucina, L. in press. Endangered ruderal communities of Slovakia and their preservation. Phytocoenologia.

Mucina, L. in press. Synchorology of weed vegetation of Europe. Tasks Veget. Sci.

Mucina, L. & Brandes, D. 1985. Communities of *Berteroa incana* in Europe and their geographical differentiation. Vegetatio 59: 125–136.

Müller, M.J. 1982. Selected climatic data for a global set of standard stations for vegetation science. Junk, The Hague.

Müller-Schneider, P. 1986. Verbreitungsbiologie der Blütenpflanzen Graubündens. Veröff. Geobot. Inst. Rübel 85: 1–263.

Navarro Andrés, F. & Valle Gutiérrez, C.J., 1984. Vegetación herbacea del centre-occidente Zamorano. Stvd. Bot. Salamanca, 3: 63–177.

Oberdorfer, E. 1953/1954. Über Unkrautgesellschaften der Balkanhalbinsel. Vegetatio 4: 379–411.

Podani, J. 1984. SYN-TAX II. Computer program for data analysis in ecology and systematics. Abstr. Bot. 8: 73–94.

Raabe, U. & Brandes, D. 1988. Flora und Vegetation der Dörfer im nördlichen Burgenland. Phytocoenologia 16: 225–258.

Ripley, B. 1981. Spatial statistics. J. Wiley & Sons, N.Y.

Rivas-Goday, S. 1964. Vegetación y florula de la cuenca extremeña del Guadiana. Publ. Excma. Diputac. Prov. Bajadoz, Madrid.

Rivas-Martínez, S. 1975. Mapa de la vegetación de la provincia de Ávila. Anal. Inst. Bot. Cavanilles 32: 1493–1556.

Rivas-Martínez, S., Peñas, A. & Diaz Gonzales, T.E. 1986. Datos sobre vegetación terofitica y nitrófila leonesa. Nota. II. Acta Bot. Malacit., Malaga, 11: 273–288.

Schubert, R. & Mahn, E.-G. 1959. Vegetationskundliche Untersuchungen in der mitteldeutschen Ackerlandschaft. I. Die Pflanzengesellschaften der Gemarkung Friedeburg (Saale). Wiss. Z. Martin-Luther-Univ. Halle-Wittemberg, Math.-Nat. R. 8: 865–1012.

Seybold, S. & Müller, T. 1972. Beitrag zur Kenntnis der Schwarznessel (*Ballota nigra* agg.) und ihre Vergesellschaftung. Veröff. Landest. Natur. Baden-Württ. 40: 51–126.

Sissingh, G. 1950. Onkruid-associaties in Nederland. Thesis Wageningen. Staatsdrukkerij, 's-Gravenhage.

Slavnić, Z. 1951. Pregled nitrofilne vegetacije Vojvodine. Nauc. Zborn. Matice Srpske, Novi Sad, 1: 84–169.

Solomakha, T.D., Solomakha, V.A. & Shelyag-Sosonko, Yu.R. 1986. Osnovni asotsiatsii ruderal'noi roslinnosti livobrezhnogo lisostepu Ukraini. Ukr. Bot. Zh., Kiiv, 43: 70–75.

Soó, R. 1947. Revue systématique des associations végétales des environs de Kolozsvár. Acta Geobot. Hung. 6: 3–50.

van Gils, H. & Keysers, E. 1977. Die *Geranion sanguinei*-Arten in verschiedenen Klimagebieten. In: Dierschke, H. (ed.), Vegetation und Klima. pp. 199–315. J. Cramer, Vaduz.

Westhoff, V. & den Held, A.J. 1969. Planten-Gemeenschappen in Nederland. Thieme & Cie, Zutphen.

Westhoff, V. & van der Maarel, E. 1978. The Braun-Blanquet approach. In: Whittaker, R.H. (ed.), Classification of plant communities. pp. 287–399. Dr. W. Junk, The Hague.

Vegetatio **81**: 117–125, 1989.

A coenocline of the high-ranked syntaxa of ruderal vegetation

Ladislav Mucina[1] & Onno F.R. van Tongeren[2]
Department of Geobotany, University of Nijmegen, Toernooiveld 1, NL-6525 ED Nijmegen, the Netherlands; [1]Present address: Department of Vegetation Ecology and Nature Conservancy, Vienna University, Althanstrasse 14, A-1091 Vienna, Austria; [2]Present address: Limnological Institute, Vijverhoflaboratorium, Rijksstraatweg 6, NL – 3631 AC Nieuwersluis, the Netherlands

Accepted 14.12.1988

Keywords: Correspondence analysis, Numerical syntaxonomy, Order, Ordination, Slovakia, Synanthropic vegetation

Abstract

Two sets comprising 1419 and 1350 phytosociological relevés of ruderal vegetation classified into 9 and 7 orders, respectively, from an area in the Podunajská nížina Lowland, western Slovakia were ordinated using correspondence and detrended correspondence analyses. The paper describes a coenocline of the high-ranked syntaxa contained in the data sets, and discusses some issues of the classification of the syntaxa involved.

The *Bidentetalia* and *Potentillo-Polygonetalia* should be considered a special category of synanthropic vegetation as habitat moisture (flooding and waterlogging) play the controlling role in the formation of structural and dynamical patterns within these communities. The latter factor complex is responsible for the clear separation of these orders from the other syntaxa included in the ordinations. The coenocline of terrestrial ruderal vegetation units has the following sequence along the CA axis 1: *Poo-Polygonetalia, Sisymbrietalia, Eragostrietalia, Onopordetalia, Agropyretalia repentis, Artemisietalia vulgaris* and *Glechometalia hederaceae.*

Various complexes of soil factors and anthropogenic disturbance are operational along the *concatenation* (a series of portions composing the coenocline). The factor complex includes soil compaction, trampling disturbance, nutrient status, soil texture and moisture, and solar irradiation. Step-by-step ordination and interpretation of concatenated portions of the coenocline proved to be useful in the analysis of complex data sets.

Abbreviations: CA = correspondence analysis, DCA = detrended correspondence analysis, PCA = principal components analysis

Nomenclature: for syntaxa: Mucina & Maglocký (1985) and Mucina (in press).

Introduction

Ruderal, synanthropic (man-made or man-controlled) vegetation comprises a varied spectrum of communities, notably summer-therophyte communities on disturbed shores *(Bidentetalia)*, temporarily flooded, heavily grazed swards *(Potentillo-Polygonetalia)*, communities on friable, loamy and sandy tips *(Sisymbrietalia* and *Eragrostietalia),* and abandoned land and rubbish tips *(Onopordetalia, Agropyretalia repentis* and *Artemisietalia vulgaris),* and nitrophilous woodland and hedgerow fringes *(Glechometalia hederaceae).* It became a very popular object of geobotanical studies especially in East European countries and West Germany as reflected in numerous vegetation surveys (for extensive lists of references and the characteristics of the units see for instance Gutte 1972; Hejný *et al.* 1979; Oberdorfer & Müller 1983; Mucina in press).

The high-ranked syntaxa of ruderal vegetation were mainly defined in the period 1950–1967. The hierarchical system built up during that period is still followed by many workers as it includes units that are well-defined both floristically and ecologically.

This paper presents some new insights into the systematic relationships of the high-ranked units (particularly orders and classes), and describes a coenocline of the syntaxa. The floristic-sociological classification of these data was elaborated by Mucina (in press).

Material and methods

The data set is from the western part of the Podunajská nížina Lowland, Slovakia (Czechoslovakia). The area is bordered by the Váh River on the east, the Danube on the south, and the Malé Karpaty Mts. to the west and north-west. The region is characterized by a slightly subcontinental climate (for climatic accounts see Mucina 1982 in press). Most of the area is built up of Quaternary sediments, including the alluvial gravels of the Podunajská rovina Plain and the loess of the Trnavská pahorkatina Hill-land. The population density is ca 250 inhabitants per km^{-2}. The lowland area is used mainly for agricultural purposes, and woodlands cover only small areas along rivers. More extensive woodlands are found on the slopes of the mountain ranges adjacent to the lowland. (Mucina 1982 in press).

Over the period 1975–1984, 1419 relevés (with 771 taxa) of ruderal vegetation were collected in this region. The vegetation was classified into 90 plant communities using the syntaxonomical procedures of the Braun-Blanquet (1964) approach. Some of the vegetation types (e.g. different subassociations within the same association) were represented by several local tables comprising relevés from different parts of the study area. Thus, there are 121 local tables of the communities belonging to the *Bidentetalia tripartitae, Potentillo-Polygonetalia* (syn. *Agrostietalia stoloniferae*), *Poo-Polygonetalia* (syn. *Plantaginetalia majoris*), *Sisymbrietalia, Eragrostietalia, Onopordetalia, Artemisietalia vulgaris, Agropyretalia repentis* and *Glechometalia hederaceae* (syn. *Galio-Alliarietalia*). For the characteristics of the units see Mucina (in press).

The relevé data set was ordinated in two steps. In the first step all relevés were included; later the *Bidentetalia* and *Potentillo-Polygonetalia* were excluded and the remainder (1350 relevés) were reordinated. The ordinations were carried out using correspondence analysis (CA; Hill 1973) and detrended correspondence analysis (DCA; Hill & Gauch 1980). No downweighting of rare species and no subjective rescaling of axes were adopted. The programme DECORANA (Hill 1979) was used for computations.

Centroids of the clusters in the ordination plane were computed. To obtain an indication of the within-cluster variation in the ordination plane, the first and second axes scores of the correspondence analysis for each cluster syntaxonomically corresponding to the orders were used as input for a centered PCA (see e.g. Orlóci 1978; Jongman *et al.* 1987). The programme PRINCOMP of Podani (1984) was used for the PCA. Standard deviations in two directions were computed from a partial sum of squares (derived from the PCA) according to the following formula:

$$\mathrm{SD} = \mathrm{SQRT}\ (\text{partial sum of squares})/(n - 1)\,, \qquad (1)$$

where

n is the number of relevés in the cluster, and the partial sum of squares is a product of an eigenvalue and the particular total sum of squares.

Orientation of the axes of the constructed

ellipses was determined by the loadings of the CA axes. Both centroid position and standard deviations were used to draw ellipses of equal concentration on the CA ordination plane of axes 1 and 2. This procedure was similar to that of Lagonegro & Feoli (1985).

Results

The *Bidentetalia* and *Potentillo-Polygonetalia* are clearly separated from the remaining orders along CA axis 1. Both orders are 'semiterrestrial' units (comprising shore communities) while the others are 'terrestrial' ones. Axis 1 in this ordination (Fig. 1) can be explained as an ecocline of water dynamics.

In the next step the *Bidentetalia* and *Potentillo-Polygonetalia* relevés were removed. The ordination of this reduced data set (Fig. 2) unfolded a 'wedge-like' arch, detecting a clear coenocline

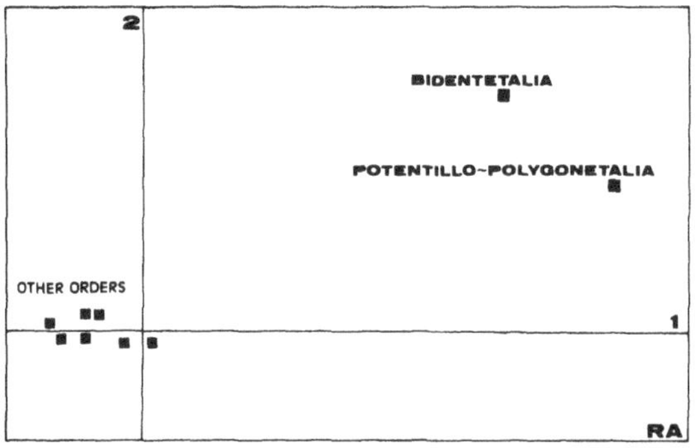

Fig. 1. First-step ordination (CA) (axes 1 and 2). Centroids of syntaxonomic orders are plotted.

with well-distinguished, compact and more or less isolated clusters of centroids of the local tables which represented 7 orders in this particular case. The first arm of the wedge was formed by the *Poo-Polygonetalia* and *Sisymbrietalia* clusters. The other arm was composed of the *Agropyretalia repentis, Artemisietalia vulgaris* and *Glechometalia*

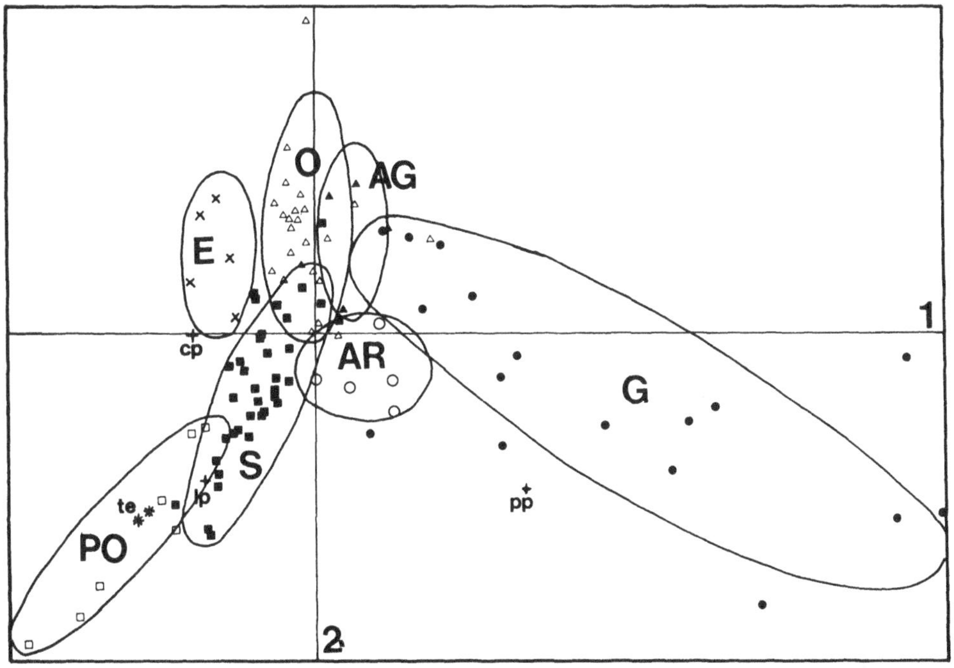

Fig. 2. Second-step ordination (reduced data set after the removal of the *Bidentetalia* and *Potentillo-Polygonetalia*); projection of axes 1 and 2. Centroids of syntaxonomic orders and concentration ellipses are plotted. Legend: PO – *Poo-Polygonetalia* (empty squares), E – *Eragrostietalia* (crosses), S – *Sisymbrietalia* (fulll squares), O – *Onopordetalia* (empty triangles), AG – *Agropyretalia repentis* (full triangles), AR – *Artemisietalia vulgaris* (empty circles), G – *Glechometalia hederaceae* (full circles); + – hemicryptophytic trampled communities: *Prunello-Plantaginetum* (pp), *Lolio-Plantaginetum* (lp) and *Cynodonto-Plantaginetum* (cp); te (*) – trampled communities belonging to the *Eragrostietalia*.

hederaceae. The *Eragrostietalia* and *Onopordetalia* were found at the joining of the arms and, therefore, were difficult to assign to either. At the same time, the *Onopordetalia* and *Agropyretalia repentis* were separable along axis 2. The *Eragrostietalia* communities were scattered over a contact zone between the *Sisymbrietalia* and *Poo-Polygonetalia* clusters in the discussed ordination scheme. The position of the *Artemisietalia vulgaris* close to the *Glechometalia hederaceae* as well as the *Onopordetalia* is indicative of a systematic similarity among the respective orders. Some of the communities, for instance the *Artemisio-Tanacetetum (Dauco-Melilotion)* or *Conio-Chaerophylletum bulbosi (Galio-Alliarion)*, exemplify typical transitions between the *Artemisietalia* and *Onopordetalia*, or between the *Artemisietalia* and *Glechometalia*, respectively.

Xero-thermophilous character-species of the *Sedo-Scleranthetea, Festuco-Brometea* and *Onopordetalia* were concentrated on the negative pole of axis 3 (not reproduced). The opposite pole was characterized by woodland species, which penetrate the nitrophilous fringe communities of the *Glechometalia hederaceae* from adjacent woods. The coenocline detected along axis 3 can be explained by intensity of solar irradiation and perhaps also by soil moisture, because the species groups characteristic of the coenocline poles are composed of either heliophilous and (sub)xerophilous or (sub)sciophilous and mesophilous species.

The coenocline detected on the ordination plot of axes 1 and 2 (Fig. 2) was largely identical with the data configuration found on the projection of axes 1 and 3 in the first step of the ordination (when also the *Bidentetalia* and *Potentillo-Polygonetalia* were included).

It is improbable that the detected coenocline is a reflection of a single, though a very complex, ecological gradient because of both the complexity of the data sets and the span of the coenocline. Instead a partial interpretation of the coenocline fractions should be considered, that is, the coenocline should be interpreted as series of concatenated portions, each portion reflecting the operation of a distant collection of environmental variables. At first sight, the coenocline appears to be rather continuous (Fig. 2). However, local discontinuities within the coenocline can be observed by a more profound inspection. These can be seen as alternation points of the controlling ecological factors operating along the coenocline. Soil compactness is a striking factor which discriminates the *Poo-Polygonetalia* from the *Sisymbrietalia* (Fig. 3; see also Krippelová 1972; Mucina in press). Within the *Poo-Polygonetalia* cluster itself there is a contrast between the xerophilous *Sclerochloo-Polygonetum* and mesophilous *Prunello-Plantaginetum*. The internal structure of the *Poo-Polygonetalia* is interpreted in terms of a soil-moisture gradient. The *Sisymbrietalia* cluster is structured according to the mechanical composition of the soil (Mucina 1982). The overlap of the *Potentillo-Polygonetalia* and *Sisymbrietalia* clusters is located in the position of the *Malvion neglectae* (Fig. 3). The sequence of the therophyte-rich ruderal communities along the coenocline is completed by the *Eragrostietalia*. The pattern within the coenocline fraction starting with the *Malvion neglectae*, through the *Atriplici-Sisymbrion* towards the *Sisymbrion officinalis* and *Salsolion ruthenicae (Eragrostietalia)* indicates also that the nutrient status of the soil is of importance in structuring this portion of the coenocline. Skeleton content is also rising along this coenocline fraction. Soil texture, particularly skeleton content, operates also within the *Onopordetalia* (Mucina 1982). This coenocline portion spans between the *Onopordion acanthii* communities (loam and nitrogen-rich soils) and the *Dauco-Melilotion*. The *Erysimo-Epilobion dodonaei (Thlaspietea rotundifolii)* communities occupy the extreme end of this fraction (Fig. 3). These are typical of skeletal habitats on riverin gravel banks where the *Dauco-Melilotion* and *Thlaspietea rotundifolii* species form peculiar assemblages.

The communities of the right arm of the wedge rank from the xerophilous *Onopordetalia* and *Agropyretalia repentis* to the mesophilous *Glechometalia hederaceae*. The latter unit includes both xerophilous/heliophilous and mesophilous/subsciophilous communities showing also features of semiruderal character. The intermediate

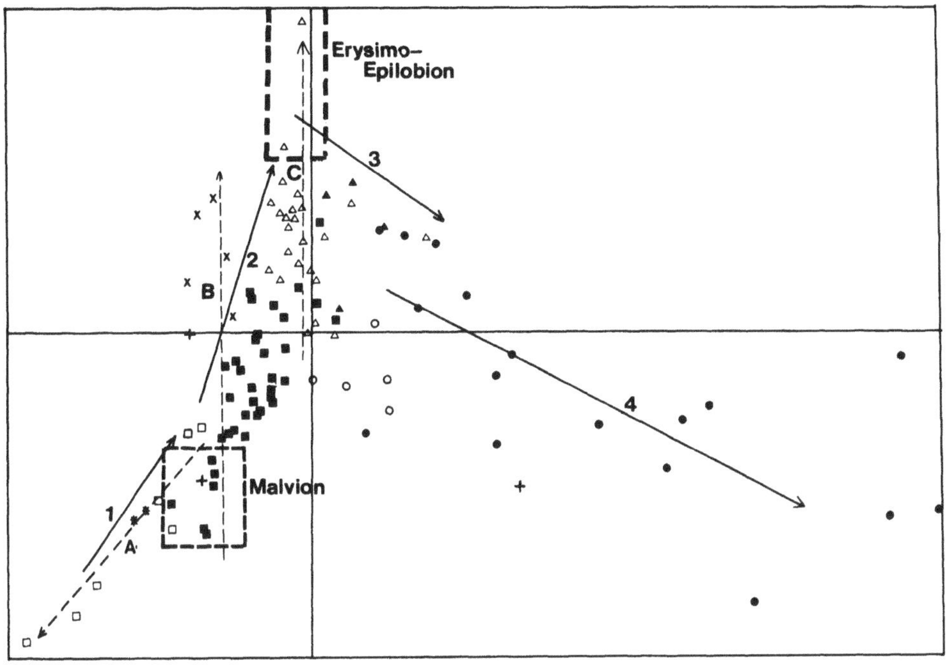

Fig. 3. Second-step ordination (see Fig. 2). Concatenated portions of the coenocline are indicated: 1–4 operational spheres of environmental factors (or their complexes): 1 – trampling (decreasing), 2 – mechanical composition of soil (from less to more skeletal), 3 – contents of nitrogen (increasing), 4 – habitat insolation (decreasing); A–C: within-cluster gradients: A: from mesic *(Poetum annuae)* to dry *(Schlerochloo-Polygonetum)*, B – from mesic and loamy *(Malvion)* to dry and sandy soils *(Salsolion)*, C – from less *(Onopordion)* to more skeletal *(Erysimo-Epilobion)*.

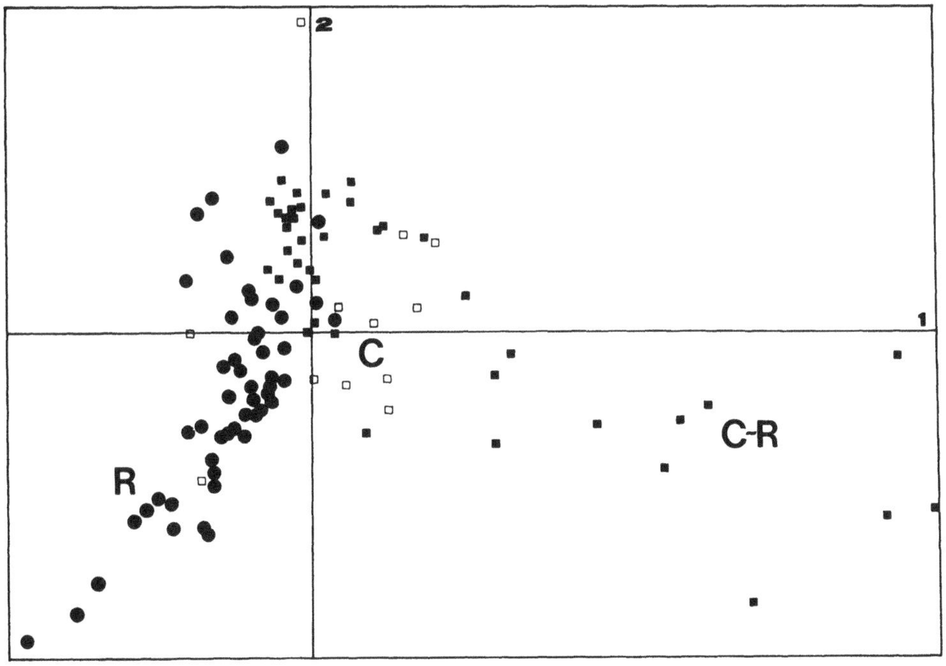

Fig. 4. Pattern of life-strategy types along the observed coenocline (see Fig. 2). The life-strategy was determined for dominants of the particular communities according to Grime (1984, p. 31). Legend: R – ruderals (full circles), C – competitors (empty squares), C–R – competitive ruderals (full squares).

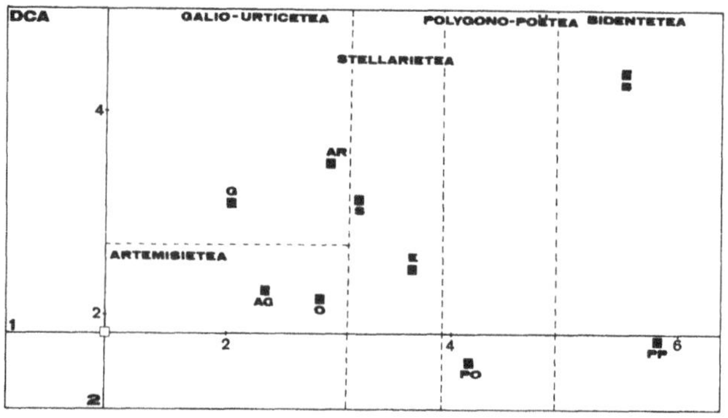

Fig. 5. Ordination (DCA) of the entire data set (including the *Bidentetalia* and *Potentillo-Polygonetalia).* Broken lines separate syntaxonomic classes for which centroids are plotted. Legend: B – *Bidentetalia*, PP – *Potentillo-Polygonetalia;* for other abbreviations see Fig. 2.

position of the *Agropyretalia repentis* between the *Onopordetalia* and *Glechometalia hederaceae* is a corollary of the pronounced floristic similarity of the respective units.

The importance of perennials rises along the described coenocline from the *Poo-Polygonetalia* towards the *Glechometalia hederaceae*. Accordingly, ruderals (R-strategists sensu Grime 1979) are replaced by C-strategists and these by C-R secondary strategists along the discussed coenocline (Fig. 4). As a rule, both intensity and frequency of human interference also decrease along the coenocline.

DCA was performed only with the entire data set (Fig. 5). The *Bidentetalia* and *Potentillo-Polygonetalia* were separated from the rest of the orders on the positive pole of the axis 1. The ranking of the other orders along axis 1 corresponded to the sequence detected by the CA in the second step of ordination (Fig. 2). The 4 orders originally classified within the *Artemisietea vulgaris* (Mucina & Maglocký 1985) were distributed along axis 2 in such a manner that two groups of orders could be recognized *(Agropyretalia/Onopordetalia,* and *Artemisietalia/Glechometalia).*

Discussion

The results of the CA show the so-called 'horseshoe', 'arch' or 'wedge' pattern (Hill & Gauch 1980; van der Maarel 1980; ter Braak 1987). The ordinated items follow a curvilinear trace in the ordination scheme, and this is claimed to be an outcome of non-linear relations between the species importance values and ecological factors (Hill & Gauch 1980; Gauch 1982; Dargie 1986). The horseshoe was considered a failure, as the method failed to reveal ordination pattern related in a linear manner to an external variable (i.e. an ordination axis). The horseshoe produced by CA is perhaps better regarding as a redescription of replacement-series data (e.g. Wartenberg *et al.* 1987) reflecting a coenocline if any is present. The pattern of the ordinated entities along axis 2 is usually difficult (or impossible) to interpret satisfactorily due to the functional dependence of the axis on axis 1. The interpretation is then supposed to be limited to the coenocline presented as an arch in the ordination scheme (van der Maarel 1980; Feoli & Feoli Chiapella 1980). An ecocline is supposed to underlie the observed coenocline (Noy-Meir & Whittaker 1977). When the data are relatively homogeneous the assumption of a complex ecocline is realistic. Van der Maarel (1980) suggested that direct or indirect ecological factors (sensu Austin 1980; Austin *et al.* 1984) can alternatively dominate fractions of a seemingly homogeneous ecocline. This implies that an ecologic factor may not be equally responsible for pattern generation along the entire coenocline. Peet (1980) also suggested that fractions of a large data set should be analysed, and that interpretation can be made with the partial ordinations of the data subsets. A combination of these approaches is particularly appealing when complex and large data sets are handled. Nevertheless, the stepwise interpretation of the observed coenocline does not exclude the possibility of interpreting the data structure along the ordination axes themselves as shown by this paper. This duality of approach towards interpretation of data structures on the ordination plane is, however, more the exception than the rule.

The recovery of an ordination pattern on the plane delimited by axes 1 and 2 of the reduced data set (Fig. 2) similar to that encountered on the projection of axes 1 and 3 in the first step of the ordination can be ascribed to the outlying position of the *Bidentetalia* and *Potentillo-Polygonetalia* clusters. Axis 1 (Fig. 1) is then an axis of outliers, whereas the ecologically interpretable information is depicted by projection of high-order axes. This observation supports the idea of discarding outliers or outlying groups, thus removing group structure, when continuous patterns are sought.

The clusters of *Glechometalia hederaceae* and *Poo-Polygonetalia* show conspicuous dispersion on the ordination plane (Fig. 2) despite being less diverse than the clusters of the *Sisymbrietalia* and *Onopordetalia* lying in the centre of the plane and showing low dispersion in terms of standard deviations and space occupied by the concentration ellipses (Lagonegro & Feoli 1985). This phenomenon is rather strange in view of one of the properties of CA, namely the contraction of a coenocline at the poles (Hill 1973, 1974; Hill & Gauch 1980). One of the reasons for the dispersion of the clusters might be a species-richness (alpha-diversity) effect (Dargie 1986) or an outcome of the size of submatrices (incl. number of relevés per community type). There are certainly many local (rare) species found in communities of the *Glechometalia hederaceae;* these are mainly woodland species not shared with the other ruderal communities. Another source of this dispersion is syntaxonomic heterogeneity, again rather high with the *Glechometalia hederaceae* (Mucina in press). The influence of syntaxonomic heterogeneity could be documented with the *Poo-Polygonetalia* (Fig. 2). When also the *Lolio-Plantaginetum, Prunello-Plantaginetum* and *Eragrostio-Polygonetum* (classified within the *Lolio-Plantaginion, Alchemillo-Ranunculion* and *Eragrostio-Amaranthion crispi,* respectively) are considered to belong to the *Poo-Polygonetalia* cluster, in order to reconstruct the broadly-conceived *Plantaginetalia* (see for instance Mucina & Maglocký 1985), the standard deviations and area occupied by the concentration ellipse would increase markedly.

Classification of ruderal plant communities on the order level is generally free from problems. These arise largely when the type is to be assigned to suballiance or alliance (Kopecký & Hejný 1974). Classification of orders into classes can also present problems as a formalized classification procedure is still not agreed (Mucina in press). The separation of the *Bidentetalia* and *Potentillo-Polygonetalia* from the other orders of ruderal vegetation highlights this particular problem. The ordinations (Figs 1 and 5) show that the *Potentillo-Polygonetalia* is more similar to the *Bidentetalia* than to the *Poo-Polygonetalia,* suggesting that the traditional classification of both units into the *Plantaginetea majoris* (Tüxen 1950; Hejný *et al.* 1979 etc.) is not acceptable for the data analysed here. It is still not clear whether the classification of the *Potentillo-Polygonetalia* (syn. *Agrostietalia stoloniferae)* in a separate class (*Agrostietea stoloniferae;* Görs 1968; Oberdorfer 1983), the inclusion of the *Bidentetalia* and *Potentillo-Polygonetalia* in a common superior unit (*Bidentetea tripartitae;* Mucina in press), or the classification of the *Potentillo-Polygonetalia ('Trifolio-Agrostietalia stoloniferae')* within the *Molinio-Arrhenatheretea* (Tüxen 1970) would be the most plausible solution for East-European phytogeographic conditions, as suggested by the present ordination analyses. However, the classification of the *Potentillo-Polygonetalia* and *Poo-Polygonetalia ('Plantaginetalia')* into a common class *(Plantaginetea majoris)* might be preferable in west and northwest Europe (Sýkora 1983; Mucina & Jarolímek 1986). If the *Plantaginetea majoris* is not considered in its original concept (Tüxen 1950), the trampled communities of the *Poo-Polygone-*

talia must be referred to the *Polygono-Poetea annuae* (Rivas-Martinéz 1975; Mucina in press).

The juxtaposition of the *Sisymbrietalia* and *Eragrostietalia* in the ordinations (Figs 2 and 5) supports their classification into one common class, the *Stellarietea mediae* (Krippelová & Mucina 1988). The overlap between the *Sisymbrietalia* and *Onopordetalia* reflects the effect of the warm and dry climate whereby many diagnostic species of both orders possess broader coenological valence (Mucina in press).

The most diverse class in this context is the *Artemisietea vulgaris*, which might include up to 5 orders, ranging from the xerophilous *Onopordetalia* to the semi-natural and natural wateredge *Convolvuletalia sepium* (for broadly-conceived concepts of the *Artemisietea vulgaris* consult Müller 1983 and Mucina & Maglocký 1985). The separation of the two groups of orders along axis 2 in DCA (Fig. 5) supports the concept of either two subclasses (Müller 1983) or separate classes, the *Artemisietea vulgaris* (approximately sensu the original concept of Tüxen 1950) and the *Galio-Urticetea* (Passarge 1967; Kopecký 1969).

Acknowledgements

The senior author was a recipient of the Research Fellowship in Biology to the University of Nijmegen 1980/1981. During that period, at the Department of Geobotany, this paper was initiated. Part of the computations was done in the Computer Centre of the Slovak Academy of Sciences in Bratislava, Czechoslovakia with the assistance of Gabo Schlosser, Vilo Čík and Pal'o Ďuriš. Gerry Doyle (Dublin), Enrico Feoli (Trieste), and Mike Dale (St. Lucia, Australia) and Jo Hughes (Trieste) kindly commented on the manuscript and corrected the English. Their help is appreciated.

References

Austin, M.P. 1980. Searching for a model for use in vegetation analysis. Vegetatio 42: 11–21.

Austin, M.P., Cunningham, R.B. & Fleming, P.M. 1984. New approaches to direct gradient analysis using environmental scalars and statistical curve-fitting procedures. Vegetatio 55: 11–27

Braun-Blanquet, J. 1964. Pflanzensoziologie. Grundzüge der Vegetationskunde. 3.Aufl. Springer-Verlag, Wien, New York.

Dargie, T.C.D. 1986. Species richness and distortion in reciprocal averaging and detrended correspondence analysis. Vegetatio 65: 95–98.

Feoli, E. & Feoli Chiapella, L. 1980. Evaluation of ordination methods through simulated coenoclines: some comments. Vegetatio 42: 35–41.

Gauch Jr H.G. 1982. Multivariate analysis in community ecology. Cambridge Univ. Press, Cambridge.

Görs, S. 1968. Der Wandel der Vegetation im Naturschutzgebiet Schwenninger Moos unter dem Einfluß des Menschen in zwei Jahrhunderten. Natur. Landschaftsschutzgeb. Baden-Württ., Ludwigsburg, 5: 190–284.

Grime, J.P. 1979. Plant strategies and vegetation processes. John Wiley and Sons, Chichester.

Grime, J.P. 1984. The ecology of species, families and communities of the contemporary British flora. New Phytol. 98: 15–33.

Gutte, P. 1972. Ruderalpflanzengesellschaften West- und Mittelsachsens. Feddes Repert. 83: 11–122.

Hejný, S., Kopecký, K., Jehlík, V. & Krippelová, T. 1979. Přehled ruderálních rostlinných společenstev Československa. Rozpr. Čs. Akad. Ved., Ser. Math.-Nat., Praha, 89/2: 1–100.

Hill, M.O. 1973. Reciprocal averaging: an eigenvector method of ordination. J. Ecol. 61: 237–249.

Hill, M.O. 1974. Correspondence analysis: a neglected multivariate method. J. Roy. Stat. Soc.; Ser. C, 23: 340–354.

Hill, M.O., 1979. DECORANA. A Fortran program for detrended correspondence analysis and reciprocal averaging. Ecology and Systematics, Cornell University, Ithaca.

Hill, M.O. & Gauch Jr H.G. 1980. Detrended correspondence analysis: an improved ordination technique. Vegetatio 42: 47–58.

Jongman, R.H., ter Braak, C.J.F. & van Tongeren, O.F.R. 1987. Data analysis in community and landscape ecology. Pudoc, Wageningen.

Kopecký, K. 1969. Zur Syntaxonomie der natürlichen nitrophilen Saumgesellschaften in der Tschechoslowakei und zur Gliederung der Klasse *Galio-Urticetea*. Fol. Geobot. Phytotax. 4: 235–259.

Kopecký, K. & Hejný, S. 1974. A new approach to the classification of anthropogenic plant communities. Vegetatio 29: 17–20.

Krippelová, T. 1972. Ruderálne spoločenstvá mesta Malaciek. Biol. Práce 18/1: 1–116.

Krippelová, T. & Mucina, L. 1988. Charakteristika vyšších syntaxónov triedy *Stellarietea mediae* na Slovensku. Preslia 60: 41–58.

Lagonegro, M. & Feoli, E. 1985. The use of ellipses of equal concentration to analyse ordination vegetation patterns. Stud. Geobot. 5: 143–165.

Mucina, L. 1982. Numerical classification and ordination of ruderal plant communities *(Sisymbrietalia, Onopordetalia)* in the western part of Slovakia. Vegetatio 48: 267–275.

Mucina, L. in press. Syntaxonomie und Ökologie der Ruderalvegetation der westlichen Donau-Tiefebene in der Slowakei. Braun-Blanquetia.

Mucina, L. & Jarolímek, I. 1986. On the syntaxonomic position of the *Plantaginetea majoris* and *Agrostietalia stoloniferae*. Preslia 58: 349–352.

Mucina, L. & Maglocký, S. (eds) 1985. A list of vegetation units of Slovakia. Doc. Phytosoc. N.S. 9: 175–220.

Müller, T. 1983. Klasse: *Artemisietea vulgaris*. In Oberdorfer, E. (ed.), Süddeutsche Pflanzengesellschaften. Teil III. 2. Aufl. Pflanzensoziologie 10: 135–277. Gustaf Fischer Verlag, Jena.

Noy-Meir, I. & Whittaker, R.H. 1977. Continuous multivariate methods in community analysis: some problems and developments. Vegetatio 33: 79–98.

Oberdorfer, E., 1983. Klasse: *Agrostietea stoloniferae*. In: Oberdorfer, E. & Müller, T. (eds), Süddeutsche Pflanzengesellschaften. Teil III. 2. Aufl. Pflanzensoziologie 10: 316–345. Gustav Fischer Verlag, Jena.

Oberdorfer, E. & Müller, T. (eds) 1983. Süddeutsche Pflanzengesellschaften. Teil III. 2. Aufl. Pflanzensoziologie 10. Gustav Fischer Verlag, Jena.

Orlóci, L. 1978. Multivariate analysis in vegetation research. 2nd ed. Junk, The Hague.

Passarge, H. 1967. Über Saumgesellschaften in nordostdeutschen Flachland. Feddes Repert. 74: 145–158.

Peet, R.K. 1980. Ordination as a tool for analyzing complex data sets. Vegetatio 42: 171–174.

Podani, J. 1984. SYN-TAX II. Computer programs for data analysis in ecology and systematics. Abstr. Bot. 8: 73–94.

Rivas-Martínez, S. 1975. Sobre la nueva classe *Polygono-Poetea annuae*. Phytocoenologia 2: 123–140.

Sýkora, K.V. 1983. The *Lolio-Potentillion anserinae* R. Tüxen 1947 in the northern part of the Atlantic domain. Thesis Nijmegen.

ter Braak, C.J.F. 1987. The analysis of vegetation-environment relationships by canonical correspondence analysis. Vegetatio 69: 69–77.

Tüxen, R. 1950. Grundriß einer Systematik der nitrophilen Unkrautgesellschaften in der eurosibirischen Region Europas. Mitt. Florist.-Soziol. Arbeitsgem. N.F. 2: 94–175.

Tüxen, R. 1970. Zur Syntaxonomie des europäischen Wirtschafts-Grünlandes (Wiesen, Weiden, Tritt- und Flutrasen). Ber. Naturhist. Ges. Hannover 114: 77–84.

van der Maarel, E. 1980. On the interpretability of ordination diagrams. Vegetatio 42: 43–45.

Wartenberg, D., Ferson, S. & Rohlf, F.J. 1987. Putting things in order: a critique of detrended correspondence analysis. Amer. Natur. 129: 434–448.

Vegetatio **81**: 127–135, 1989.

Ecological species groups in forest communities in South Belgium

Marie-Françoise Godart
Laboratoire de Botanique Systématique et de Phytosociologie, CP. 169, Université Libre de Bruxelles, Av. P. Héger 28, B-1050 Bruxelles, Belgium

Accepted 8.12.1988

Keywords: Cluster analysis, Correspondence analysis, Ecological profile, Soil property

Abstract

Cluster analysis, correspondence analysis and the method of ecological profiles were used to determine and characterize ecological species groups in woodlands in South Belgium. Two data sets were used, one composed of 260 relevés, the other of 130 relevés with soil analyses (A1). Correspondence analysis points out the most important factors influencing the floristic composition of the stations.

Nomenclature: De Langhe J.-E. *et al.* 1983. Nouvelle Flore de la Belgique, du Grand-Duché de Luxembourg, du Nord de la France et des régions voisines. 3e ed. Jard. Bot. Belg., Meise.

Introduction

A long tradition of ecological studies in southern Belgian forests at the 'Laboratoire de Botanique systématique et de Phytosociologie', yielded a high number of relevés stored in field notes. This material has been studied here mainly to check the disagreement between local botanists about the composition of ecological species groups characterizing Belgian forests (Duvigneaud & Denaeyer 1970; Noirfalise 1984; Stieperaere & Fransen 1982; Tanghe 1967, 1968, 1971).

The purpose of this paper is to objectively define ecological groups of species and elaborate their characteristics and the prevailing conditions indicated.

Study areas

Three distinct regions of South Belgium are involved:

1) Brabant, southeast of Brussels, gently hilly and mostly cultivated, with some parts still covered with forests. The climate is atlantic.
2) the Ardennes, SE Belgium, with acid bedrock and a cooler and wetter climate, still mainly covered by forest. Most relevés are from the Semois valley.
3) the Lorraine, southernmost Belgium, more hilly and forested than Brabant, with a slightly more continental climate (Poncelet & Martin 1947).

Two different data sets were used, one (PHYTO2) containing 260 relevés and the other (PHYTO5), containing 130 relevés with soil data.

Methods

Three different numerical methods were used for the detection and definition of ecological groups.

a. Agglomerative minimum-variance **cluster analysis**, with the CLTR program (Wildi & Orlóci 1983; similar to average linkage clustering, cf. Gauch 1982), with the data transformed according to an angular transformation (van der Maarel 1979), and with van der Maarel's Similarity Ratio as a similarity coefficient (this coefficient is similar to Jaccard's index, which is recommended by Roux & Roux 1980). Clustering was performed on species as well as relevés.

b. **Correspondence analysis**, after presence-absence transformation and elimination of species present in less than five relevés. The analysis was of the reciprocal averaging type (RA) and was performed with the ANFACOR program (for PHYTO2), based on binary correspondence, as developed by Trinquart (Unité de Mathématiques, Sciences sociales, Université de Grenoble II) and the MULTM program (for PHYTO5), using multiple correspondences (Lebart 1979).

c. **Ecological profiles**. This method (e.g. Guillerm 1971; Daget *et al.* 1972; Daget & Godron 1982) is based on the frequency distribution of a species in the different classes of a factor (the ecological profile of a species for this factor) and the mutual information between the species and the factor, which is the quantity of information involved in the presence or absence of this species for that factor (Bottliková *et al.* 1976; Lepart & Debussche 1980). Daget (1976, 1977) also proposed the use of g, the abscissa of the barycentre of the ecological profile, in order to classify the indicator species.

$$g = \sum m_i \cdot d_i / \sum m_i$$

where m_i is the frequency of species i and d_i the number of the considered class of the factor.

Results

Cluster analysis

The relevés clustering of the PHYTO2 data set resulted in two main groups: 136 relevés formed a cluster of rich-soil-relevés (PHYTO2A) and 124 relevés a cluster of poor-soil-relevés (PHYTO2B). The PHYTO2 species clustering also showed two obvious groupings of species characterizing poor and fertile soils respectively. PHYTO2A and B were subdivided separately, with nine species groups in each (at a fusion level of $> 20\%$). These groups will not be discussed further.

Clustering of the PHYTO5 data set yielded nine species groups, three of them including species growing on poor soils. At this stage of the analysis we compared the composition of each species group produced by the four clustering approaches by an empirical method, which is very much like tabular comparison (Mueller-Dombois & Ellenberg 1974). The results will be included in the next sections.

Correspondence analysis

Fig. 1 shows the ordination diagram for the PHYTO2 data. We notice the 'arch effect' described, e.g., by Gauch (1982). This effect is not considered as a problem by Volle (1985) who interpreted the second axis just as if the species were ordinated following their coordinates on that axis (cf. van der Maarel 1980). As in the cluster analysis, the first axis shows a clear dichotomy with the species known from poor soils at the negative side, while the axis as a whole reflects a soil richness gradient. The first two axes account for only 11% and 6% extracted variance respectively.

The second axis shows first a segregation between calcicolous and other species. The calcicolous plants are also heliophytes, like most of the other species with positive coordinate values on the second axis. The species with negative coordinate values are sciophytes. So, the second axis can be interpreted as to reflect a light intensity gradient. The major factors determining the first

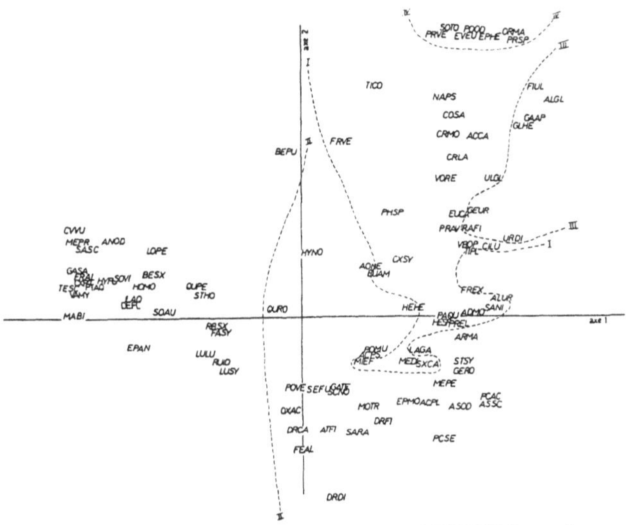

Fig. 1. Two-dimensional RA ordination of PHYTO2 species (first two axes). Symbols for species usually consist of the first two letters of the genus and the first two letters of the species. For complete names, see Fig. 3. The dotted lines indicate divisions resulting from the cluster analysis.

two axes are the same as those found for woodlands in North Belgium by Hermy & Stieperaere (1981). The main clusters as derived from the cluster analysis have been indicated by division lines in Fig. 1. They are reasonably coherent with the ordination pattern.

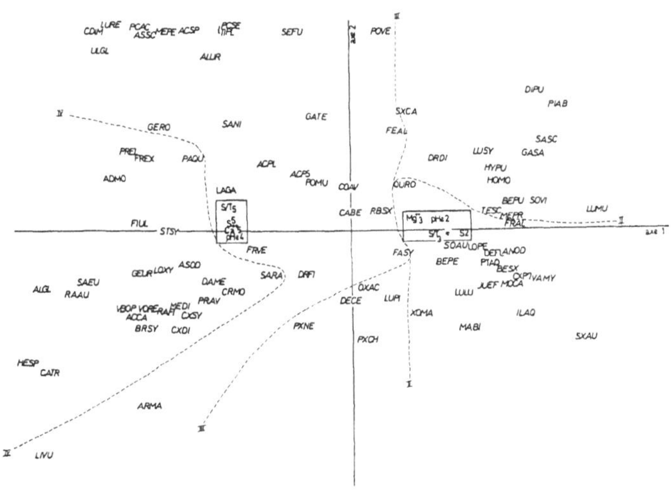

Fig. 2. Two-dimensional RA ordination of PHYTO5 species (first two axes) with the position of soil vectors indicated. See further Fig. 1. * = QUPE.

Fig. 2 shows the ordination diagram for the PHYTO5 data including the soil data. The first two axes account in this case for 29% and 12% extracted variance. The first axis is even more easy to interpret than in the case of PHYTO2. Regarding the position of the soil vectors, we conclude:

(1) High chemical values of pH, exchangeable cations and base saturation rate at the negative side of axis 1.

(2) Comparatively low values of pH, exchangeable magnesium and sum of cations at the opposite pole of the axis. This confirms that soil richness is the determining factor for the first axis. The second axis is strongly influenced by a few species at its positive side; the environmental factor correlated with those species is a topographical one: bottom of steep slopes, on raw colluvial material: *Aceri-Fraxinetum* type (Godart *et al.* 1984).

The main clusters, as indicated by their division lines in Fig. 2, fit better in the ordination pattern than the PHYTO2 material. From the comparison of the cluster structures in both figures a system of 12 ecological groups was obtained (Appendix 1).

Ecological profiles

Two examples of ecological profiles of the species are presented in Figs 3 and 4. The species are grouped according to the results from the previous analyses. 10 of the 12 groups are included. The groups are ordinated along a richness gradient and within the group, the species are ordinated with respect to their *g* values. The same approach was followed for the other soil factors, but the best results were obtained with the factors which are 'efficient' along the first axis of the correspondence analysis (base saturation rate and sum of cations).

From Fig. 3 it can be seen that:

(1) The mutual information between the species and the factor (see last column) may give useful information: if its value is low (say under 100), caution must be taken in inter-

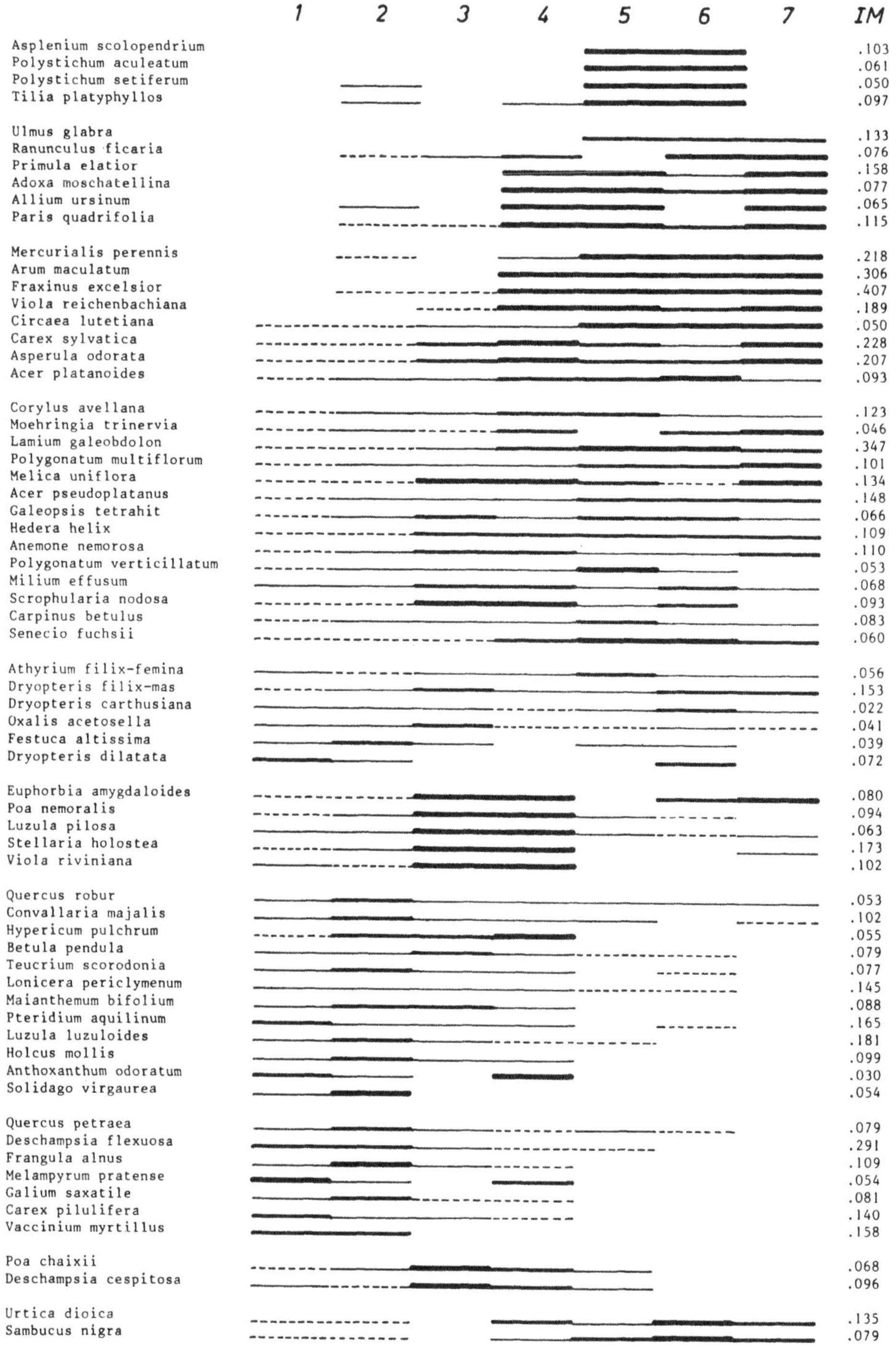

Fig. 3. Ecological profiles: ordination following the base saturation rate of the soil.
Class value of the describer (%):
1 = 0–15; 2 = 15.1–30; 3 = 30.1–45; 4 = 45.1–60;
5 = 60.1–80; 6 = 80.1–100; 7> = 100.

------ corrected frequency <0.700
——— corrected frequency between 0.700 and 1.400
——— corrected frequency between 1.400 and 2.100
▬▬▬ corrected frequency >2.100

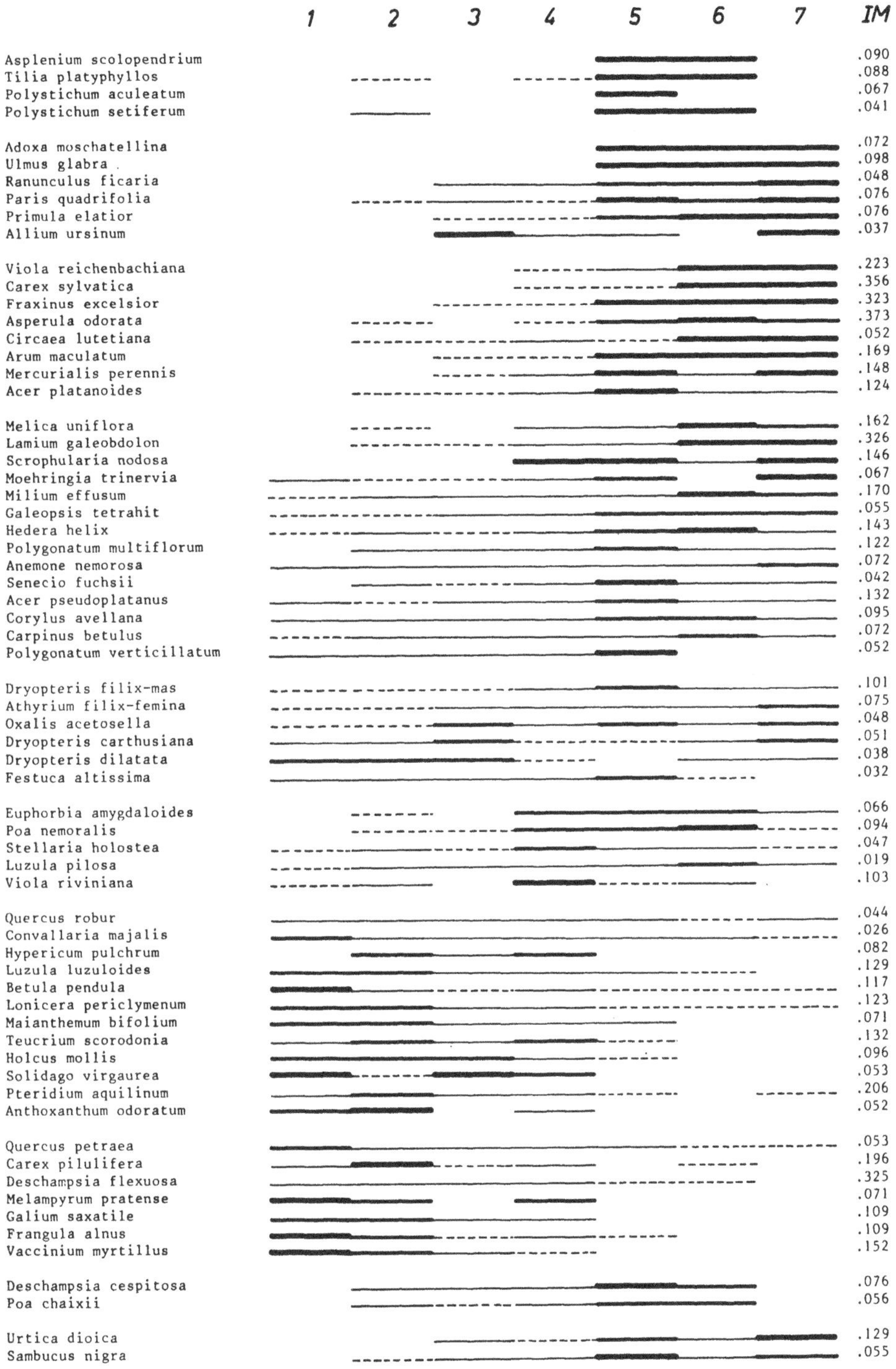

Fig. 4. Ecological profiles: ordination following the sum of cations. Class value of the describer (meq-100 g): 1 = 0–1; 2 = 1.01–2; 3 = 2.01–3; 4 = 3.01–5; 5 = 5.01–9; 6 = 9.01–15; 7> = 15. See further Fig. 3

preting the results. For example, if a species seems to diverge from the others within a group, it does not necessarily mean that it is different in its phytosociological behavior but maybe that the factor used is not appropriate to describe its ecology.

(2) A gradual decrease in the *g* value may be observed from groups in top of Fig. 3 to those towards the bottom. There are groups with most of their species showing a high mutual information (indicating that the factor used is well suited to describe those groups), e.g., group 3, while others have mainly low mutual information values, such as group 7, indicating that such species are not well differentiated towards the factor included.

(3) Some species were not included in the profiles because of their rarity.

On the whole the groups seem to be well characterized by the method, either on the basis of the *g* values or on the general aspect of their profiles.

With another important factor, sum of cations, more or less the same pattern has been obtained (Fig. 4).

However, the *g* values do not follow the decreasing gradient to the same extent as for the base saturation rate.

The two centrally positioned groups for instance, are difficult to ordinate. Apparently these groups do not respond well to the cation sum factor and do better towards other factors, not included in this study, such as humidity. The mutual information values are indeed low for nearly all species of such difficult groups.

Fig. 4 shows also that tree species usually have a more ubiquitous distribution in relation to the ecological factor than the herbaceous species.

Ecological groups: integration of the three methods

First the coherence between the clustering results and the ordination results will be discussed. The cluster division lines as plotted on the ordination diagrams (Figs 1 and 2) show the following relations:

Table 1. Ecological characteristics for 12 species groups (See App. 1).

Group name	pH	C/N	S/T	F (1)	R (1)	N (1)	Other characters
1: Scree	± 5	<12	>60	6	7	7	Steep slope (>35°) colluvial material–shadow
2: Ranufica	6–7	<12	>50	6	7	7	–
3: Arummacu	6–7	9–15	>50	5	7	7	–
4: Lamigale	5–7	9–15	<80	5	6	6–7	–
5: Sciophytes	4–5	12–18	50	6	4	6 (2)	>50% silt flat and fresh stands
6: Stelholo	4–5	12–15	30–50	5	5 (2)	5	–
7: Moder	3–5	12–18	<50	5	3 (2)	3 (2)	Flat stands Al 3 + exch. > 5 meq/100 g
8: Descflex	± 4	15–25	<35	5	2 (2)	3 (2)	Light canopy
9: Nitrophytes	5–6	9–15	>50	6	7	8	–
10: Filiulma	5–6	12	>50	8	6	5 (2)	Wet stands (*) river bank (*)
11: Desccesp	4–5	12–18	<60	6–7	3	4	Damp stands (*) variable water level (*)
12: Calcicolous	6–7	12–15	satur.	4	8	4 (2)	Calcareous soils (*) dry places

(*) Field observations – not mathematically analysed.
(1) Median of the group's species values.
(2) Means were computed and equaled always medians. Percentage standard deviation is in this case comprised between 25 and 40%, in any other case it is less than 25% (usually nearer 10%).

(1) The first division in the PHYTO2 material does more or less coincide with the fully positive first quadrant of the graph, including species characteristic for rich soils and ample light conditions. On the whole the second division separates the species along the second axis, separating typically the Quercion robori-petraeae species. The third and fourth divisions separate small groups of species with particular habitat conditions, i.e. moist soils (III) and calcareous soils (IV).
(2) Regarding Fig. 2, the first division (not represented) concerns the calcicolous species. The second division line coincides with the second quadrant including species of the poorer soils. The third division follows the second axis.

Following guidelines for the interpretation of correspondence analyses (Felzines 1979), we may achieve a better ecological understanding of the species groups. For example, if all (or nearly all) the species of a group have similar coordinate values on one axis, we must look for an environmental vector with the same value and which reasonably accounts for this axis. This will express the interaction between factors and groups. In this way some of the broad groups could be refined.

An exhaustive list of the 12 ecological species groups is given in Appendix 1. Each group has been named either after the most important component species or after the particular ecological factor. The ecological indication of each group is summarized in Table 1. In addition to the soil properties, Ellenberg's (1979) indicator values for moisture (F), reaction (R) and nitrogen status (N) are included (as averages for the species per group).

The species grouping presented here is still provisional. Some groups are well-defined because of their steady appearance in all the methods and samples. This is the case for the scree group, for instance. Other groups vary somewhat depending on the data set and the method used.

Conclusions and prospects

This paper presents a first attempt to classify objectively southern Belgian forest species according to the main ecological factors. Ecological groups may be preferred to typify forest communities because they provide a more direct interpretation of the local eco-geographical factors (e.g. Sebald 1964; Guillemin *et al.* 1978; Becker 1979a, b).

A comparison of our groups with those already published in Belgium (see references in the introduction) shows that the groups presented here:
(1) do not deviate substantially from other systems presented
(2) contain fewer species, however
(3) are ecologically better defined
(4) have smaller ecological amplitudes

It is not considered a disadvantage that rare species were omitted, as the system of ecological groups is intended, among other things, to help foresters to characterize forests by an easy and rapid survey method, where the most frequent species are best fitted for the purpose.

In this context, it is interesting to note that the ecological groups published in Weissen *et al.* (1988) in the framework of a Belgian interuniversity project to define the 'aptitude of forest stations' in Wallonia, are based on the groups defined and described here!

Acknowledgements

Thanks are due to Drs J.-C. Deroubaix and E. Crivisqui (Laboratoire de méthodologie du traitement des données, Institut de sociologie, Université Libre de Bruxelles), who kindly performed the RA analyses.

Appendix

Ecological species group

1: **Scree group**
Tilia platyphyllos, Asplenium scolopendrium, Polystichum aculeatum, Polystichum setiferum, Actaea spicata, Lunaria rediviva, Cardamine impatiens.

2: ***Ranunculus ficaria*** **group**
Ulmus glabra, Ranunculus ficaria, Paris quadrifolia, Allium ursinum, Adoxa moschatellina, Primula elatior, Sanicula europaea, Ranunculus auricomus, (Circaea lutetiana), (Mercurialis perennis).

3: ***Arum maculatum*** **group**
Fraxinus excelsior, Acer platanoides, Crataegus laevigata, Cornus sanguinea, Mercurialis perennis, Arum maculatum, Asperula odorata, Viola reichenbachiana, Circaea lutetiana, Carex sylvatica, Ajuga reptans, Brachypodium sylvaticum, Campanula trachelium.

4: ***Lamium galeobdolon*** **group**
Acer pseudoplatanus, Quercus robur, Carpinus betulus, Corylus avellana, (Fraxinus excelsior), Lamium galeobdolon, Anemone nemorosa, Milium effusum, Hedera helix, Polygonatum multiflorum, Polygonatum verticillatum, Galeopsis tetrahit, Melica uniflora, Scrophularia nodosa, Senecio fuchsii, Epilobium montanum, Moehringia trinervia.

5: **Sciophyte group**
(Fagus sylvatica), Athyrium filix-femina, Oxalis acetosella, Dryopteris carthusiana, Dryopteris dilatata, (Festuca altissima), (Dryopteris filix-mas).

6: ***Stellaria holostea*** **group**
Stellaria holostea, Viola riviniana, Fragaria vesca, Potentilla sterilis, Phyteuma nigrum, Phyteuma spicatum, Euphorbia amygdaloides, Luzula pilosa, Poa nemoralis, Carex digitata, (Melica uniflora), (Solidago virgaurea), (Stellaria nemorosa).

7: **Moder group**
Quercus robur, Betula pendula, Sorbus aucuparia, (Quercus petraea), Sarothamnus scoparius, (Ilex aquifolium), Lonicera periclymenum, Holcus mollis, Teucrium scorodonia, Luzula luzuloides, Hypericum pulchrum, Solidago virgaurea, Convallaria majalis, Maianthemum bifolium, Anthoxanthum odoratum, Hieracium murorum, Pteridium aquilinum, (Digitalis purpurea).

8: ***Deschampsia flexuosa*** **group**
Quercus petraea, (Fagus sylvatica), Frangula alnus, Calluna vulgaris, Vaccinium myrtillus, Deschampsia flexuosa, Carex pilulifera, Galium saxatile, Melampyrum pratense, (Pteridium aquilinum).

9: **Nitrophyte group**
Sambucus nigra, Rubus caesius, Urtica dioica, Glechoma hederacea, Galium aparine, Veronica hederifolia, Alliaria petiolata, Geum urbanum, Aegopodium podagraria, (Moehringia trinervia), (Melandrium dioicum).

10: ***Filipendula ulmaria*** **group**
Alnus glutinosa, Filipendula ulmaria, Polygonum bistorta, Cirsium oleraceum, Valeriana repens, (Heracleum sphondylium).

11: ***Deschampsia cespitosa*** **group**
Deschampsia cespitosa, Juncus effusus, Carex pallescens, Poa chaixii.

12: **Calcicole group**
Sorbus torminalis, Pyrus pyraster, Cornus mas, Viburnum lantana, Evonymus europaeus, Orchis mascula, Polygonatum odoratum, Primula veris, Carex flacca, Clematis vitalba.

References

Becker, M. 1979a. Une étude phyto-écologique sur les plateaux calcaires du Nord-Est. Utilisation de l'analyse des correspondances dans la typologie des stations. Relations avec la productivité et la qualité du hêtre et du chêne. Ann. Sci. Forest. 36: 93–124.

Becker, M. 1979b. Influence du traitement sylvicole sur la flore forestière: cas de la futaie et du taillis-sous-futaie. Vegetatio 40: 155–161.

Bottliková, A., Daget, P., Drdoš, J., Guillerm, J.L., Romane, F. & Ružičková, H. 1976. Quelques résultats obtenus par l'analyse factorielle et les profils écologiques sur des observations phyto-écologiques recueillies dans la vallée de Liptov (Tchécoslovaquie). Vegetatio 31: 79–91.

Daget, P. 1976. Ordination des profils écologiques. Natur. Monspel. Ser. Bot., 26: 109–128.

Daget, P. 1977. Espèces indicatrices et leur valeur caractérisante vis-à-vis du milieu. Natur. Monspel. Sér. Bot., 27: 93–108.

Daget, P., Godron, M., Guillerm, J.L., Drdoš, J., Ružičková, H. & Urvichiarova, E. 1972. Profils écologiques et information mutuelle entre espèces et facteurs écologiques. In: van der Maarel, E. & Tüxen, R. (eds), Grundfragen und Methoden in der Pflanzensoziologie. pp. 121–149. Junk, Den Haag.

Daget, P. & Godron, M. 1982. Analyse fréquentielle de l'écologie des espèces. Masson, Paris.

Duvigneaud, P. & Denaeyer, S. 1970. Phytogéochimie des groupes écologiques forestiers de Haute-Belgique. Oecol. Plant. 5: 1–32.

Ellenberg, H. 1979. Zeigerwerte der Gefässpflanzen Mitteleuropas. 2nd ed. Scipta Geobotanica, 9.

Felzines, J-C. 1979. L'analyse des correspondances et l'information mutuelle entre les espèces et les facteurs du milieu: application à l'écologie des macrophytes aquatiques et palustres. Bull. Soc. Bot. N. France 32: 39–63.

Gauch, H.G. 1982. Multivariate analysis in community ecology. Cambridge Univ. Press. Cambridge.

Godart, M-F., Herbauts. J. & Tanghe, M. 1984. Relations sols-composition floristique dans les forêts de la Semois ardennaise. Bull. Soc. Roy. Bot. Belg. 117: 289–304.

Guillerm, J.L. 1971. Calcul de l'information fournie par un profil écologique et valeur indicatrice des espèces. Oecol. Plant. 6: 229–225.

Guillemin, M., Rouiller, J. & Bruckert, S. 1978. Relations

entre groupes écologiques et propriétés physico-chimiques des sols brunifiés des Vosges cristallines. Ann. Sci. Forest. 35: 139–150.

Hermy, M. & Stieperaere, H. 1981. An indirect gradient analysis of the ecological relationships between ancient and recent riverine woodlands of the south of Bruges (Flanders, Belgium). Vegetatio 44: 43–49.

Lebart, L. 1979. Traitement des données statistiques. Dunod, Paris.

Lepart, J. & Debussche, M. 1980. Information efficiency and regional constellation of environmental variables. Vegetatio 42: 85–91.

Mueller-Dombois, D. & Ellenberg, H. 1974. Aims and methods of vegetation ecology. John Wiley & Sons, New York.

Noirfalise, A. 1984. Forêts et stations forestières en Belgique. Les presses agronomiques, Gembloux.

Poncelet, L. & Martin, H. 1947. Esquisse climatographique de la Belgique. I.R.M.B., Mémoires, 27.

Roux, G. & Roux, M. 1980. A propos de quelques méthodes de classification en phytosociologie. In: Benzecri, J.P. (ed.), L'analyse des données: II. L'analyse des correspondances. Dunod, Paris.

Sebald, O. 1964. Ökologische Artengruppen für den Wuchsbezirk 'Oberer Neckar'. Mitt. Ver. Forstl. Standortskunde Forstpflanzenzüchtung 14.

Stieperaere, H. & Fransen, K. 1982. Standaardlijst van de Belgische vaatplanten, met aanduiding van zeldzaamheid en socio-ecologische groep. Dumortiera 22: 1–41.

Tanghe, M. 1967. Les groupes écologiques forestiers de la Gaume. Lejeunia N.S. 43: 1–64.

Tanghe, M. 1968. Recherches sur l'écosystème forêt. Série E: Forêts de Haute Belgique. Contribution no 3: La végétation forestière de la vallée de la Semois ardennaise; 1e partie: Les groupes écologiques. Bull. Inst. Roy. Sci. Nat. Belg. 44: 1–55.

Tanghe, M. 1971. Recherches sur l'écosystème forêt. Série E: Forêts de Haute Belgique. Contribution no 13: Étude d'un transect topo-lithologique de la région d'Eprave-Rochefort (vallée de la Lomme) pour la délimitation des groupes écologiques forestiers de la Calestienne. Bull. Soc. Roy. Bot. Belg. 104: 333–371.

van der Maarel, E. 1979. Transformation of cover-abundance values in phytosociology and its effects on community similarity. Vegetatio 39: 97–114.

van der Maarel, E. 1980. On the interpretability of ordination diagrams. Vegetatio 42: 43–45.

Volle, M. 1985. L'analyse des données. Collect. 'économie et statistiques avancées'. Economica, Paris.

Weissen, F. *et al.* 1988 (in press). Le fichier écologique des essences – Texte explicatif (définition de l'aptitude des stations forestières). Ministère de la Région Wallonne.

Wildi, O. & Orlóci, L. 1983. Management and multivariate analysis of vegetation data. 2nd. ed., Swiss Fed. Inst. of Forest. Res., Birmensdorf.

Vegetatio **81**: 137–144, 1989

On the delimitation of the *Mesobromion* and *Xerobromion* in Belgium and French Lorraine

D. Van Speybroeck, N. Koedam & P. Büscher
Vrije Universiteit Brussel, Laboratorium voor Algemene Plantkunde en Natuurbeheer, Pleinlaan 2, B-1050 Brussels, Belgium

Accepted 10.7.1988

Keywords: *Brometalia,* Canonical correlation, Classification, Discriminant analysis, Ordination, Soil parameter

Abstract

Within a broader study of the chalk and limestone grasslands in Belgium, Luxemburg and northern France, the limestone grasslands of the Meuse valley in Belgium and in French Lorraine were investigated. Floristic data as well as environmental variables were taken into account for the delimitation of species and relevé groups. Discriminant analysis and canonical correlation analysis proved to be very helpful in the interpretation of the detected groups as well as for the reconsideration of interrelationships between recognised syntaxa. The present results are consistent with the hypothesis that the limestone grasslands in French Lorraine, even those previously assigned to the *Xerobromion,* belong to the *Mesobromion,* and that the limestone grasslands in Belgium belong either to the *Mesobromion* or to the *Xerobromion* in an impoverished form.

Nomenclature: De Langhe, J.-E., Delvosalle, L., Duvigneaud, J., Lambinon, J. & Vanden Berghen, C. 1983.

Introduction

The chalk and limestone grasslands of western Europe are either assigned to the *Mesobromion* or to the *Xerobromion* alliance (*Brometalia erecti, Festuco-Brometea*). *Mesobromion* grasslands are widely distributed on mesophilous calcareous soils from southern Sweden to northern Spain and from Ireland to the Alps and the Brandenburg region in East Germany. *Xerobromion* grasslands on the other hand are less widely distributed in Europe, being mainly found in the north-western mediterranean calcareous formations and extending in fairly impoverished forms in localised areas up to southern England, northern France, western Switzerland and the southern parts of West and East Germany (Wolkinger & Plank 1981; Willems 1982; Royer 1984, 1985, 1987).

Both *Mesobromion* and *Xerobromion* grasslands have been described for Belgium and northern France. Duvigneaud *et al.* (1982) give an annotated bibliographical list of studies dealing with chalk and limestone grasslands in Belgium and northern France. Within a broader syntaxonomical study of the *Mesobromion* grasslands in Belgium, northern France and Luxemburg, we investigated the limestone grasslands of the Meuse valley in Belgium and in French Lorraine. One hundred thirty relevés were recorded from limestone grasslands assigned to the *Mesobromion* and *Xerobromion* in Belgium as described by Lebrun *et al.* (1949), Vanden Berghen (1955), Duvigneaud (1981a), Noirfalise & Dethioux (1984) and others and from *Mesobromion* and *Xerobromion* grasslands in French Lorraine as described by Duvigneaud *et al.* (1970), Mullenders *et al.* (1972), Duvigneaud (1981b) and others.

Following the foregoing authors, one may conclude that *Xerobromion* and *Mesobromion* grasslands in the studied area not only have a different floristic composition, but also appear in different habitats. *Xerobromion* grasslands colonise shallow soils on steep south facing slopes and generally have a total cover between 40 and 80%. *Mesobromion* grasslands on the other hand develop on deeper soils, on moderate slopes without exposure preference and with a total cover always approaching 100%.

Multivariate analysis was applied to detect the pattern of both *Mesobromion* and *Xerobromion* in northern France and Belgium. Floristic as well as environmental data were taken into account to yield relevé groups (syntaxa) and species groups.

Two phytogeographical districts are distinguished within the study area: the Lorraine and the Meuse district (De Langhe *et al.* 1983). French Lorraine and Belgian Lorraine belong to the Lorraine district (only Jurassic limestone), whilst the remaining grasslands recorded in Belgium and the Givet region in the French Ardennes belong to the Meuse district (Devonian and Carboniferous limestone). We will refer to the former as the 'Lorraine district' and to the latter as the 'Meuse district'.

In the present study it will be shown that multivariate methods can generate new insights into the syntaxonomy and synecology of vegetation types formerly described in the traditional way (Westhoff & van der Maarel 1978).

Material and methods

The 130 relevés discussed were recorded during the period 1981–1984 in Belgium (provinces of Namur, Luxembourg and Liège) and in French Lorraine (France, departments of Meuse, Ardennes and Meurthe-et-Moselle). Every plot was observed at least both in early and late summer. For the description of the grasslands we followed the Braun-Blanquet approach (Westhoff & van der Maarel 1978).

In each plot a number of subsamples from the A-horizon were pooled to give an average soil sample which was analysed in the laboratory after air drying and sieving (Ø 2 mm). Soil pH was measured electrometrically in the supernatant of a suspension of dried soil in $CaCl_2$ (0.02 N) in a weight ratio 1/5 after 30 minutes equilibration. Carbonate in the soil was determined by the volumetric calcimeter method (Allison & Moodie 1965) and expressed as % dry weight of soil. Organic carbon was determined according to the method Anne (Bonneau & Souchier 1979). Percentage total cover of the vegetation and average slope of the plot were estimated. The following notations correspond to the environmental variables: pH, CO_3^{--}, organic C, % cover and slope.

Since each multivariate method has both specific advantages and limitations (Orlóci 1978; Greig-Smith 1983; Gauch 1984; Orlóci & Kenkel 1985), none of the existing methods completely fits the 'floristic-sociological' approach of the present study. But by simultaneously using two conceptually different methods for the same data, the difficulties of the respective methods can be largely overcome. Here we applied a hierarchical classification algorithm (minimum variance clustering of a chord distance matrix) and an ordination algorithm (correspondence analysis starting from a covariance matrix). Algorithms are explained by Orlóci (1978). All 250 species occurring at least twice in all the relevés are included in the computations. For the analysis we implemented the programme package of Wildi & Orlóci (1983, version 1985) on the CDC-855 main frame of the Free Universities of Brussels. The box plot presentation (Chambers *et al.* 1983) is used to display the distribution of a parameter from a set of data showing the median value (the innermost horizontal line), the 10 and 90th percentile (the outermost horizontal lines) and the 25 and 75th percentile (the in between horizontal lines). The values above and below the 10 and 90th percentile are individually represented (see Fig. 4).

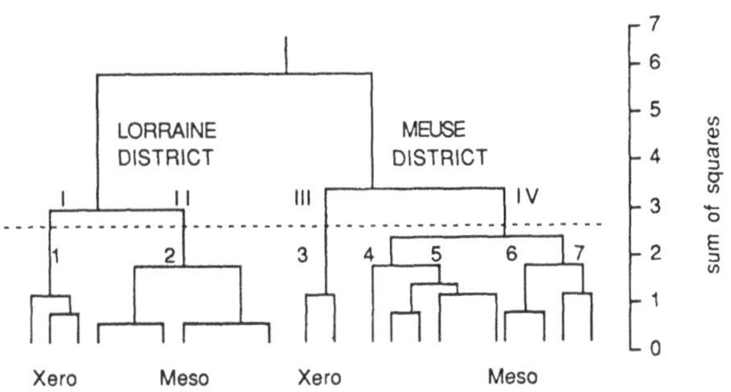

Fig. 1. Dendrogram resulting from minimum variance clustering of a chord distance matrix of 130 relevés of limestone grasslands in Belgium and French Lorraine. The similarity range (0–7) is given by the within-group sum of squares. 'Meso' and 'Xero': grasslands described in the literature as belonging to *Mesobromion* respectively *Xerobromion*. The numbers 1–7 refer to the relevé groups.

Results and discussion

Figure 1 shows the dendrogram of the relevés after minimum variance clustering (sum of squares method) of a chord distance matrix (Orlóci 1978; Wildi & Orlóci 1983; Orlóci & Kenkel 1985). There are seven clusters, three more than at the optimal level according to Feoli & Orlóci (1979) and Popma *et al.* (1983). Visual inspection of the dendrogram shows the subdivision of the two geographical regions into four main groups: clusters I and II represent the xerophilous and the mesophilous grasslands originating from the 'Lorraine district' respectively, whilst clusters III and IV represent xerophilous and mesophilous grasslands from the 'Meuse district'. Note that the xerophilous grasslands assigned to the *Xerobromion* (clusters I and III) show more affinity to the mesophilous (*Mesobromion*) grasslands in the same region (cluster II for the 'Lorraine district' and cluster IV for the 'Meuse district') than to each other. Floristic affinity of vegetation types within a particular region, being higher than between similar syntaxonomic units from different regions, is a recognised phenomenon and may affect the concept of a syntaxon. Therefore other parameters than floristic affinity should also be taken into consideration.

Discriminant analysis estimates the discriminating power of the measured environmental variables versus the relevé groups as defined by the classification (Wildi & Orlóci 1983). Table 1 shows that the seven main clusters are well separated along the first two discriminant axes ($\lambda_1 = 1.58$; $\lambda_2 = 0.53$). The first axis is highly correlated with pH (discriminant coefficient = -0.80), whilst the second axis is highly correlated with % cover of the vegetation (discriminant coefficient = -0.82). In other words, the relevé set can first be divided into two groups on the basis of a gradient in pH (geological and pedological substrates); within each 'pH group' a further subdivision can be made on the basis of a difference in % cover of the vegetation.

On the basis of the relevé cluster dendrogram (Fig. 1) and the species cluster dendrogram (not shown), it was decided to construct a 7 (relevé groups) × 25 (species groups) contingency table. Analysis of concentration (Feoli & Orlóci 1979) was performed on the 7×25 contingency table. Inspection of the χ^2 value confirmed that the block structure in the vegetation table deviates significantly from those determined on the basis of random distribution expectations (Orlóci & Kenkel 1985). Canonical correlations and scores for the relevé and species groups are computed (Table 2) and their joint scatter plot is given in Fig. 2 (Wildi & Orlóci 1983). It is apparent that the 'Lorraine district' grasslands (F1 and F2) fall in the right hand side of the diagram, whilst the 'Meuse district' grasslands (F3 to F7) fall in the left hand side. The mesophilous grasslands (F2, F4, F5, F6 and F7) are plotted in the middle of the diagram, whilst the xerophilous groups (F1 and F3) show affinity to the respective mesophilous grass-

Table 1. Discriminant analysis of the seven relevé groups defined by classification (Fig. 1). Discriminant coefficients for the environmental variables along the four discriminant axes are shown.

Eigenvalues:	0.158E+01	0.532E+00	0.128E+00	0.100E+00
Percentages:	65.156	22.018	5.278	4.154
Discriminant coefficients:				
pH	−0.79909	−0.05462	−0.95440	0.91933
CO_3^{--}	−0.29647	0.20154	0.06029	−0.10755
Org. C	0.13706	0.51188	−0.01383	−0.05300
% Cover	−0.49937	−0.82240	0.27469	0.33645
Slope	0.07358	0.13427	0.09920	0.16512

Table 2. Canonical correlations and scores for the contingency table (7 relevé groups × 25 species groups).

Canonical correlations:					
	0.46605	0.40869	0.29988	0.26793	0.19898
Chi square			= 2369.237		
Mean square contingency coeff.			= 0.098		
Canonical scores of species groups					
1	0.38378	0.18820	−0.30151	−0.38236	−3.0736
2	1.57359	−0.19784	0.52508	0.42894	1.04468
3	−1.63250	−3.47908	2.21183	−0.34985	1.70426
4	−0.63036	−1.56880	−0.39594	−0.12242	−0.34140
5	−1.84687	−2.90634	1.88176	−1.31279	1.35483
6	−1.37762	−0.63659	1.29658	0.64255	−0.72573
7	−1.28600	0.06548	2.10110	1.94625	−0.43748
8	0.41740	0.93352	0.70977	1.43526	0.29096
9	−0.20287	−0.58695	−0.55260	0.36124	−0.78037
10	−0.46667	0.08931	0.15295	−0.17107	−1.93541
11	−0.72803	1.11467	1.63511	2.44722	−0.29584
12	−1.11778	−0.30410	−2.76789	1.58423	0.25786
13	−0.88339	0.11525	−3.38556	1.12680	0.83368
14	1.37755	−0.15932	−0.41633	−0.03390	−0.11504
15	0.56535	−0.11136	0.28257	1.17395	1.70960
16	3.19191	−0.98324	0.61163	0.72660	1.16407
17	2.97251	−1.41408	0.09774	−0.75548	−3.16846
18	2.78920	−0.41344	0.71282	0.55616	2.79432
19	−0.09128	1.31020	0.44925	0.35049	0.59710
20	−0.09274	1.97833	−0.57563	−3.67092	2.96157
21	−0.92415	1.59781	1.79011	−1.27559	−1.17769
22	−0.42127	1.77315	0.41228	1.16758	−1.46744
23	−0.78136	2.13748	2.05120	2.07091	−2.29829
24	−0.63132	0.51846	−0.29244	−0.63109	−0.20488
25	−0.34163	1.41167	0.63100	−2.68275	0.68203
Canonical scores of relevé groups					
1	1.93512	−1.09345	−0.01794	−0.10398	−1.10028
2	1.09892	0.30638	0.46012	0.77577	1.82910
3	−1.23749	−1.62293	0.75832	−0.17541	0.35714
4	−0.58456	0.04670	−2.01068	0.95413	0.11132
5	−0.09977	0.99478	−0.10210	−1.90445	0.62826
6	−0.09396	0.30001	−0.44608	−0.82628	−1.15374
7	−0.50546	1.25030	1.23400	1.06127	−0.98310

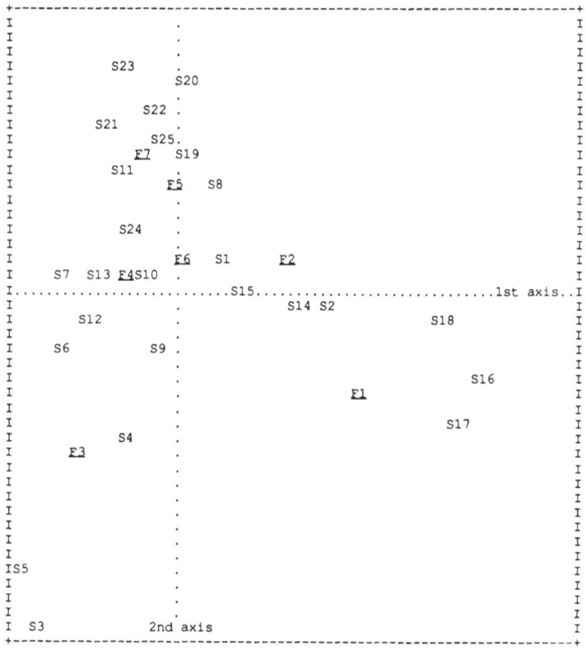

Fig. 2. Joint scattergram of the canonical variables for 7 relevé groups (F1–F7) defined in Fig. 1 and 25 species groups (S1–S25) of limestone grasslands in Belgium and French Lorraine. 'Meso' in 'Meuse district': F4, F5, F6, F7; 'Xero' in 'Meuse district': F3; 'Meso' in 'Lorraine district': F2; 'Xero' in 'Lorraine district': F1. Species group composition is specified in Appendix 1.

lands rather than to each other. It is clear from Fig. 2 that relevé group F1 (Xero 'Lorraine district') is characterised by species groups S16, S17 and S18 and that the other relevé groups are likewise characterized by certain species groups.

In a recent overview of the *Festuco-Brometea* grasslands in the French Jura (Royer 1987), *Carex humilis, Genista pilosa, Globularia punctata, Linum tenuifolium, Potentilla neumanniana, Seseli montanum, Sesleria albicans, Stachys recta, Teucrium chamaedrys* and *Teucrium montanum* are considered as xerophilous species characteristic for the *Xerobromion* as well as for the drier types of the *Mesobromion,* and no longer as species faithful to the *Xerobromion.* Differentiating species of the *Xerobromion* according to Royer (1984) in the present study are: *Allium sphaerocephalon, Anthericum liliago, Fumana procumbens, Helianthemum apenninum* and locally, *Anthericum ramosum, Anemone pulsatilla, Dianthus carthusianorum* and *Seseli montanum.* Inspection of the floristic composition of relevé group F1 (Xero 'Lorraine district') and F3 (Xero 'Meuse district') shows that F1 is extremely poor in these differentiating species of the *Xerobromion* (only *A. ramosum* and *S. montanum*), whilst F3 includes *A. sphaerocephalon, A. liliago, D. carthusianorum, F. procumbens* and *H. apenninum.*

Ordination results are shown in Fig. 3. Correspondence analysis (Wildi & Orlóci 1983) of the initial 130×250 data set was performed based on a

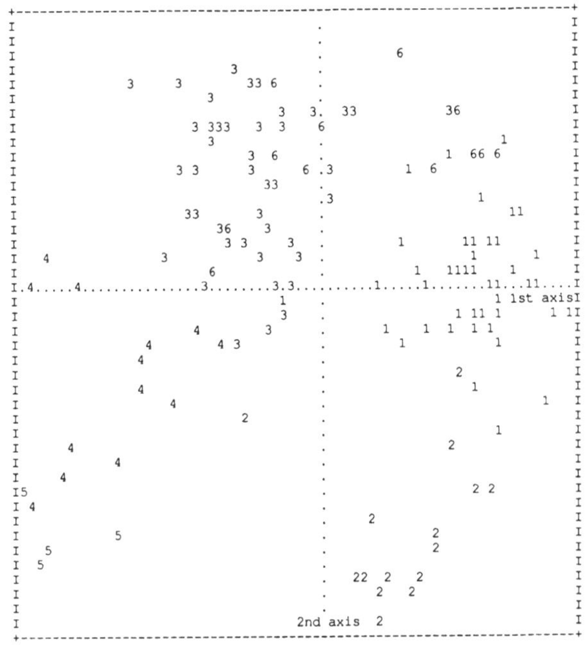

Fig. 3. Ordination of 130 relevés of limestone grasslands in Belgium and French Lorraine on the first two axes derived by correspondence analysis (variance explained by the first axis: 12%, by the second axis: 9%). The 6 groups represented resulted from a space density analysis in the three dimensional space (axes 1, 2 and 3). In the two dimensional space, 4 relevé groups are apparent: 'Meso' in 'Meuse district' (3), 'Meso' in 'Lorraine district' (1), 'Xero' in 'Meuse district' (4 and 5) and 'Xero' in 'Lorraine district' (2).

covariance matrix between the relevés. The relevés are plotted in the two-dimensional space (defined by axes 1 and 2) in 6 more or less distinct groups. These groups are the result of a space density analysis of the relevé point cloud in the three dimensional space defined by the three first ordination axes (programme Grid, Wildi 1979; Orlóci & Kenkel 1985). Visual inspection of Fig. 3 clearly shows two main relevé groups which can be further subdivided into two groups. Along the first ordination axis, the 'Lorraine district' grasslands (relevé groups 1 and 2) are separated from the 'Meuse district' grasslands (relevé groups 3, 4 and 5). The second ordination axis separates the more mesophilous grasslands in the 'Meuse district' (relevé group 3) and the 'Lorraine district' (relevé group 1) from the more xerophilous grasslands in the 'Meuse district' (relevé groups 4 and 5) and the 'Lorraine district' (relevé group 2).

Correlations of the environmental variables with

Table 3. Correlation coefficients between co-ordinate sets of environmental variables (adjusted to unit length) and ordination axes 1 and 2 as represented in Fig. 3.

	pH	CO_3	Org. C	% Cover	Slope
Axis No. 1	0.40153	0.50503	−0.38247	0.25072	−0.11948
Axis No. 2	−0.36300	−0.48753	−0.05235	0.47597	−0.30619
Axis No. 3	0.05406	−0.00776	−0.09449	−0.16192	−0.10998
Axis No. 4	−0.01048	0.10404	0.06241	−0.10029	0.16765
Axis No. 5	0.15364	−0.06020	0.03522	−0.18694	0.19025
Axis No. 6	0.25295	0.11852	−0.11000	−0.09962	0.01698

the ordination axes are given in Table 3. The first ordination axis shows the relatively highest correlation with % CO_3^{--} in the soil (correlation coefficient = 0.51) and with pH (0.40). The importance of the substrate parameters pH and % CO_3^{--}, reflecting acidity and carbonate content, confirms the overriding role of geological and pedological properties of the sampling sites in determining floristic composition and physiognomy of the grasslands. Correlation of organic C with the first axis (−0.38) illustrates that more acid soils, obviously having low amounts of carbonates, also tend to accumulate more organic matter in the A horizon. The second ordination axis shows a high correlation with % vegetation cover (0.48) and with slope (−0.31). This explains the differentiation along the second axis of the mesophilous grasslands (closed grassland vegetation on moderate slopes) from the xerophilous ones (more open grassland vegetation on steeper slopes).

Figure 4 shows the box plots of pH and % CO_3^{--} for grasslands occurring on the three main geological substrates within the study area.

Discriminant analysis shows also the discriminating power of the environmental variables with regard to the detected relevé groups in the ordination axes. Table 4 gives the relative importance of the discriminating axes and the discriminating coefficients for the environmental variables along those axes. Taking into account that intercorrelation between parameters influences the discriminant coefficients (Levine 1977), inspection of the eigenvalues in Table 4 shows that the groups are separated in the first place along the first axis (λ_1 = 1.02) and secondly along the second axis (λ_2 = 0.44). The first axis is

Table 4. Discriminant analysis of the 6 relevé groups as represented in Fig. 3. Discriminant coefficients for the environmental variables with the 4 discriminant axes are shown.

Eigenvalues:	.102E+01	.441E+00	.800E-01	.122E-01
Percentages:	65.595	28.307	5.134	.781
Discriminant coefficients:				
pH	−.86033	.22588	.97099	.62734
CO_3^{--}	−.32450	.14322	−.07297	−.10513
Org. C	.11142	.43786	.04244	−.27098
% Cover	−.35520	−.84770	.21991	−.71281
Slope	.12633	.13470	.04122	−.11775

highly correlated with pH (discriminant coefficient = −0.86), whilst the second axis shows high correlation with % vegetation cover (discriminant coefficient = −0.85). This confirms the results discussed in previous paragraphs: the relevé groups are first divided into two substrate dependent main groups, i.e. the limestone grasslands of the 'Meuse district' and the 'Lorraine district', whilst further division corresponds to a difference in % cover of the vegetation.

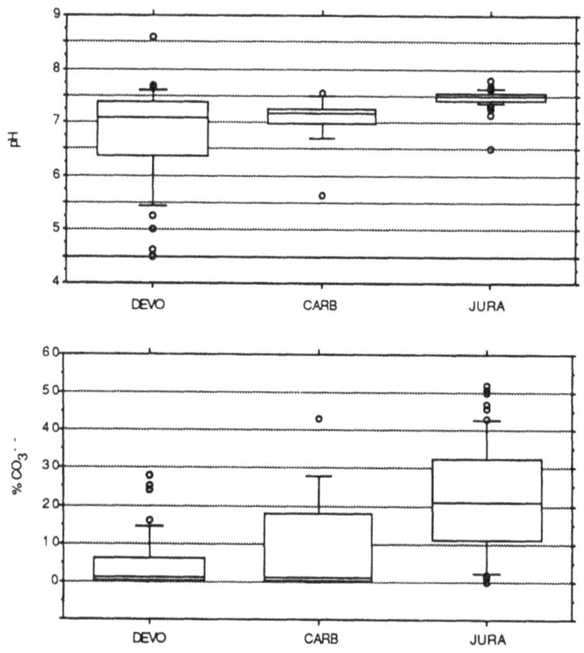

Fig. 4. Box plots of pH and % CO_3^{--} for each of the three main geological substrates in the study area. Devonian (Devo) and Carboniferous (Carb) limestone are dominant in the 'Meuse district', whilst Jurassic (Jura) limestone only outcrops in the 'Lorraine district'.

Conclusion

Two conceptually different multivariate methods (a hierarchical classification and an eigenanalysis ordination) revealed the same group structure in a 130 relevé × 250 species matrix obtained from limestone grasslands in the Meuse valley in Belgium and northern France. Four clearly defined relevé groups were detected. According to practice, we described them as two mesophilous and two xerophilous groups. The mesophilous groups belong to the *Mesobromion* whilst the xerophilous groups are currently assigned to the *Xerobromion* (Mullenders *et al.* 1972; Noirfalise & Dethioux 1984). Bearing this in mind and knowing that *'the assignment of an association to an alliance (and other higher units) is primarily based on comparison of floristic relationships'* (Westhoff & van der Maarel 1978, p. 332), the two xerophilous groups (*Xerobromion*) should express higher affinity for each other than for the two mesophilous groups (*Mesobromion*). But in contrast with this, the two xerophilous groups show more affinity to the respective mesophilous groups. These results were fully supported by the analysis of the environmental parameters. The difference between xerophilous and mesophilous grasslands (% cover of the vegetation and slope) was not the main difference found. More important was a substrate-depending aspect of the grasslands (pH and % CO_3^{--}). Both a mesophilous and a xerophilous group were found on Jurassic limestone ('Lorraine district') as well as on Devonian and Carboniferous limestone ('Meuse district').

This supports our hypothesis that the two xerophilous groups should not be pooled together in the *Xerobromion* alliance. All limestone grasslands recorded in the 'Lorraine district' (F1 and F2) should rather be assigned to the *Mesobromion,* whilst the grasslands from the 'Meuse district' either should be assigned to the *Mesobromion* (F4, F5, F6 and F7) or to an impoverished form of the *Xerobromion* (F3), reaching its northern limit of distribution in Belgium.

Though it cannot be excluded that a general tendency to 'less xeric conditions' during the last few decades could at least partly explain our results as compared to those in previous reports (Mullenders *et al.* 1972), it appears that our conclusions frame in a more general view regarding limestone grasslands, as exemplified by the work on the Bourgogne region and the French Jura of Royer (1984, 1987).

Acknowledgements

Financial support was provided by the Belgian research funds IWONL and NFWO and is gratefully acknowledged. We thank Dr M. D. Castle for correcting the English.

References

Allison, L. E. & Moodie, C. D. 1965. Carbonate. In: Black, C. A. (ed.), Methods of soil analysis. Part 2. Chemical and microbiological properties: pp. 1379-1396. American Society of Agronomy, Madison.

Bonneau, M. & Souchier, B. 1979. Pédologie. 2. Constituants et propriétés du sol. Masson, Paris.

Chambers, J. M., Cleveland, W. S., Kleiner, B. & Tukey, P. A. 1983. Graphical methods for data analysis. The Wadsworth statistics/probability series. Duxbury Press, Boston.

De Langhe, J.-E., Delvosalle, L., Duvigneaud, J., Lambinon, J. & Vanden Berghen, C. 1983. Flora van België, het Groothertogdom Luxemburg, Noord-Frankrijk en de aangrenzende gebieden. Patrimonium van de Nationale Plantentuin van België, Meise.

Duvigneaud, J. 1981a. Une nouvelle réserve d'Ardenne et Gaume: le tienne Breumont à Nismes (province de Namur, Belgique). Natura Mosana 34: 57-79.

Duvigneaud, J. 1981b. L'herborisation générale de la Société Royale de Botanique de Belgique dans la partie septentrionale de la Lorraine française les 6 et 7 septembre 1980. Bull. Soc. Roy. Bot. Belg. 114: 140-154.

Duvigneaud, J., Durin, L. & Mullenders, W. 1970. La végétation des éboulis de Pagny-la-Blanche-Côte (Meuse, France). Vegetatio 21: 48-73.

Duvigneaud, J., Mériaux, J. L. & Van Speybroeck, D. 1982. La conservation des pelouses calcaires de la Belgique et du nord de la France. Nécessité de leur protection, proposition d'intervention et méthodes de gestion. Institut européen d'écologie, Metz.

Feoli, E. & Orlóci, L. 1979. Analysis of concentration and detection of underlying factors in structured tables. Vegetatio 40: 49-54.

Gauch, H. G. 1984. Multivariate analysis in community ecology. Cambridge University Press, Cambridge.

Greig-Smith, P. 1983. Quantitative plant ecology. (3rd ed.). Blackwell, Oxford.

Lebrun, J., Noirfalise, A., Heinemann, P. & Vanden Berghen, C. 1949. Les associations végétales de Belgique. Bull. Soc. Roy. Bot. Belg. 82: 105-207.

Levine, M. S. 1977. Canonical analysis and factor comparison. Sage Univ. Pap. Ser. Quant. Appl. Soc. Sci. Sage publications, London.

Mullenders, W., Durin, L., Duvigneaud, J. & Vanden Berghen, C. 1972. Les pelouses calcaires de la Lorraine mosane et du Châtillonnais. Docum. Phytosoc. 2: 45-52.

Noirfalise, A. & Dethioux, M. 1984. Les pelouses calcaires de la Belgique et leur protection. Coll. Phytosoc. 11: 201-218.

Orlóci, L. 1978. Multivariate analysis in vegetation research (2nd ed.). Junk, The Hague.

Orlóci, L. & Kenkel, N. C. 1985. Introduction to data analysis. Statistical Ecology Monographs, vol. 1. International Cooperative Publishing House, University Park, MD, USA.

Popma, J., Mucina, L., van Tongeren, O. & van der Maarel, E. 1983. On the determination of optimal levels in phytosociological classification. Vegetatio 52: 65-75.

Royer, J. M. 1984. Caractérisation, répartition et origine du Xerobromion. Coll. Phytosoc. 11: 243-267.

Royer, J. M. 1985. Liens entre chorologie et différenciation de quelques associations du Mesobromion erecti d'Europe occidentale et centrale. Vegetatio 59: 85-96.

Royer, J. M. 1987. Les pelouses des Festuco-Brometea. D'un exemple régional à une vision eurosibérienne. Etude phytosociologique et phytogéographique. Thesis, Besançon.

Vanden Berghen, C. 1955. Etude sur les irradiations de plantes méridionales dans la vallée de la Meuse wallonne. Bull. Soc. Roy. Bot. Belg. 87: 29-54.

Westhoff, V. & van der Maarel, E. 1978. The Braun-Blanquet approach. In: Whittaker, R. H. (ed.), Classification of plant communities 2nd ed. pp. 287-399. Junk, The Hague.

Wildi, O. 1979. Grid - a space density analysis for recognition of noda in vegetation samples. Vegetatio 41: 95-100.

Wildi, O. & Orlóci, L. 1983. Management and multivariate analysis of vegetation data (2nd ed.). Swiss Federal Institute of Forestry Research, Birmensdorf.

Willems, J. H. 1982. Phytosociological and geographical survey of Mesobromion communities in Western Europe. Vegetatio 48: 227-240.

Wolkinger, F. & Plank, S. 1981. Les pelouses sèches en Europe. Collection Sauvegarde de la Nature 21. Conseil de l'Europe, Strasbourg.

Appendix 1

Limestone grasslands of Belgium and French Lorraine. Constancy table of the seven relevé groups and the 25 species groups as presented in Fig. 2. The species within the species groups are listed in alphabetical order. "Xero" in 'Lorraine district': F1; "Meso" in 'Lorraine district': F2; "Xero" in 'Meuse district': F3; "Meso" in 'Meuse district': F4, F5, F6, F7. (1=0 – 20%, 2 = 21 – 40%, 3 = 41 – 60 %, 4 = 61 – 80%, 5 = 81 – 100%).

RELEVE GROUP	F1	F2	F3	F4	F5	F6	F7
number of relevés	14	49	14	6	26	11	10
Species group 1							
ANTHYLLIS VULNERARIA	3	3	3	5	4	3	1
BRACHYPODIUM PINNATUM	5	5	4	.	4	5	5
BRIZA MEDIA	3	5	2	5	5	3	3
BROMUS ERECTUS	5	5	3	5	5	.	3
CAMPANULA ROTUNDIFOLIA	5	4	4	5	5	5	3
CAREX CARYOPHYLLEA	2	3	2	5	3	5	1
CAREX FLACCA	4	5	2	4	5	5	3
CARLINA VULGARIS	4	4	.	4	2	4	1
CENTAUREA SCABIOSA	4	3	.	3	3	3	2
CENTAUREA SEROTINA	3	3	.	3	5	4	4
CIRSIUM ACAULE	5	4	.	2	4	5	1
FESTUCA OVINA	4	5	5	3	5	5	3
GYMNADENIA CONOPSEA	4	4	1	.	4	4	3
HIERACIUM PILOSELLA	4	4	4	5	3	4	2
HYPERICUM PERFORATUM	2	3	4	4	3	5	4
KNAUTIA ARVENSIS	2	3	2	2	4	3	3
LEONTODON HISPIDUS	5	4	1	1	4	3	1
LINUM CATHARTICUM	3	5	4	5	5	5	4
LOTUS CORNICULATUS	2	5	3	5	5	5	5
PIMPINELLA SAXIFRAGA	5	4	1	5	5	4	3
POTENTILLA NEUMANNIANA	3	4	5	5	5	5	2
SANGUISORBA MINOR	5	5	5	5	5	5	4
SCABIOSA COLUMBARIA	4	5	5	5	4	3	2
TEUCRIUM CHAMAEDRYS	5	3	5	.	4	5	2
Species group 2							
ACERAS ANTHROPOPHORUM	2	3	.	.	.	1	.
ANEMONE PULSATILLA	3	3	.	.	.	.	.
ASPERULA CYNANCHICA	5	5	2	.	2	2	1
AVENULA PRATENSIS	.	2	.	.	1	.	.
BUPLEURUM FALCATUM	5	4	2	.	2	3	2
CAREX TOMENTOSA	.	1	.	.	.	.	.
EUPHORBIA CYPARISSIAS	5	5	3	.	4	2	2
HIPPOCREPIS COMOSA	4	4	4	3	1	1	3
KOELERIA PYRAMIDATA	4	5	1	.	1	.	1
LEUCANTHEMUM VULGARE	1	3	2	2	3	1	3
LISTERA OVATA	.	2	.	.	2	1	1
ONONIS REPENS	.	4	.	3	3	.	1
OPHRYS FUCIFLORA	3	3	.	.	.	2	.
ORCHIS MILITARIS	5	4	1	.	1	.	1
PLATANTHERA CHLORANTHA	2	3	.	2	1	1	2
POLYGALA CALCAREA	.	2	1	.	.	1	.
PRUNELLA VULGARIS	1	1	.	.	1	.	.
SESELI MONTANUM	4	2	.	.	.	.	.
THYMUS PRAECOX	5	5	.	.	1	1	.
Species group 3							
GERANIUM SANGUINEUM	.	.	2	.	.	.	.
ODONTITES VERNA	.	.	1	.	.	.	.
PHLEUM PHLEOIDES	1	1	2	.	.	.	.
Species group 4							
ACINOS ARVENSIS	1	1	5	2	.	2	1
ARABIS HIRSUTA	3	1	4	5	1	.	1
ARENARIA SERPYLLIFOLIA	1	.	3	2	1	1	.
CAREX HUMILIS	3	1	4	.	1	.	1
DIANTHUS CARTHUSIANORUM	.	1	4	.	1	.	1
ECHIUM VULGARE	2	.	3	5	1	1	.
GLOBULARIA PUNCTATA	5	1	4	.	2	4	.
HELIANTHEMUM APENNINUM	.	.	3	.	1	.	.
POA COMPRESSA	1	1	3	3	1	1	.
SEDUM ACRE	1	.	3	3	.	1	.
SEDUM ALBUM	.	.	5	2	1	1	.
SEDUM REFLEXUM	1	.	3	.	1	.	.
SESELI LIBANOTIS	3	1	4	2	3	1	2
SESLERIA ALBICANS	4	1	5	5	3	3	1
Species group 5							
ALLIUM SPHAEROCEPHALON	1	1	4	.	.	.	.
ANTHERICUM LILIAGO	.	.	2	.	.	1	.
ASPLENIUM RUTA-MURARIA	.	.	3	.	1	.	.
ASPLENIUM TRICHOMANES	.	.	2	.	.	.	.
ASTER LINOSYRIS	.	.	2	.	1	.	.
EROPHILA VERNA	.	.	1	.	.	.	.
MELICA CILIATA	.	.	3	.	1	.	.
OROBANCHE CARYOPHYLLACEA	.	1	1	.	.	.	.
POLYGONATUM ODORATUM	1	1	1	.	1	1	.
RUMEX SCUTATUS	.	.	1	.	1	.	.
TEUCRIUM BOTRYS	.	.	3	.	1	1	.
Species group 6							
BUXUS SEMPERVIRENS	.	1	2	.	.	.	1
CARDUUS NUTANS	.	.	2	.	.	.	.
LIGUSTRUM VULGARE	1	1	2	2	1	.	1
ROSA CANINA	.	.	2	.	.	2	2
TRIFOLIUM DUBIUM	.	1	2	.	1	.	1
Species group 7							
CARPINUS BETULUS	.	.	.	1	1	.	1
HYPERICUM TETRAPTERUM	.	.	1	.	.	.	1
RHAMNUS CATHARTICUS	.	1	.	.	.	.	1
RUBUS div. sp.	.	1	1	.	.	.	1
VIBURNUM LANTANA	.	1	2	.	.	.	1
Species group 8							
CIRSIUM ARVENSE	.	.	.	.	.	.	1
DIGITALIS LUTEA	.	1	.	.	1	1	1
HIERACIUM MACULATUM	1	1	.	.	1	.	.
OROBANCHE div. sp.	1	1	1	.	.	1	1
POA PRATENSIS subsp. ANGUSTIFOLIA	.	3	1	3	1	.	4
PRUNUS MAHALEB	2	.	.	.	.	.	1
SENECIO ERUCIFOLIUS	.	1	.	.	1	.	.
TRAGOPOGON PRATENSIS	1	2	.	1	1	.	1
TRIFOLIUM RUBENS	.	1	.	.	.	.	1
VALERIANA WALLROTHII	.	1	.	.	.	.	.
VICIA CRACCA	.	1	.	.	1	.	.
Species group 9							
ANACAMPTIS PYRAMIDALIS	1	1	.	.	.	.	1
HELLEBORUS FOETIDUS	1	1	2	.	.	2	1
HIMANTOGLOSSUM HIRCINUM	1	1	.	.	1	1	.
PICRIS HIERACIOIDES	2	1	1	4	1	1	1
VINCETOXICUM HIRUNDINARIA	2	1	5	4	1	2	2
Species group 10							
AQUILEGIA VULGARIS	1	1	.	.	1	.	1
CENTAURIUM ERYTHRAEA	.	1	1	1	1	.	1
CERASTIUM HOLOSTEOIDES	.	1	2	.	1	2	.
COLCHICUM AUTUMNALE	.	1	.	.	.	.	1
CONVOLVULUS ARVENSIS	1	1	2	2	1	1	1
CUSCUTA EPITHYMUM	2	1	.	.	1	1	1
EPIPACTIS ATRORUBENS	1	1	.	1	1	3	1
FRAGARIA VIRIDIS	1	1	2	1	1	3	3
LACTUCA PERENNIS	1	1	2	.	.	1	.
OROBANCHE TEUCRII	1	1	1	.	.	1	.
PHLEUM PRATENSE	.	1	1	.	.	.	.
POLYGALA COMOSA	.	.	2	.	.	1	1
ROSA RUBIGINOSA	.	.	.	.	.	1	.
RUBUS CAESIUS	.	.	.	1	.	.	1
Species group 11							
CALAMINTHA CLINOPODIUM	.	1	.	1	1	.	1
CLEMATIS VITALBA	.	1	2	1	.	.	2
EUPHORBIA AMYGDALOIDES	.	.	.	.	.	.	2
VICIA HIRSUTA	.	1	.		1	.	1
VICIA SATIVA	.	1	.	.	.	.	1
Species group 12							
LEPIDIUM CAMPESTRE	.	.	2	2	.	.	.
SILENE OTITES	.	.	.	2	.	.	.
SILENE VULGARIS	1	1	1	2	1	1	1
ULMUS MINOR	.	1	.	1	.	.	.
Species group 13							
COTONEASTER INTEGERRIMUS	.	.	.	2	.	.	.
FRAXINUS EXCELSIOR	.	.	.	1	1	1	.
FUMANA PROCUMBENS	.	.	1	.	.	1	.
ROSA div. sp.	1	2	1	4	2	1	1
SAROTHAMNUS SCOPARIUS	.	1	1	2	1	.	1
SILENE NUTANS	.	.	1	1	.	.	.
TARAXACUM div. sp.	.	1	.	1	2	.	.
TEUCRIUM SCORODONIA	.	1	1	4	.	.	.
VERBASCUM LYCHNITIS	.	1	1	2	.	.	.
Species group 14							
ASTRAGALUS GLYCYPHYLLOS	1	1	.	.	1	.	1
DAUCUS CAROTA	1	2	1	3	1	.	1
EUPHRASIA STRICTA	3	2	.	1	2	1	1
FAGUS SYLVATICA	1	1	.	1	.	1	.
GENTIANELLA CILIATA	.	2	.	.	1	.	.
GENTIANELLA GERMANICA	3	2	.	.	2	3	1
HIERACIUM LACHENALII	1	1	.	.	1	.	.
HIERACIUM LAEVIGATUM	1	2	.	1	1	.	1
IBERIS AMARA	1	.	.	.	.	.	.
LATHYRUS APHACA	.	1	.	.	.	.	.
LINUM TENUIFOLIUM	2	1	.	.	1	.	1
ONONIS NATRIX	1	1	.	.	.	.	.
OPHRYS APIFERA	.	1	1	.	.	1	.
OROBANCHE ALBA	2	1	.	.	.	1	.
PINUS SYLVESTRIS	2	1	.	2	2	1	.
POLYGALA AMARELLA	2	1	.	.	.	.	.
RESEDA LUTEA	2	1	.	.	.	.	.
SENECIO JACOBAEA	.	2	2	2	1	.	1
SORBUS ARIA	.	1	.	.	.	.	.
THESIUM HUMIFUSUM	3	1	.	.	1	.	.
THLASPI PERFOLIATUM	1	1	1	.	.	.	.
Species group 15							
AGROSTIS GIGANTEA	1	3	.	.	1	.	2
CAMPANULA GLOMERATA	1	2	1	.	.	.	1
CORNUS SANGUINEA	1	2	2	4	1	1	2
CORONILLA VARIA	1	3	.	.	.	.	.
GALIUM MOLLUGO	2	3	4	1	1	1	3
INULA CONYZA	2	2	3	3	1	1	1
LATHYRUS PRATENSIS	.	1	.	.	1	.	.
LINARIA REPENS	1	1	.	.	.	.	.
MEDICAGO FALCATA	2	3	.	.	.	.	.
ORIGANUM VULGARE	2	3	3	5	2	1	4
PHLEUM BERTOLONII	.	1	1	.	.	.	.
RANUNCULUS BULBOSUS	2	3	1	1	2	1	.
SOLIDAGO VIRGAUREA	2	3	.	.	1	.	.
SONCHUS ASPER	.	1	1	.	1	.	.
STACHYS RECTA	2	3	2	.	.	.	.
Species group 16							
BELLIS PERENNIS	.	1	.	.	.	.	.
CIRSIUM ERIOPHORUM	.	1	.	.	.	.	.
CYTISUS DECUMBENS	2	1	.	.	.	.	.
ERYNGIUM CAMPESTRE	1	1	.	.	1	.	.
OPHRYS SPHEGODES	.	1	.	.	.	.	.
PHYTEUMA TENERUM	1	1	.	.	.	.	.
PRUNELLA GRANDIFLORA	2	1	.	.	.	.	.
SALVIA PRATENSIS	1	1	.	.	.	.	.
TETRAGONOLOBUS MARITIMUS	1	1	.	.	.	.	.
Species group 17							
ANTHERICUM RAMOSUM	4	1	.	.	.	.	.
BETULA PENDULA	1	.	.	.	1	1	1
CENTAUREA APPROXIMATA	1	.	.	.	1	1	.
CERASTIUM ARVENSE	.	.	1	.	1	1	.
GENISTA PILOSA	5	1	.	.	1	.	.
IBERIS VIOLLETII	2	.	.	.	.	.	.
JUNIPERUS COMMUNIS	3	1	.	.	1	.	.
LEONTODON HYOSEROIDES	2	.	.	.	.	.	.
OPHRYS INSECTIFERA	3	1	.	.	1	1	1
ORCHIS PURPUREA	1	1	.	.	.	.	.
SORBUS TORMINALIS	1	.	.	.	.	1	.
TEUCRIUM MONTANUM	5	2	.	.	.	1	.
Species group 18							
BUNIUM BULBOCASTANUM	.	1	.	.	.	.	.
HIERACIUM UMBELLATUM	1	1	.	.	1	.	.
ONONIS SPINOSA	.	1	.	.	.	.	.
RHINANTHUS ALECTOROLOPHUS	.	1	.	.	.	.	.
THALICTRUM MINUS	.	1	.	.	.	.	.
Species group 19							
ACHILLEA MILLEFOLIUM	2	3	1	2	3	1	3
AGRIMONIA EUPATORIA	1	2	.	.	2	1	4
ARRHENATHERUM ELATIUS	.	3	1	.	1	1	4
AVENULA PUBESCENS	.	2	1	.	2	2	1
CAMPANULA RAPUNCULUS	.	1	1	.	1	.	1
DACTYLIS GLOMERATA	.	2	1	1	1	1	2
MEDICAGO LUPULINA	.	2	1	3	1	1	2
MELAMPYRUM ARVENSE	.	1	.	.	1	.	.
ONOBRYCHIS VICIIFOLIA	1	1	.	1	.	1	1
PLANTAGO LANCEOLATA	.	1	1	3	4	1	1
PLANTAGO MEDIA	.	2	.	.	3	3	1
PRIMULA VERIS	.	2	1	2	1	2	2
RHINANTHUS MINOR	.	2	.	.	2	.	1
TRIFOLIUM PRATENSE	.	1	.	.	1	1	2
TRISETUM FLAVESCENS	.	1	2	2	1	1	2
VICIA TETRASPERMA	.	1	.	.	.	.	1
Species group 20							
BRACHYPODIUM SYLVATICUM	.	.	.	.	1	.	.
DACTYLORHIZA FUCHSII	.	.	.	.	2	.	.
INULA SALICINA	.	1	.	.	.	.	.
POTENTILLA ERECTA	.	.	.	.	1	.	.
SILAUM SILAUS	.	1	.	.	1	.	.
SORBUS AUCUPARIA	.	.	.	.	1	.	.
SUCCISA PRATENSIS	.	.	.	1	1	.	.
VERONICA OFFICINALIS	.	1	.	.	1	.	1
Species group 21							
ANAGALLIS ARVENSIS	.	.	.	.	1	.	.
CALLUNA VULGARIS	.	.	.	.	1	.	1
GENTIANA CRUCIATA	.	.	1	.	1	.	.
POTENTILLA REPTANS	.	.	1	.	1	1	1
Species group 22							
ALLIUM OLERACEUM	.	.	.	1	1	.	1
CYNOSURUS CRISTATUS	.	1	.	.	.	.	1
TRIFOLIUM CAMPESTRE	.	1	.	.	1	1	1
TRIFOLIUM REPENS	.	1	.	.	.	1	1
Species group 23							
ANTHOXANTHUM ODORATUM	.	1	1	.	.	1	2
LATHYRUS TUBEROSUS	.	1	.	.	.	.	1
LUZULA CAMPESTRIS	.	.	.	.	1	.	1
MYOSOTIS SYLVATICA	.	.	.	.	.	.	1
PYRUS PYRASTER	.	1	.	.	1	.	.
RUMEX ACETOSA	.	1	.	1	1	.	1
VIOLA CANINA	.	.	.	.	1	.	1
Species group 24							
CORYLUS AVELLANA	.	1	2	3	2	2	2
CRATAEGUS MONOGYNA	1	3	3	5	3	3	5
GALIUM PUMILUM	.	2	3	4	5	5	3
GALIUM VERUM	.	2	3	1	4	3	3
GENISTA TINCTORIA	.	2	2	.	4	1	2
HELIANTHEMUM NUMMULARIUM	2	2	5	5	4	5	3
KOELERIA MACRANTHA	.	2	3	5	4	4	2
POLYGALA VULGARIS	.	.	4	4	4	2	2
PRUNUS SPINOSA	2	2	3	2	3	4	4
QUERCUS ROBUR	2	3	3	4	4	3	3
THYMUS PULEGIOIDES	.	1	4	5	5	4	3
VIOLA HIRTA	1	2	3	2	4	5	5
Species group 25							
AGROSTIS TENUIS	.	1	1	1	2	.	2
FRAGARIA VESCA	.	1	.	.	3	.	1
GENISTELLA SAGITTALIS	.	1	1	.	2	1	1
PLATANTHERA BIFOLIA	1	2	.	.	2	.	.
PRUNELLA LACINIATA	.	1	1	.	2	2	1
SIEGLINGIA DECUMBENS	.	.	.	.	3	.	1
STACHYS OFFICINALIS	.	1	2	.	2	1	2
TRIFOLIUM MEDIUM	.	1	.	.	1	.	.

Vegetatio **81**: 145–157, 1989.

Classification of South Swedish Isoetid vegetation with the help of numerical methods

Urho Mäkirinta
Department of Botany, University of Oulu, SF-90570 Oulu, Finland

Accepted 1.2.1989

Keywords: Aquatic vegetation, *Littorelletea*, PCA, TABORD

Abstract

Relevés of Isoetid vegetation from 60 lakes in southern Sweden have been classified with the help on numerical methods. A community system is constructed, at the variant level by clustering with the TABORD program, and at the community and subcommunity levels by reference to traditional floristic characteristics. The diagnostic species were selected with the help of PCA. *Isoëtes lacustris*, *Lobelia dortmanna* and *Littorella uniflora* determined the community composition together with *Eleocharis acicularis*. The differentation of the syntaxa along the first three ordination axes of a PCA was clear. The complex water depth factor determines clearly the position of the small syntaxa (variants) in the PCA ordination space. At high levels of similarity the relevés were grouped effectively with the TABORD program, so that the clusters are floristically homogeneous and easy to identify on the basis of floristic-sociological criteria. At higher syntaxon levels a selective use of diagnostic species was made.

Nomenclature: for vascular plants Tutin, T.G. *et al.* (1965–1980). Flora Europaea with some updatings notably *Isoëtes echinospora* Durieu instead of *Isoëtes setacea* Lam.; for bryophytes Nyholm, E. (1954–69) Illustrated Moss Flora of Fennoscandia, II. Lund; for Charophytes Cedercreutz, E. (1933), die Characeeen Finnlands, Mem. Soc. Fauna Flora Fenn. 8: 241–254; the nomenclature of the life forms is according to Mäkirinta (1978b).

Introduction

The vegetation monograph on the Isoetid vegetation of Lake Kukkia (Mäkirinta 1978a) resulted in a new system of Isoetid plant communities, which allowed a syntaxonomical comparison with earlier studies of the *Littorelletea*. In the rich literature on the subject two relatively new syntaxonomical studies on the *Littorelletea* were found useful (Dierssen 1975; Pietsch 1977). These thorough studies have two obvious shortcomings, however. Firstly, they were based on only few real relevés from Fennoscandia, although this vegetation is very common there, and secondly, they used species lists of entire lakes, as presented by e.g. Lohammar (1938) and Maristo (1941), as relevés. Such species lists refer to large areas, from several ha up to many sq.km and do not allow a syntaxonomical treatment, due to the inherent heterogeneity. On the other hand, also very small and equally unrealistic small samples of only 1 dm^2 as reported in Vaarama (1938) were used as independent relevés. In thus seems, that the syntaxonomy of the *Littorelletea* has been built up with insufficient knowledge of the synsociological behaviour of the main *Littorelletea* species in subcontinental boreal Europe.

The present study aims at extending the Finnish classification system obtained (Mäkirinta 1978a) to southern Sweden and to interpret the entire Fennoscandian results in a syntaxonomic way (Mäkirinta 1985). Some earlier Swedish works (e.g. Thunmark 1931) contain accurate descriptions of the sociological behaviour of the Isoetid vegetation in southern Sweden, and point to clear differences with the material from the Finnish Lake district (e.g. Mäkirinta 1978a).

Methods

Study area

The study area includes southern Sweden from the northern border of Dalarna to the southern border of Småland, an area 550 km long and maximally 400 km wide (Fig. 1). The bedrock in the western part of this area consists mainly of gneiss and that of the eastern half of various granites. There are limestone and other calcareous rocks overlying the Archaean basement in restricted areas (Magnusson *et al.* 1957).

About 60 lakes in this area were chosen, partly at random, partly selectively, based on information in the phytogeographical literature (Fig. 1 and Appendix 1). Further ca 20 lakes were found not to house any Isoetid vegetation, perhaps due to too eutrophic water. Most of the Isoetid lakes are located in areas of silicate rock and not directly surrounded by intensively used agricultural or industrial areas. Isoetid vegetation was found to be quite well developed on wet calcareous substrate on Öland (Albertson 1950). Most of the lakes are small in area, typically a few km^2. Lake Sommen (132 km^2) in Östergötland and Lake Siljan (290 km^2) in Dalarna represent some of the biggest areas of well-developed Isoetid vegetation. This vegetation has also been reported from the northern shores of the very large Lake Vättern (Stålberg 1939).

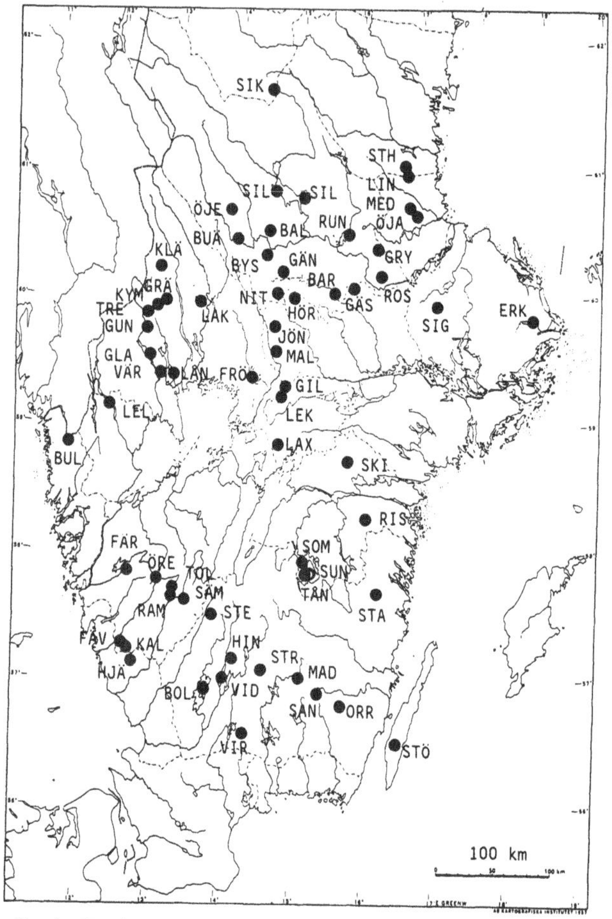

Fig. 1. Study area and location of the 60 lakes studied. The lakes are labelled with symbols; see Appendix 1 for the full names.

A special area of 1–2 ha with a shoreline of 100–200 m in which Isoetid vegetation could be found was defined in each of the 60 lakes and a sufficient number of relevés (size 10 m^2) were made. The number of relevés depended on the variation in the Isoetid vegetation belt. The mean number of relevés per lake was 4. The 220 relevés originated from belts extending from around mean water level (MW) to the outer border of the Isoetid vegetation.

The shape of the relevé plots varied. Rectangle plots, usually 2 m × 5 m, were used in shallow water, the longer side being parallel to the shoreline and perpendicular to the depth gradient. Square plots were often used in extensive, homogeneous isoetid vegetation, e.g. in deep water with *Isoëtes lacustris* stands.

A detailed analysis was made as follows: Five 1 m^2 subsidiary relevés, located at regular intervals in the 10 m^2 main relevé were used to analyse the helophyte and elodeid layers (synusia). Five small relevés of 3 dm^2 were analysed using a 'scoop' (a combination of a hoe and a spade with a 4 m long handle, Mäkirinta 1978a) to estimate the projected percentage cover of each species, especially of the isoetid layer. This analysis usually gave further information on the other layers as well. In deep water and under bad light conditions, the 3 dm^2 scoop relevés were mostly the only source of information, not only for isoetids but also for macro-algae, aquatic bryophytes, small elodeids and seedlings of helophytes. Therefore it was necessary to analyse a larger number of these units (up to 10). The mean abundance values of the species in the subsidiary relevés of 1 m^2 and 3 dm^2, expressed in projected coverage percentages, were determined for the main 10 m^2 relevé. In deep water a drag of the Maristo type (Maristo 1941) was also used especially for elodeids and aquatic bryophytes. (For more details see Mäkirinta 1978a).

Exact determination of the mean water level (MW) was possible only in a few cases. Otherwise, the location of the 'hook' (i.e. at the point where it is a clear upward step in the shore profile at the water line), was used to obtain the zero level. Observations with exact hydrographic measurements at Lake Kukkia (Mäkirinta 1978a) and at some other lakes suggested that this was located a little above the true zero on open shores and a little below it in sheltered coves. This implies that the best place to determine the 'hook level' was a half-open site. The border between geoamphibiontic and hydroamphibiontic vegetation (Du Rietz & Hannerz 1939) points to the same level. These methods give only approximate values, however.

Numerical methods

The original scores of species as estimated in the field in projected cover percentages were transformed as follows (close to the scheme of van der Maarel 1979): $<0.5\% = 1$, $1\% = 2$, $2\% = 3$, $3\% = 4$, $5\text{-}10\% = 5$, $15\text{–}25\% = 6$, $30\text{–}45\% = 7$, $50\text{–}75\% = 8$, $80\text{–}100\% = 9$. The scale is (except in its uppermost part) roughly logarithmic, like that used for the numerical establishment of associations by Dale & Webb (1975). The choice of an appropriate transformation scale of cover values is very important in the numerical classification of aquatic vegetation (Jensén 1978).

All numerical procedures were applied in two ways: a. including all species occurring in the relevé material, and b. with only the hydrophytes, i.e. excluding the helophytes (Mäkirinta 1978a, b). The clustering and classification procedure advanced in a stepwise manner:

1. An ordination of the species with the PCA program ORDINA (Roskam 1971) to detect which are of major importance for selecting the diagnostic species for the community types as appearing along the first 3 axes.
2. Construction of a dendrogram with the program GROUPAGE (by Hajdu, Uppsala; see also Andersson 1988) and its drawing program REDRAW. This program includes average linkage of relevés, with the similarity ratio SR as resemblance measure. It was decided on inspection of the dendrogram where the optimal level of the cluster structure and the optimum number of the (not too small) basic clusters should be selected.
3. To obtain optimally homogeneous basic clusters (Popma *et al.* 1983), the material was treated interactively using the iteration-based program TABORD (van der Maarel *et al.* 1978) with an initial cluster array derived from the GROUPAGE dendrogram and with some alternative arrays with slightly more or fewer cluster.
4. A fusion of basic clusters into higher units was achieved in two ways, which were compared and evaluated:

4a. A numerical fusion of the basic clusters using TABORD without iteration, to form a hierarchy of higher clusters, and

4b. A fusion of the basic clusters into higher units (subcommunities, communities and alliances) by their diagnostic species. To facilitate practical use of the system it was important to find a community type corresponding to the association of the floristic-sociological tradition of the Zürich-Montpellier-School (Westhoff & van der Maarel 1978; Barkman *et al.* 1986).

5. An ordination of the relevés with the PCA program ORDINA to evaluate the positions of the basic clusters (variants) and higher units with respect to each other and to possible ecological factors.

Results

Basic clusters and diagnostic species

A good classification result for practical field use is achieved when the basic clusters are so homogeneous that individual relevés can be identified from simple floristic characteristics, notably diagnostic species. An interpretable result in terms of this criterium was obtained in the TABORD clustering procedure with the demands: 17 clusters, minimal cluster size 4 relevés, fusion level at 0.65, automatic ICLA, 20 iterations allowed, no threshold value and only hydrophytes (species no. 1–29) included. If the fusion limit was varied between 0.5 and 0.7, the clusters still contain the same relevés.

Table 1. Isoëtes lacustris community (IlacC) with 2 subcommunitities (SC), and 2 variants (V).

Community	C -->	Ilac	
Subcomm.	SC -->	t	Lit
Variant	V -->	nu	nu
Cluster no.		01	13
Relevé no.		0000000000000000002222222222222222	0000000022
		0333344444444445500000011111111112	2222233300
		7258901234567890134578901234567890	5678901616
Lake symbol		LFVTFBFSLSSTTLSLTFMBRLRHSHBBKGGÖSS	SSSBBSÖÖBV
		EÄIOÄUÄIEIIÅÅEIEORAYUÅIITÖYYLRRJTK	OIIUUORRAÄ
		LVDLRLVLLGLNLLLLLÖDSNNSNERSSÄÄÄAHI	MLLÄLMEERR
1 *Littorella uniflora*		-2122------------222--------------	5755676665
2 *Lobelia dortmanna*		1---------------------------------	--------1-
3 *Isoëtes lacustris*		6558666787666868976666586567777657	5677755347
4 *Ranunculus reptans*		--1-------------------------------	-----22---
5 *Isoëtes echinospora*		--------------------5-------------	-----1----
6 *Eleocharis acicularis*		-32------------------5------------	-----1--54
7 *Subularia aquatica*		--------------------1-------------	----------
11 *Myriophyllum alternifl.*		-2-----------35------52-----------	---------1
12 *Juncus bulbosus*		----------------------------------	----------
13 *Elodea canadensis*		---1------------------------------	-----5-4--
14 *Potamogeton perfoliatus*		----------------------------------	------13--
15 *Potamogeton gramineus*		--4-------------------------------	----------
16 *Ranunculus peltatus*		----------------------------------	------3---
19 *Nitella flexilis*		-3--1-----------------------------	----------
20 *Utricularia intermedia*		-------------------1--------------	----------
24 *Potamogeton natans*		7---------------------------------	----------
30 *Equisetum fluviatile*		1-5-------------5-----------------	----------
31 *Phragmites australis*		--5-------------------------------	--------5-

Table 2. Isoëtes lacustris-Lobelia dortmanna community (IlacLoC) with 2 subcommunities (SC) and 4 variants (V).

Community	C -->	IlacLo			
Subcomm.	SC -->	t		Lit	
Variant	V -->	nu	Rr	Lit	Ma
Cluster no.		11	10	02	04
Relevé no.		001111111	1111	000001111111111111	000000011
		003344444	4445	011110111122222335	111113512
		453534568	1271	816895345613479022	234573788
Lake symbol		TLMGKÖÖGF	GKGF	FSSBLHGTÖGSMMSKVSL	FFTTSFTSB
		ÅAERLJJUR	ÄLUR	ÄUIUEIRRJRAAATLÄKÅ	ÄÄOÅIÄÅKY
		NKDÄÄAANÖ	NÄNÖ	RNGÄLNÄEAÄNDDEÄRIN	VRLNGRNIS
1 *Littorella uniflora*		-------22	----	566677554656665766	677366756
2 *Lobelia dortmanna*		657666735	6656	665668676785466587	36362-665
3 *Isoëtes lacustris*		376655577	4465	566563766635565755	522425-54
4 *Ranunculus reptans*		2----13-2	6356	--1--1--1--------3	--13----1
5 *Isoëtes echinospora*		----2----	---4	----1----------1-2	2--------
6 *Eleocharis acicularis*		-----1153	2562	----2----------2--	-11--1--2
7 *Subularia aquatica*		----1----	--3-	---1-----------1-2	-1---1---
10 *Pilularia globulifera*		---------	---1	-----------------3	---------
11 *Myriophyllum alternifl.*		--------3	---2	--2---------------	555565755
12 *Juncus bulbosus*		-----12--	23--	--14-1-2-22--2----	-1----14-
13 *Elodea canadensis*		---------	----	------------------	--5------
14 *Potamogeton perfoliatus*		---------	1---	------------------	---1-11--
16 *Ranunculus peltatus*		---------	----	------------------	----51---
18 *Chara fragilis*		---------	----	------------------	---2-----
19 *Nitella flexilis*		---------	----	------------------	2-1---1--
21 *Utricularia vulgaris*		------1--	----	------------------	---------
23 *Nuphar lutea*		---------	----	----4-------------	---------
24 *Potamogeton natans*		---------	----	----4--1----------	---------
25 *Polygonum amphibium*		---------	----	---------------1-2	---------
26 *Sparganium angustifol.*		---------	4---	------------------	---------
27 *Nymphaea candida*		---------	-2--	------------------	---------
29 *Glyceria fluitans*		---------	-2--	------------------	---------
30 *Equisetum fluviatile*		----5--32	2-23	---22---------2--1	----2----
31 *Phragmites australis*		42----555	5--5	--225---2--------6	----5----
32 *Eleocharis palustris*		-----6---	--4-	------------------	----2----
33 *Carex lasiocarpa*		---------	----	--5---------------	---------
34 *Schoenoplectus lacutr.*		-----14-3	---2	----5--22--------3	---------
35 *Lysimachia thyrsiflora*		---------	---4	------------------	---------
36 *Alisma plantago-aquat.*		---------	--4-	------------------	---------

A few relevés did not share the floristic characters of the cluster (as determined by the key in Table 6, see below) to which they were assigned because of the mathematical similarity. Therefore Relevés 98, 177 and 178 were moved from cluster 6 to cluster 12 and relevé 208 from cluster 12 to cluster 1. One relevé from cluster 12 (97) was considered a unit of its own.

In addition not more than 10 other relevés are transitional between two clusters or do not fit in the scheme of Table 7: Nr 9 (clusters 3 and 13), 14 (4/7?), 17 (4/unclear), 35 (4/13), 37 (4/14), 61 (7/9), 106 (3/14), 119 (16/2), 125 (5/8) and 154 (9/10?). These were not moved from their original clusters.

In this way 18 community types were derived and interpreted on three levels of similarity, called community, subcommunity and variant. Tables 1–5 present the floristic composition of the types.

The PCA diagrams (Fig. 2) show that the three main species, which have an overall dissimilarity in occurrence are *Littorella uniflora*, *Lobelia dortmanna* and *Isoëtes lacustris*. *Eleocharis acicularis*, *Ranunculus reptans*, *Juncus bulbosus*, *Myriophyllum*

Table 3. Lobelia dortmanna community (LoC) with 3 subcommunities (SC) and 4 variants (V).

Community	C -->	Lo			
Subcomm.	SC -->	Iech		Jb	Lo
Variant	C -->	Su	Rr	Rr	nu
Cluster no.		06	08	09	16
Relevé no.		011111111 945566666 990334567	011111 203358 201885	0001111111111111111111 2263334555666667788889 3466790479012890212341	00000000000000000000001111111111 00007888888888899999991388889999 12369012345678901234569467890234
Lake symbol		SVVRRRRRG IÄÄOOOOOR LRRSSSSSY	BSVHGL AUÄIRÅ LNRNÄN	LLTSSLHKSGKKTHHHTLLLGH EEÅAAAÖLARYYRÖÖÖRAAARÖ LLNNNXRÄNÄMMERRREXXXÄR	MGGSMJJJBBMMGGJJGMKGJSSSMMÖSHNNG AIIIAÖÖÖAUAAIIÖÖIAAIÖITTEEJTIIIÄ LLLKLNNNLÄLLLLNNLLLLNKAADDAANTTN
1 *Littorella uniflora*		---------	--2-11	--2--------2----2-2---	----------------------3----2----
2 *Lobelia dortmanna*		265652274	627788	5666666738667356677576	56755787667775886676857568776756
3 *Isoëtes lacustris*		-112-----	-222--	---22323--------2---12	2122------------------22-1--1---
4 *Ranunculus reptans*		566455654	345343	123-3552636664446-2---	-----1--------------------11---1
5 *Isoëtes echinospora*		5332453-5	335222	22122--221422---------	--------1-----------------------
6 *Eleocharis acicularis*		545554256	------	53----------5---------	------13--------------------1---
7 *Subularia aquatica*		22663-657	--2-2-	--1---5--1--2-12-----1	-----------------------------1--
10 *Pilularia globulifera*		-61------	------	----------------------	--------------------------------
11 *Myriophyllum alternifl.*		---66572-	------	--1-------------------	-----252------------------------
12 *Juncus bulbosus*		----32-2-	1--22-	6864565565665756665544	------1-1------------1----2221--
13 *Elodea canadensis*		---1-2---	------	----------------------	--------------------------------
14 *Potamogeton perfoliatus*		-----4---	------	----------------------	---1----------------31----------
15 *Potamogeton gramineus*		-------3-	------	--4-------------------	---1----------------------------
19 *Nitella flexilis*		-----1---	------	----------------------	--------------------------------
20 *Utricularia intermedia*		---------	------	----------------------	-----------------------------1--
23 *Nuphar lutea*		------2--	2-----	-5-1-6----------------	----------------------------2---
24 *Potamogeton natans*		---------	------	-2--------------------	--------------------------------
26 *Sparganium angustifol.*		1--------	2-----	-------2--------------	---1-1112--1--------11----------
29 *Glyceria fluitans*		---------	------	------2---12----------	-----------12------2------------
30 *Equisetum fluviatile*		-------1-	-312--	-142------------------	-1------111221--------1-1-321-1-
31 *Phragmites australis*		-------5-	-----5	-5-------5------------	--2---------------------------4-
32 *Eleocharis palustris*		-------2-	------	--1-------------------	---------------------------1----
33 *Carex lasiocarpa*		---------	------	----------------------	--------------562---------------
34 *Schoenoplectus lacustr.*		---------	-3---5	-5-----5-----------4--	-----------------32-----------1-
35 *Lysimachia thyrsiflora*		---------	----5-	----------------------	------------2-----12------------
36 *Alisma plantago-aquat.*		---2-----	---1--	----------------------	-----------1------------1-------
37 *Carec rostrata*		---------	---111	----------------------	--------------61--------1---3---

Table 4. Littorella uniflora-Lobelia dortmanna community (LitLoC) with 2 subcommunities (SC) and 4 variants (V).

Community	C -->	LitLo			
Subcomm.	SC -->	t		Jb	
Variant	V -->	nu	Rr	Rr	nu
Cluster no.		14	05	07	15
Releve no.		00000000000001111 15556677777770002 02562801234562480	0000011 2256612 0195715	000011111 666612557 013406561	0001111 5560112 3891272
Lake symbol		TSSSSRSSBFBSBSMSS ÅIOTIAIIUÄUIUTATA NLMEGMGGÄVLGLHDEN	VVTFÖBM IIOÄRAA DDLRERD	LLVBLSSHH EEIOATTÖÖ KKRLXRRRR	SHTSSSS IJOKKKA GÄLIIIN
1 *Littorella uniflora*		67875686587766755	6576562	755665676	6665555
2 *Lobelia dortmanna*		55466766774566575	3655546	516666665	6654776
3 *Isoëtes lacustris*		1------------22-2	1---1-1	----11-1-	-----22
4 *Ranunculus reptans*		----11--------1--	5446523	645646567	--1----
5 *Isoëtes echinospora*		--------2-------5	251-2-2	---2-112-	-------
6 *Eleocharis acicularis*		-1------2--------	521--2-	--1------	-------
7 *Subularia aquatica*		-1------1--------	-2-----	---211-12	-------
8 *Elatine hexandra*		-----------------	-------	---2-----	-------
11 *Myriophyllum alternifl.*		--1--22----------	--54245	---------	1-2--2-
12 *Juncus bulbosus*		---22--21------22	--1---2	676425525	5436665
13 *Elodea canadensis*		-----------------	--2----	---------	--2----
15 *Potamogeton gramineus*		-----------------	1---2--	---------	-------
23 *Nuphar lutea*		-----------------	-------	---------	----1--
25 *Polygonum amphibium*		-----------------	-------	-1-------	-------
26 *Sparganium angustifol.*		-----------------	-------	---------	----11-
29 *Glyceria fluitans*		-----------------	-------	-----3435	---1---
30 *Equisetum fluviatile*		----2-1-2-12-----	5--5-2-	---3--122	-------
31 *Phragmites australis*		----5-553-657----	6--5---	---1---5-	-------
32 *Eleocharis palustris*		----4--1-5------2	---66--	--1--55--	--2----
33 *Carex lasiocarpa*		-----3-----------	-------	-----1---	--2----
34 *Schoenoplectus lacustr.*		---------2-------	-------	---------	-------
35 *Lysimachia thyrsiflora*		---------5-------	-------	------2--	---1---
36 *Alisma plantago-aquat.*		-----------------	-------	--1------	-------
38 *Lythrum salicaria*		-----------------	-------	---1-----	-------
39 *Ranunculus flammula*		-----------------	-------	---------	---1---
43 *Juncus articulatus*		-----------------	-------	---------	---1---
46 *Lysimachia vulgaris*		-----------------	-------	-----2---	-------
48 *Veronica scutellata*		-----------------	-------	--------2	-------

Table 5. Littorella uniflora community (LitC) and *Eleocharis acicularis* community (EacC), both with 2 subcommunities (SC) and 2 variants (V).

Community	C -->	Lit		Eac	
Subcomm.	SC -->	t	Pg	t	Pg
Variant	V -->	nu	Cha	Rr	nu
Cluster no.		03	17	12	12
Releve no.		0000001111122 0335770000900 9474783679902	1111 9999 5678	011111111 977777778 834567890	0 9 7
Lake symbol		SÖSSBSVHHSOOE OROOUOÄIITRRR MEMMLMRNNERRK	SSSS TTTT ÖÖÖÖ	SGGGGLVBS ILLLLIÄAT LAAAANRRH	S Ä M
1 *Littorella uniflora*		6676877878777	7898	---------	-
2 *Lobelia dortmanna*		1--2112322---	----	-------12	-
3 *Isoëtes lacustris*		322----121--2	----	31511--11	-
4 *Ranunculus reptans*		-----3-------	----	55265794-	-
5 *Isoëtes echinospora*		-1---1-------	----	22-1-2-1-	-
6 *Eleocharis acicularis*		-1----4------	----	667587569	8
7 *Subularia aquatica*		-------------	----	5----14-3	-
9 *Elatine hydropiper*		-------------	----	-1254----	-
11 *Myriophyllum alternifl.*		--32--------2	----	-5555--4-	-
12 *Juncus bulbosus*		-2-------222-	----	---------	-
13 *Elodea canadensis*		-------------	----	---------	2
15 *Potamogeton gramineus*		-----1-------	4626	---------	5
16 *Ranunculus peltatus*		-------------	----	-3536----	-
17 *Chara aspera*		-------------	5565	---------	-
19 *Nitella flexilis*		-------------	----	---------	1
22 *Scorpidium scorpioides*		-------------	-5--	---------	-
28 *Sagittaria sagittifolia*		-------------	----	-251-----	-
29 *Glyceria fluitans*		---------2---	----	------4--	-
30 *Equisetum fluviatile*		---------2---	11-6	-------45	2
31 *Phragmites australis*		----4--------	----	---------	-
32 *Eleocharis palustris*		-2--1--------	1446	---------	-
35 *Lysimachia thyrsiflora*		---------5---	----	---------	-
37 *Carex rostrata*		-------------	----	--------2	-
38 *Lythrum salicaria*		-------------	----	------3--	-
39 *Ranunculus flammula*		-------------	-1--	---------	-
40 *Baldellia ranunculoides*		-------------	12--	---------	-
41 *Carex elata*		-------------	2---	---------	-
42 *Mentha arvensis*		-------------	2---	---------	-
43 *Juncus articulatus*		-------------	13-1	------2--	-
44 *Agrostis stolonifera*		-------------	-11-	---------	-
45 *Schoenoplectus lacustr.*		-------------	--5-	---------	-
47 *Juncus filiformis*		---------2---	----	------3--	-

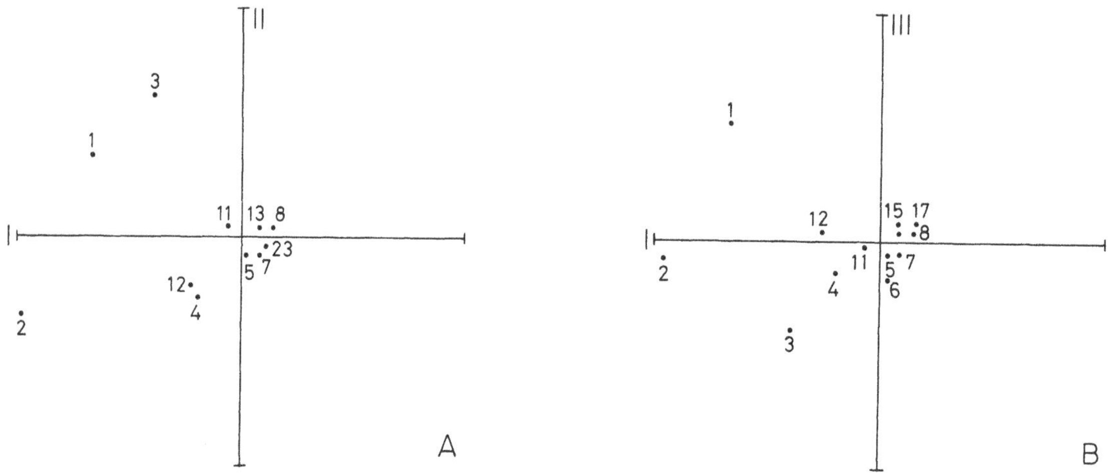

Fig. 2. Species ordination determined by the first three PCA axes. A: axes I and II; B: axes I and III. Main diagnostic species: 1. *Littorella uniflora*, 2. *Lobelia dortmanna* 3. *Isoëtes lacustris*, 4. *Ranunculus reptans*, 12. *Juncus bulbosus*. The species numbers are the same as in the relevé tables (2–6). Identical species are in 2A: ** 5,6 ** 8–10, 16–22, 24–25, 27–28 ** 13–15 ** 23, 26, 29 *; in 2B: ** 8–10, 16, 18–22, 24–25, 27–28 ** 13–14, 23, 26, 29 **.

alterniflorum and *Chara aspera* are additional differentiating species which also show clear dissimilarities. All other species are found near the origo and have no significance on the three first axes. Of course, they may be important on higher dimensions.

The floristic system obtained on the basis of the characteristic species combinations with 6 communities (syntaxa) is presented in Table 6. The numerical fusion of the 17 basic clusters into 5 major clusters (A–E) is shown in Fig. 3. A comparison of the two systems is presented in Table 7. The main difference is that of the four main diagnostic species. *Isoëtes lacustris* is considered more important than *Littorella uniflora*, for which reason the clusters 11 and 10 are united with 02 and 04. The six syntaxa of Table 6 may be considered as associations. They can easily be dis-

Table 6. Key for determining the syntaxa in South Swedish Isoetid vegetation. The names of the syntaxa are indicated with symbols, for the whole names, see Tables 1–5.

Community C -->	Ilac		IlacLo				Lo				LitLo				Lit		Eac	
Subcommunity SC -->	t	Lit	t		Lit		Iech		Jb	Lo	t		Jb		t	Pg	t	Pg
Variant V -->	nu	nu	nu	Rr	Lit	Ma	Su	Rr	Rr	nu	nu	Rr	Rr	nu	nu	Ch	Rr	nu
Basic clusters 1-17	01	13	11	10	02	04	06	08	09	16	14	05	07	15	03	17	12	12
3 *Isoëtes lacustris*	X	X	X	X	X	X	.	.	.	.	.	.	.	.	.	.	.	.
2 *Lobelia dortmanna*	.	.	X	X	X	X	X	X	X	X	X	X	X	X	.	.	.	.
1 *Littorella uniflora*	.	X	.	.	X	X	.	.	.	.	X	X	X	X	X	X	.	.
6 *Eleocharis acicularis*	.	.	.	X	.	.	X	.	.	.	.	.	.	.	.	.	X	X
5 *Isoëtes echinospora*	.	.	.	.	.	.	X	X	.	.	.	.	.	.	.	.	.	.
7 *Subularia aquatica*	.	.	.	.	.	.	X	.	.	.	.	.	.	.	.	.	.	.
4 *Ranunculus reptans*	.	.	.	X	.	.	X	X	X	.	.	X	X	.	.	.	X	.
12 *Juncus bulbosus*	.	.	.	.	.	.	.	.	X	.	.	.	X	X	.	.	.	.
11 *Myriophyllum alterniflorum*	.	.	.	.	.	X	.	.	.	.	.	.	.	.	.	.	z	.
15 *Potamogeton gramineus*	.	.	.	.	.	.	.	.	.	.	.	.	.	.	.	X	.	X
17 *Chara aspera*	.	.	.	.	.	.	.	.	.	.	.	.	.	.	.	X	.	.
10 *Pilularia globulifera*	.	.	.	.	z	.	z	.	.	.	.	.	.	.	.	.	.	.

X = a constant species with cover more than 1%
z = not a constant species with varying cover
· = species absent or its cover very low, not more than 1%

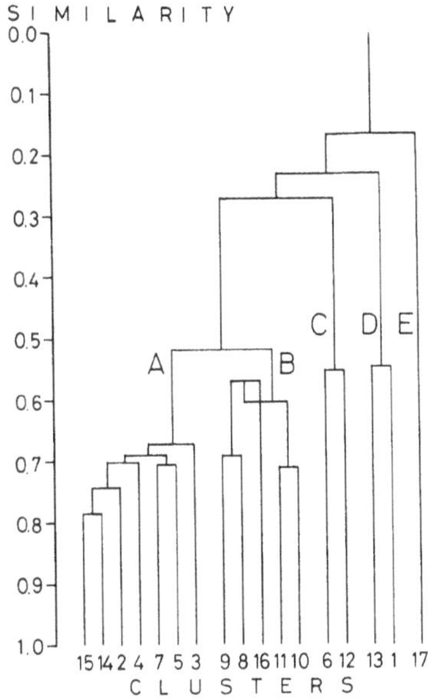

Fig. 3. Fusion of the 17 basic clusters with the program TABORD to 5 major clusters A–E at a level of 0.5–0.6 similarity and without iterations.

tinguished from each other by reference to the three main species, *Littorella uniflora*, *Lobelia dortmanna* and *Isoëtes lacustris* and their absence or scarcity in the case of the *Eleocharis acicularis* community, which is dominated by *Eleocharis acicularis*.

In the nomenclature of the syntaxa (Table 6) the term 'nudum' is used for variants without additional diagnostic species as compared with the next higher subcommunity level, while the variant is poor in species at the same time. This idea is borrowed from Raspopov (1971), who used the term 'subpurum' in comparable cases.

Ordination of the relevés

The PCA ordination was first run with all 220 relevés and second on a selection of 68 relevés (by taking 4 typical relevés from each of the 17 clusters). In both cases the same trend is shown for the clusters, i.e. to be clearly separated in the space formed by the first three PCA–axes. Fig. 4 presents the PCA diagrams for the 68 typical relevés.

The positions of the 17 clusters and the community types in the PCA space (Fig. 4A) shows that the *Isoëtes lacustris-Lobelia dortmanna* community (IlacLoC) occupies a central position in the system while the other 5 communities are situated towards the periphery. Contrasts are formed by a. *Lobelia* community (LoC) and the *Isoëtes lacustris* community (IlacC), and b. the *Littorella uniflora-Lobelia dortmanna* community

Table 7. Comparision between the numerical system of clusters A–E (See Fig. 3) and the floristic community types. The 17 basic clusters (variants) are the same in both systems. For symbols, see Tables 1–6.

Major clusters A-E	A							B					C		D		E
Basic clusters 1-17	02	04	15	14	07	05	03	11	10	09	08	16	06	12	13	01	17
Community C -->	IlacLo		LitLo				Lit	IlacLo		Lo			Lo	Eac	Ilac		Lit
Subcommunity SC -->	Lit		Jb	t	Jb	t	t	t		Jb	Iech	Lo	Iech	t	Lit	t	Pg
Variant V -->	Lit	Ma	nu	nu	Rr	Rr	nu	Rr	Rr	Rr	Rr	nu	Su	Rr	nu	nu	Ch
3 *Isoëtes lacustris*	X	X	.	.	.	.	.	X	X	.	.	.	.	.	X	X	.
2 *Lobelia dortmanna*	X	X	X	X	X	X	.	X	X	X	X	X	X	.	.	.	.
1 *Littorella uniflora*	X	X	X	X	X	X	X	.	.	.	.	.	.	.	X	.	X
6 *Eleocharis acicularis*	.	.	.	.	.	.	.	.	X	.	.	.	X	X	.	.	.
5 *Isoëtes echinospora*	.	.	.	.	.	.	.	.	.	.	X	.	X	.	.	.	.
7 *Subularia aquatica*	.	.	.	.	.	.	.	.	.	.	.	.	X	.	.	.	.
4 *Ranunculus reptans*	.	.	.	.	X	X	.	.	X	X	X	.	X	X	.	.	.
12 *Juncus bulbosus*	.	.	X	.	X	.	.	.	.	X	.	.	.	.	.	.	.
11 *Myriophyllum alternifl.*	.	X	.	.	.	.	.	.	.	.	.	.	.	z	.	.	.
15 *Potamogeton gramineus*	.	.	.	.	.	.	.	.	.	.	.	.	.	.	.	.	X
17 *Chara aspera*	.	.	.	.	.	.	.	.	.	.	.	.	.	.	.	.	X
10 *Pilularia globulifera*	z	.	.	.	.	.	.	.	.	.	.	.	z	.	.	.	.

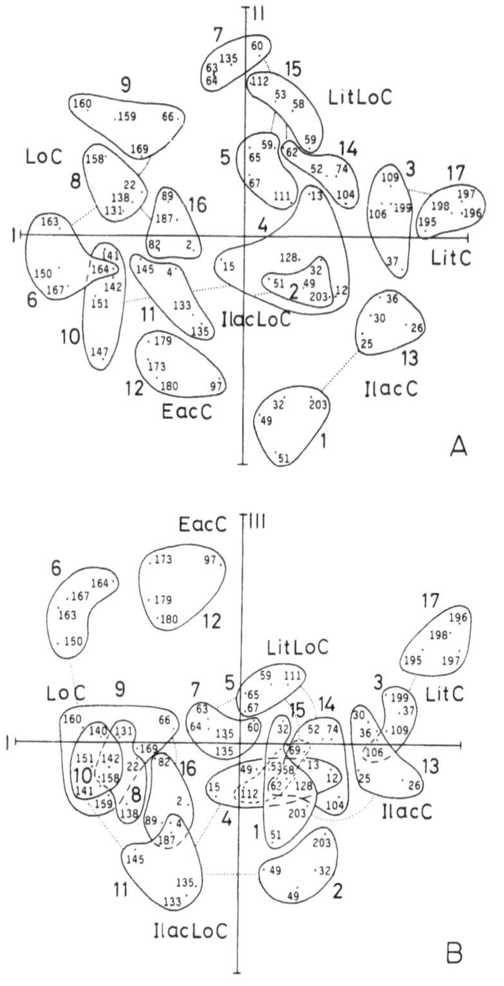

Fig. 4. Ordination of 68 typical relevés (4 relevés from each of the 17 basic clusters) with the short names of the 6 communities indicated. A: axes I and II, B: axes I and III. The numbers of relevés and basic clusters are the same as in the relevé tables (1–5).

(LitLoC) and the *Eleocharis acicularis* community (EacC). These contrasts show at the same time the floristic dissimilarities between the types.

Relevé 97, which was separated from cluster 12 to become a syntaxon of its own (PgSC, nuV), differs clearly from the other relevés of the same basic cluster (tSC, RrV; Table 5): it is characterized by the diagnostic species *Potamogeton gramineus*, which is lacking in the other relevés of the cluster. Also in the ordination space this relevé is clearly separated. (Fig. 4).

Ecological interpretation

In the ordination space of the two first axes (Fig. 4A) the clusters are ordened as a compositional gradient which is correlated with water depth (Fig. 5): the deep water vegetation (IlacC) is located right down and the shallow water vegetation (LoC) left up, while the vegetation of medium depths (IlacLoC and a part of LitLoC) is situated inbetween. The shallow-deep water gradient is running along the diagonal; clearly both floristic axes are correlated with this environmental factor and axes 1 and 2 must be rotated ca 45 degrees to the right in order to obtain water depth as the first vector (Carleton 1980; van der Maarel 1980; Abbott *et al.* 1985; Mäkirinta 1986). One anomaly occurs in the scheme: cluster 17 is situated at the 'deep' end of the 1st axis, but it originates from very shallow waters. The effective ecological factors (e.g. the light climate), are correlated with the depth-gradient, but not in a perfect manner. The ecological factors to be correlated with the 2nd and 3rd dimensions (Fig. 4A, B) are unclear, as may happen in PCA (Beals 1984).

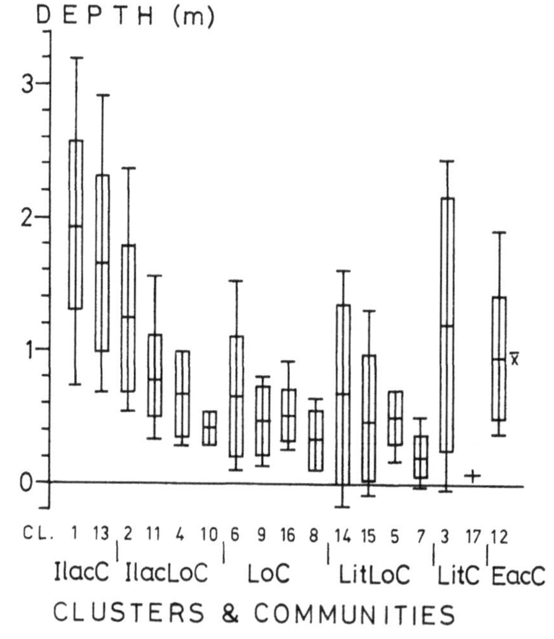

Fig. 5. Water depth in the sites, arranged according to the clusters (CL) and communities (C) of the material (n = 220). The mean, standard deviation and extreme values are given.

Role of the helophytes

If all the species, also shore plants, were included in the TABORD analysis, the clusters were less homogeneous in terms of their real aquatic species composition. The helophytes (species nrs 30–48) never appear in abundance in this kind of vegetation, but a cover of 0–5% is quite common. The most common helophyte species in this material are *Eleocharis palustris*, *Equisetum fluviatile* and *Phragmites australis*, which do not have a very clear correlation with any isoetid community type (Mäkirinta 1978a). Helophytes are terrestrial rather than aquatic species, i.e. they are not of a hydrophyte life form (see the systems in Braun-Blanquet 1964 and Mäkirinta 1978a, b). Thus their ecological behaviour does not correlate very well with that of the true aquatic macrophytes. Removal of the helophytes from the numerical analysis reduces the heterogeneity of the material and makes the result more meaningful, both floristically and ecologically. Deletion of species is widely used in numerical methods (e.g. Dale & Williams 1978; van der Maarel 1982; Kissling *et al.* 1988), although this usually concerns rare species.

Connecting the communities into other systems

On the whole the South-Swedish system presented here and the Finnish system (Mäkirinta 1978a) are similar, but some remarkable differences exist. The Finnish 'community types' are 'communities' here and the subcommunity level is new in the Swedish system. The other main changes are that the IechVar of ILoT is now placed in the Lo community as Iech subcommunity and the IlacVar of ILoT is now raised to the range of community (IlacC). The few relevés with *Pilularia globulifera* are spread over different syntaxa.

According to the system of Pietsch (1977) the syntaxa represent three alliances: the *Isoëtes lacustris* community fits in the *Subulario-Isoëtion* (Nordhagen 1936) Pietsch 1977; the *Eleocharis acicularis community* and the *Littorella uniflora* community in the *Eu-Littorellion uniflorae* Koch 1926 s.str., suballiance *Elatini-Eleocharition acicularis* (Pietsch 1965) Pietsch 1977; and the other three communities in the *Lobelio-Isoëtion* Pietsch 1965. The alliances distinguished by Dierssen (1975) have different names and partly different contents.

The limits of our communities are in many cases somewhat different from those of the associations described in the literature, but it is still possible to assign most communities and subunits to a Braun-Blanquet syntaxon. Two new subassociations and 4 variants in the association system may be suggested (with their diagnostic species presented in Table 6):

- IlacC/tSC/nuV (cluster 1) represents the *Isoëtetum echinosporae* W. Koch 1926, typical subass. Pietsch 1977, typical variant, var. nova;
- IlacC/LitSC/nuV (cluster 13) represents the *Isoëtetum echinosporae* W. Koch 1926, typical subass. Pietsch 1977, *Littorella uniflora* variant, var. nova;
- LoC/JbSC/RrV (cluster 9) represents the *Isoëto-Lobelietum* (W. Koch 1926) Tx. 1937, subass. of *Juncus bulbosus*, subass. nova, *Ranunculus reptans* variant, var. nova;
- LitC/PgSC/PgV (cluster 17) represents the *Subulario-Eleocharitetum* acicularis (Sørensen 1942) Pietsch 1977, subass. of *Potamogeton gramineus*, subass. nova, *Chara aspera* variant, var. nova.

Discussion

Sampling design

The relevé size was reduced from 25 m^2 (Mäkirinta 1978a) to 10 m^2, because this seemed to be enough even for the higher but poorly represented strata (synusia) in the Swedish lakes. A still smaller size (e.g. 1–2 m^2, see Pietsch 1977) could have been enough for the isoetid synusia, but not for the higher ones. The small sample plots of 3 dm^2 are very effective in analysing this kind of vegetation (Mäkirinta 1978a) if they are numerous and scattered over the large relevé, and if the mean values of the scores in the small plots are used to describe the large relevé (Mäkirinta 1978a). The advantage of many small scattered relevés against one large in a stand to be analysed is well documented (e.g., Downing & Anderson, 1985).

A series of separated, large (10 m^2) but still homogeneous relevés along the depth gradient was preferred to a continuous belt transect, as used by Jensén (1979, also Jensén & van der Maarel 1980).

The structural type concept

The role of helophytes in *Littorelletea* vegetation is usually disregarded (e.g., Tüxen 1937; Dierssen 1975; Pietsch 1977). However, dense stands of *Phragmites* or other helophytes can be of considerable importance in this kind of lake vegetation, and attention must be paid to them in some way. The solution recommended and adopted here is to add the names of any relevant helophyte to the community type, after this has been established without the helophytes. The solution is a modification of the 'structural type' concept (Mäkirinta 1978a). Thus one should speak of an *Isoëtes lacustris-Lobelia dortmanna* community (association) with *Phragmites*.

Some remarks on the TABORD program

The program TABORD (van der Maarel *et al.* 1978) has been used successfully on both aquatic (e.g., Jensén 1979; Jensén & van der Maarel 1980) and terrestrial vegetation (e.g., Klokk 1981; Krahulec *et al.* 1986, Mäkirinta 1986). From the optimality point of view, TABORD is not perfect because in the iteration procedure only one relevé (not two or more) is relocated at a time, which may cause some suboptimal location of relevés (M.B. Dale, pers. comm.). It is not possible to say what this means in terms of optimality, but for a floristic classification where characteristic species may be weighted, relevés have to be moved anyway. Another problem which cannot be solved by a numerical approach is that some relevés are transitional and are difficult to place in any cluster or syntaxon. In the present case 95% of all relevés were grouped in a floristically satisfactory way.

The present study suggests that the TABORD program can be effectively used on higher levels of similarity, i.e. for forming lower units of vegetation (syntaxa). Indeed, as the authors (van der Maarel *et al.* 1978) say: 'TABORD can be a guide in creating some order in an apparently chaotic ensemble of phytosociological data'. In its basic form it is no equal effective for constructing higher units, because it does not selectively use the diagnostic species, but takes all species into account. In the florisitc-sociological tradition the classification procedure is generally based on the individual capability and experience of the scientist himself (Ellenberg 1956).

Conclusion

The classification of *Littorelletea* vegetation on the association level, is mostly based on differential species only, while character species are usually lacking (Pietsch 1977). The subassociations and variants must be differentiated by species which are typical of the *Littorelletea* or other classes (Pietsch 1977). When an attempt is made to apply the classification to the whole distribution area of the class, the system will become complicated and will often given an impression of an arbitrary selection of diagnostic species.

The Swedish Isoetid vegetation community system is an extension of the Finnish system (Mäkirinta 1978a) to the Swedish area, making use of new approaches: 1. Selection of the diagnostic species is based on a PCA ordination of the species in the material; 2. Establishment of the lowest syntaxa by clustering with the TABORD program produces quite small compact groups of relevés (basic clusters); 3. The use of the selective floristic principle for the combination of the basic clusters (variants) into larger units makes the system reliable in practical identification.

The classification system now obtained for an

area considered optimal for the *Littorelletea* (the southern half of Sweden), may be used for at least the whole southern half of Fennoscandia. It will provide a reliable basis for the final goal, a total syntaxonomical system of the *Littorelletea.*

Acknowledgements

I express my sincere gratitude to Prof. Dr Eddy van der Maarel for having taught me the use of numerical methods and providing me with working space in the Institute of Ecological Botany, University of Uppsala, during my Nordic Council for Ministers research scholarship 1982–1983. I am very grateful to Dr Lajos Hajdu and Dr Hans Olofsson for practical help in computer work and especially to Docent Håkan Hytteborn and other kind people at 'Växtbio' in Uppsala for other practical help and friendship. The English test of the manuscript was corrected by Mr Malcolm Hicks, MA (Oulu). For a very valuable review of the manuscript I am thankful to Prof. E. van der Maarel and two anonymous reviewers.

Appendix 1.

List of the 60 lakes and the special areas studied. Information from left to right: 3-letter symbol for the lake, date of study, name of lake and part of the shore with the relevés: locality (town, community, parish, village etc.). See Fig. 1.

BAL 780712 BALLSEN, W shore; Dala-Järn, 9 km NE from the church
BAR 830728 BARKEN, N., SE end; Söderbärke, near the church
BOL 780722 BOLMEN, E part; Tannåker, S of the bridge to Bolmsö
BUÄ 780713 BUSJÖN, ÄPP, E shore; Äppelbo, Nordibyn
BUL 780717 BULLAREN, N., NE shore; Fagerhult, Vassbotten
BYS 830731 BYSJÖN, N shore; Nås, Närsjö, Närsjöbadet
ERK 830711 ERKEN, E end; Svanberga
FÄV 780720 FÄVREN, E shore; Kungsäter, Stackenäs
FÄR 780719 FÄRGEN, ST., S shore; Hemsjö, Edsås
FRÖ 830808 FRÖVETTERN, E shore; Käxsund
GÄN 830730 GÄNSEN, W shore; Gänsen, 0,5 km N from the village
GIL 820918,19 GILSÅSSJÖN, NE shore; Garphyttan
GLA 830805 GLAFSFJORDEN, E shore; Högerund, Västra Hungvik
GRÄ 830803 GRÄSSJÖ, ENE shore; Gräsmark sn, Grinnemo
GRY 830824 GRYCKEN, SSE end; Horndal, bathing beach
GUN 830805 GUNNERN, E shore; Gunnerskog, shore between Årbotten and Byn
HIN 830720 HINDSEN, W shore; Värnamo, Hindsekind
HJÄ 780720 HJÄRTAREDSSJÖN, SE shore; Källsjö, Öv. Hjärtared
HÖR 830728,29 HÖRKEN, N., SW shore; Grängesberg, Hörken
JÖN 820921 JÖNSSJÖN, E shore; Hällefors, 6 km ENE from Gillershöjden
KAL 780720 KALSAREDSJÖN, S shore; Karl Gustav, a small L betw. Å and Kalsarebo
KLÄ 830802 KLÄGGEN, NNW shore; Östmark, Finnbron
KYM 830804 KYMMEN, N shore; Gräsmark, 1 km SE of Timbonäs
LAK 780714 LAKENE, NE shore; Lakene, 1 km N from Åleby
LÅN 830807 LÅNGEN, S end; Grums, 3 km S of Lång estate
LAX 830723 LAXSJÖN, ÖSTRA., NE shore; Stenboda
LEK 820917 LEKEN, NE shore; Lekhyttan
LEL 780715 LELÅNGEN, E shore; Gustavsfors, Bön, Kesviken
LIN 830827 LINGAN, N shore; Lingbo
MAD 830718 MADKROKEN, S shore; Braås, Madkroksbaden
MAL 820920 MALEN, W shore; Kärvingeborn, 1,5 km S from the village
MED 830827,31 MEDSKOGSSJÖN, NW shore; Medskog
NIT 830729 NITTEN, ST., E shore; Nittkvarn, 5 km N from the village
ÖJA 830827,31 ÖJAREN, SW shore; Högbo, Badet

ÖJE 830801 ÖJESJÖN, SE shore; (Dalarna) Öje
ÖRE 780724 ÖRESJÖN, NNE shore; Fristad, Skalle
ORR 830717 ORRENÄSSJÖN, W shore; Hälleberga, 5 km SE of the church
RAM 821019 RAMSJÖN, W shore; Dannike
RIS 830714 RISTELN, N shore; Björsäter, Byrum, Byrumsbadet
ROS 830823 ROSSEN, SSE shore; Horndal
RUN 830824 RUNN, NE shore; Korsnäs, near Staberg
SÄM 821018 SÄMSJÖN, N end; Vegby, 1 km E of the village
SAN 830717 SANDSJÖN, NE shore; Vägershult
SIG 820910,29 SIGGEFORASJÖN, E shore; Järlåsa, near Östfora
SIK 780709 SIKSJÖN, W shore; Sveg, Siksjön, near the border of Dalarna
SIL 780711 SILJAN, W part; Sollerön, SW shore of the island
SIL 820912 SILJAN, E part; Tällberg, shore 2 km ESE of Laknäs
SKI 830714 SKIREN, ESE shore; Lotorp, Yxviken
SOM 821015 SOMMEM, SSW shore; Torpa, Dragsnäs, SE of the bridge
STA 830715 STORSJÖN, ANK., S shore; Ankarsrum, betw. Fagersand and Hökhult
STE 830721 STENGÅRDHULTSJÖN, NNW shore; Stengårdshult, Badplats
STH 830828,31 STORSJÖN, HÄR. (in Härjedalen), S shore; Långbo
STÖ 830716 STORSJÖN, S shore; Öland, St. Alvaret betw. Resmo and Brunneby
STR 830719 STRÅKEN, N end; Aneboda, 2 km NNE from the church
SUN 821018 SUNDSJÖN, ST., S shore; Sund, Badstrand
TÅN 821017 TÅNGEN, N shore; Asby, Asbysand
TOL 821021 TOLKEN, SW shore; Äspered, Sjötorp
TRE 830804 TREEN, NE shore; Gräsmark, 2 km S from Fredros
VÄR 830806 VÄRMELN, S part; Värmskog, 1,5 km W of the church
VID 780721 VIDÖSTERN, E shore; Tånnö, Gravfält
VIR 780723 VIRESTADSJÖN, E shore; Virestad, 2 km S of the church

References

Abbott, L.A., Bisby, F.A. & Rogers, D.J. 1985. Taxonomic analysis in biology. Columbia University Press, New York.

Albertson, N. 1950. Das grosse südliche Alvar der Insel Öland. Eine pflanzensoziologische übersicht. Svensk Bot. Tidsskr. 44: 269–331.

Andersson, P.-A. 1988. Ordination and classification of operational geographic units in Southwest Sweden. Vegetatio 74: 95–106.

Barkman, J.J., Moravec, J. & Rauschert, S. 1986. Code of phytosociological nomenclature. 2nd ed. Vegetatio 67: 145–195.

Beals, E.W. 1984. Bray-Curtis ordination: an effective strategy for analysis of multivariate ecological data. Adv. Ecol. Res. 14: 1–55.

Braun-Blanquet, J. 1964. Pflanzensoziologie, Grundlagen der Vegetationskunde. 3. Aufl. Springer Verlag. Wien, New York.

Carleton, T.J. 1980. Non-centred component analysis of vegetation data: a comparison of orthogonal and oblique rotation. Vegetatio 42: 59–66.

Dale, M.B. & Webb, L.J. 1975. Numerical methods for the establishment of associations. Vegetatio 30: 77–87.

Dale, M.B. & Williams, W.I. 1978. A new method of species reduction for ecological data. Austr. J. Ecol. 3: 1–5.

Dierssen, K. 1975. *Littorelletea uniflorae* Br.-Bl. et Tw. 1943. J. Cramer, Vaduz.

Downing, J.A. & Anderson, M.R. 1985. Estimating the standing biomass of aquatic macrophytes. Can. J. Fish. Aquat. Sci. 42: 1860–1869.

Du Rietz, G.E. & Hannerz, A.G. 1939. Gefässpflanzenvegetation. In: Du Rietz, G.E., Hannerz, A.G., Lohammar, G., Santesson, R. & Wärn, M. Zur Kenntnis der Vegetation des Sees Tåkern. Acta Phytogeogr. Suec. 12: 1–41.

Ellenberg, H. 1956. Aufgaben und Methoden der Vegetationskunde. In: Walter, H. (ed.) Einführung in die Phytologie IV. Grundlagen der Vegetationsgliederung 1: 1–136. Eugen Ulmer, Stuttgart.

Jensén, S. 1978. Influences of transformation of cover values on classification and ordination of lake vegetation. Vegetatio 34: 19–31.

Jensén, S. 1979. Classification of lakes in Southern Sweden on the basis of their macrophyte composition by means of multivariate methods. Vegetatio 39: 129–146.

Jensén, S. & van der Maarel, E. 1980. Numerical approaches to lake classification with special reference to macrophyte communities. Vegetatio 42: 117–128.

Kissling, P., Kuhn, N. & Wildi, O. 1988. The merocoenotic sample and its use in investigating floristic changes in forests. Bot. Helv. 98: 39–75.

Klokk, T. 1981. Classification and ordination of river bank

vegetation from middle and upper parts of the river Gaula, Central Norway. K. Norske Vidensk. Selsk. Skr. 1980 (2): 1–43.

Krahulec, F., Rosén, E. & van der Maarel, E. 1986. Preliminary classification and ecology of dry grassland communities on Ölands Stora Alvar (Sweden). Nord. J. Bot 6: 797–809.

Lohammar, G. 1938. Wasserchemie und höhere Vegetation schwedischer Seen. Symb. Bot. Ups. 3 (1): 1–252.

Magnusson, N.H. *et al.* 1957. Karta över Sveriges berggrund. Pre-Quaternary rocks of Sweden. Utarbetad vid Sveriges Geologiska Undersökning 1957. Skala 1 : 1 000 000. Stockholm.

Mäkirinta, U. 1978a. Die pflanzensoziologische Gliederung der Wasservegetation im See Kukkia, Südfinnland. Acta Univ. Oulu A75: 1–157.

Mäkirinta, U. 1978b. Ein neues ökomorphologisches Lebenformensystem der aquatischen Makrophyten. Phytocoenologia 4: 446–470.

Mäkirinta, U. 1985. Classification and ecology of the Isoetid vegetation of lakes in Southern Sweden. In: Hytteborn, H. van der Maarel, E. & Wallström, K. (eds), Research in Progress at the Institute of Ecological Botany, Uppsala University 1982-84. Medd. Växtbiol. Inst., Uppsala, 1985 (1): 55.

Mäkirinta, U. 1986. Vegetation types and exposure on acid rocks in South Häme, South Finland. Coll. Phytosoc. 13: 469–484.

Maristo, L. 1941. Die Seetypen Finnlands auf floristischer und vegetationsfysiognomischer Grundlage. Ann. Bot. Soc. Vanamo 6 (4): 1–55.

Pietsch, W. 1977. Beitrag zur Soziologie und Ökologie der europäischen *Littorelletea-* und *Utricularietea*-Gesellschaften. Feddes Repert. 88: 141–245.

Popma, J., Mucina, L., van Tongeren, O. & van der Maarel, E. 1983. On the determination of optimal levels in phytosociological classification. Vegetatio 52: 65–75.

Raspopov, I.M. 1971. Makrofiti Onezkogo ozera. – In: Raspopov, I.M. (ed.), Rastitel'nyi mir' Onezhskogo ozera: 21–87. Izd. Akad. Nauk SSSR, Leningrad.

Roskam, E. 1971. Program ORDINA: Multidimensional ordination of observation vectors. Progr. Bull. Psych. Lab. Nijmegen 16: 1–8.

Stålberg, N. 1939. Lake Vättern. Outlines of its natural history, especially its vegetation. Acta Phytogeogr. Suec. 11: 1–52.

Thunmark, S. 1931. Der See Fiolen und seine Vegetation. Acta Phytogeogr. Suec. 2: 1–198.

Tüxen, R. 1937. Die Pflanzengesellschaften Nordwestdeutschlands. Mitt. Florist.-soziol. Arbeitsgem. Niedersachsen 3: 1–170.

Vaarama, A. 1938. Wasservegetationsstudien am Grossee Kallavesi. Ann. Bot. Soc. Vanamo 13 (1) 1–318.

van der Maarel, E. 1979. Transformation of cover-abundance values in phytosociology and its effects on community similarity. Vegetatio 39: 97–114.

van der Maarel, E. 1980. On the interpretability of ordination diagrams. Vegetatio 42: 43–45.

van der Maarel, E. 1982. On the manipulation and editing of phytosociological and ecological data. Vegetatio 50: 71–76.

van der Maarel, E., Janssen, J. & Louppen, J. 1978. TABORD, a program for structuring phytosociological tables. Vegetatio 38: 143–156.

Westhoff, V. & van der Maarel, E. 1978. the Braun-Blanquet approach. In: Whittaker, R.H. (ed.) Classification of plant communities: 287–399. Junk, The Hague.

Vegetatio **81**: 159–168, 1989

Syntaxonomy of the Australian mangal refined through iterative ordinations

P. B. Bridgewater
Bureau of Flora & Fauna, GPO Box 1383, Canberra, ACT 2607 Australia

Accepted 21.10.1988

Keywords: Avicennia, Coastal vegetation, Mangrove, Numerical syntaxonomy, Ordination, Phytosociology

Abstract

Mangal (tidal forest) communities vary with respect to position in the littoral sequence, and latitude. A previous classification of mangal from the southern and western coastlines of Australia has been subjected to a series of ordinations, in an attempt to produce a robust syntaxonomy. An iterative ordination process allowed for the progressive refinement of the syntaxonomic units. This process allows an ecologist to use critically a numerical technique to refine ecological assumptions developed from field work, and/or non-numerical analyses. Ordination analysis suggested a need for the classification to be refined, and allowed the construction of a final syntaxonomy. In conjunction with broad environmental data collected during the investigation, the ordinations allowed the development of some environmental perspectives. These suggested that littoral position is a stronger environmental influence on mangal communities than balance between salt and freshwater.

Nomenclature: Green, J. W. 1985. Census of the vascular plants of Western Australia. Western Australian Herbarium. Perth.

The problem

The Australian mangal occurs over a wide spectrum of climatic environments, from the temperate southern coast to dry tropical in the north-west and wet tropical in the north-east. The north-eastern and northern coastlines have the greatest species richness (Wells 1982; Busby & Bridgewater 1986; Hutchings & Saenger 1987). This paper will deal with the mangal of the southern and western coastlines.

Studies in Australia from MacNae (1968) through Saenger *et al.* (1977) to Semeniuk & Wurm (1987) have described the major zones of mangal as 'communities' without a clear definition of such 'communities' as repeatable units in space and/or time. Semeniuk (1983) and Tomlinson (1986) strongly emphasised a direct environmental relationship to zonation, although Buckley (1982) suggests an alterative view: that groupings of species are related to repeatable environmental variables. Semeniuk & Wurm (1987) have developed a terminology largely based on geomorphological attributes, and they give heavy emphasis to their 'community', which they term an 'assemblage', as a basic phytosociological unit.

Other workers (e.g. Dowling & McDonald 1982) have utilised the schemata of Specht (1970, 1981) which impose a largely structural framework as the determinant for community pattern. This has the unfortunate effect of creating an anomalous and confusing terminology (e.g. *Avicennia* heath). Bunt & Williams (1980) and Semeniuk & Wurm (1987) have noted that the major environmental effects on mangal pattern are tidal influence and salinity. Biogeographic studies such as Busby & Bridgewater (1986), Hutchings & Saenger (1987) and Tomlinson (1986) show a reduction in species richness associated with an increasing southerly latitude. Any consideration of mangal communities must therefore take account of biogeographic variation, variation imposed by position in the littoral sequence and variation imposed by the balance between freshwater and seawater. In addition, many mangrove communities are composed of only a single species of vascular plant. Development of a suitable syntaxonomy poses thus some difficulties.

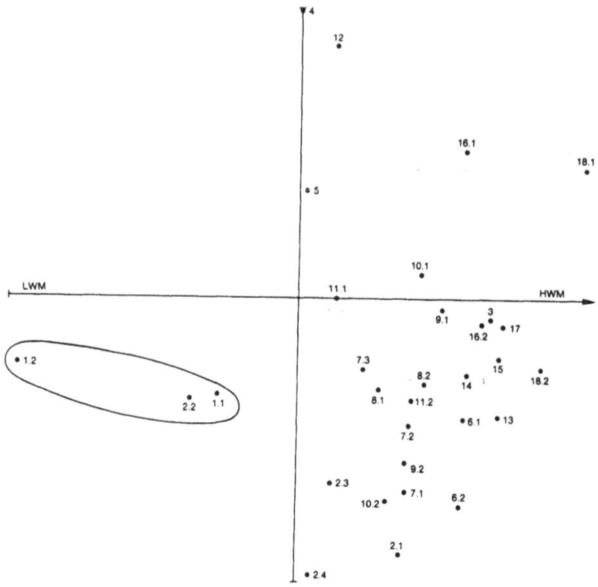

Fig. 1. Ordination of original data set for all mangrove communities from Bridgewater (1985).

Previous work

Syntaxonomic studies of the Australian mangal have been attempted previously by Chapman (1977) and Bridgewater (1987). In both cases the authors created *nomina nuda,* relying on field observation and experience to establish the boundaries of the higher syntaxonomic units (terms after Barkman *et al.* 1986). Although creation of syntaxonomic terminology in this *ad hoc* way is tempting, it inevitably leads to confusion and an undesirable proliferation of names in the literature.

In order to provide a sound syntaxonomy for Australian mangal, and potentially the Asian-Pacific region, the present study utilised an available data set of mangrove communities published in Bridgewater (1985). This data set consisted of 160 relevés from the southern and western coasts of Australia (Fig. 1). It built on the studies reported in Bridgewater (1986a). Environmental data gathered as part of the exercise included littoral position, salinity, substrate and tidal flux.

Table 1 is a summary table of the mangrove communities published in Bridgewater (1985) and Appendix 1 the conspectus of nomenclature used in that publication. The terminology used for these communities was informal, but was the basis of the syntaxonomy published as *nomina nuda* in Bridgewater (1987).

Present study

The present paper tests and refines the syntaxonomy of Bridgewater (1987) by using the ordination technique of Principal Axes Ordination (PAO) (van der Maarel 1969). In essence a range of ordination techniques could have been used. PAO was used (Bridgewater 1980) in a syntaxonomic analysis of heathland in Britain. Results from that study demonstrated that this ordination technique was particularly helpful for syntaxonomic work. The success of that analysis coupled with the simple nature of the data set, suggested PAO (using the Sørensen coefficient) as an appropriate technique. PAO combines the simplicity of reference set ordination techniques (Anderson 1971) with a particularly clear display of the relevés in the ordination diagrams. The presence class values in Table 1 were used in developing the similarity and distance matrices for the PAO, which is included in the computer program VEGCLASS.

Use of the full data set produced the ordination as shown in Fig. 2. The three communities containing *Sonneratia alba* are separated at one end of the x-axis, with all other communities arranged against the x and y-axes in a pattern which does not reflect

Table 1. Synoptic table of presence values for mangrove communities described in Bridgewater (1985). Appendix 1 lists the classification used.

Community	1		2				3	4	5	6		7			8		9		10		11		12	13	14	15	16		17	18	
Sub-community	1	2	1	2	3	4				1	2	1	2	3	1	2	1	2	1	2	1	2					1	2		1	2
Sonneratia alba	5	5		5																											
Rhizophora stylosa		1	5	5	5	5			2			2		2				3	1	5		1									
Aegialitis annulata				3	1		1	5	1	2							2		2				5				5			5	
Aegiceras corniculatum									5	3	4																	2			
Camptostemon schultzii						5				5	5	5	5	4	1		1			1											
Bruguiera exaristata													5		4	5	5	4		1											
Xylocarpus australasicus														5	5	5						1									
Excoecaria agallocha											2		1	1						1		5							4	2	
Lumnitzera racemosa															4					1	5	2									
Osbornia octodonta										2								4		5	1										
Sarcocornia quinqueflora																									5						
Halosarcia halocnemoides																									1	5					2
Suaeda arbusculoides							1				5								1							5	5	2	3	3	
Sporobolus virginicus																												5	3		
Batis argillicola																												2		5	5
Halosarcia indica																											1		4	1	5
Ceriops tagal										1	2	5	5	5	2		5	5	5	5	5	4	5			2		3	2	2	1
Avicennia marina	5	5		5	2	2	5		2	5	5	5	3	3	2	4	4	5	5	4		5		5	5	5	5	5	3	5	5
Suaeda australis																									2						
Samolus repens																									1						
Sclerostegia arbuscula																									1						
Spartina townsendii																									1						
Rhizoclonium sp.																									1						
Muellerolimon salicorniaceum																										2					
Hemichroa diandra																										2				1	
Halosarcia pterygosperma																										2					
Xerochloa imberbis																												2			
Hibiscus tiliaceus																						1									
Acrostichum speciosum															3																
Diospyros ferrea															1																

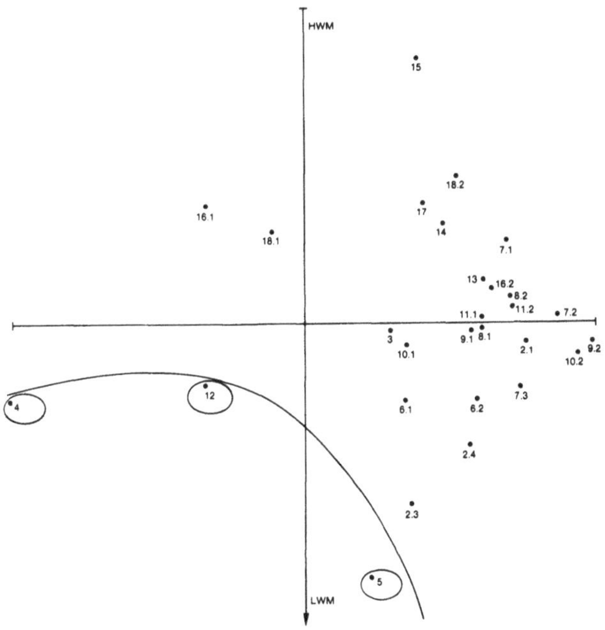

Fig. 2. Ordination of original data set reduced by removal of vegetations units 1.1, 1.2, 2.2.

the classification. In general the communities are arranged along the x-axis corresponding to positions on the littoral gradient, with no obvious environmental gradient along the y-axis. The three *Sonneratia* communities were removed from the data set, and the ordination re-run. A referee commented that the diagrams would be clearer if dominant species were noted by symbols and the units containing them joined up. I have not followed this course, because for mangal vegetation over-emphasis on dominance has caused much of the syntaxonomic confusion.

The second ordination had the *Aegialitis annulata* comm., *Aegiceras corniculatum* comm. and *Ceriops tagal* comm. as outliers, with all other communities contracted together (Fig. 3). No obvious environmental factor appeared to separate communities on the x-axis, although the sequence of communities in the littoral zone appeared strongly reflected on the y-axis. The three outlier communities were removed from the data set and the ordination run again. It is possible to suggest that the *Avicennia marina – Suaeda arbusculoides* typical comm. and the

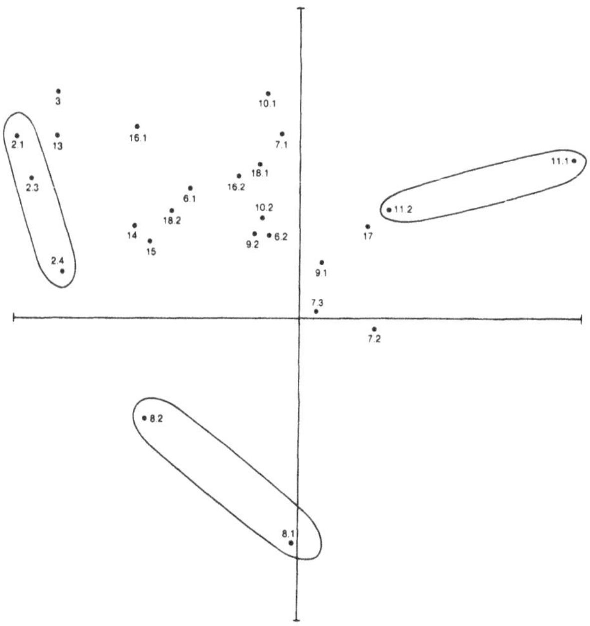

Fig. 3. Ordination of original data set reduced by removal of vegetations units 1.1, 1.2, 2.2, 4, 5, 12.

Avicennia marina – Batis argillicola typical comm. should be removed. However, they were retained to see the results of a new comparative ordination with their other sub-communities.

The third ordination (Fig. 4) shows the *Rhizophora stylosa* comm. (except the already separated *Sonneratia alba* sub-comm.), the *Bruguiera exaristata – Xylocarpus australasicus* comm. and the *Excoecaria agallocha* comm. as outliers on the edge of the ordination, with other communities again arranged in a 'confused' pattern. These outlier communities represent respectively the most tidal, most fresh-water influenced and most landward tropical mangal. They were removed and the data set run again.

The fourth ordination (Fig. 5) has communities arranged from tidal (lower left) to supra-littoral (upper right). Two communities are in anomalous positions – the *Avicennia marina* comm. (included in the supra-littoral group although strongly eulittoral) and the *Avicennia marina – Sarcocornia quinqueflora* comm. This latter community occurs only along the southern coastline of Australia where mangal abuts saltmarsh (Bridgewater 1982b). [Maps

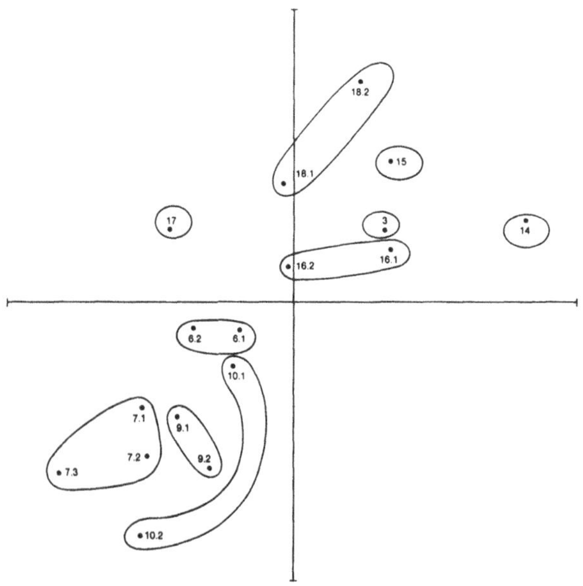

Fig. 4. Ordination of original data set reduced by removal of vegetations units 1.1, 1.2, 2.1, 2.2, 2.3, 2.4, 4, 5, 8.1, 8.2, 11.1, 11.2, 12.

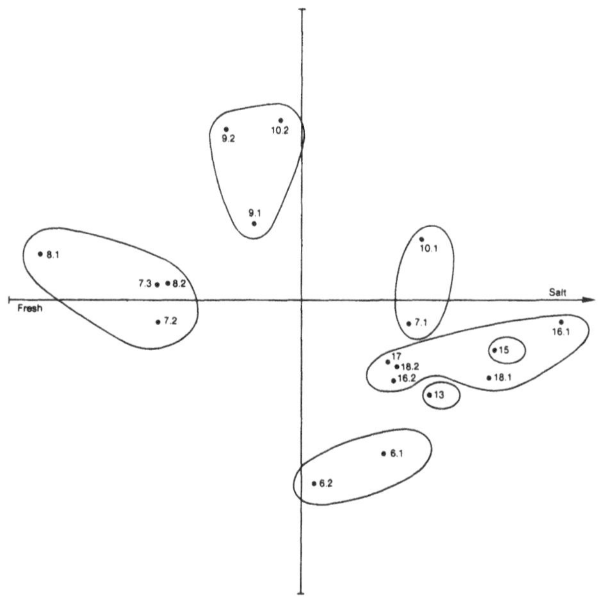

Fig. 5. Ordination of original data set reduced by removal of vegetations units 1.1, 1.2, 2.1, 2.2, 2.3, 2.4, 3, 4, 5, 8.1, 8.2, 11.1, 11.2, 12, 13.

of all associations described in the final syntaxonomy appear as Figs. 8–14].

These communities were accordingly removed and a fifth ordination run. This ordination (Fig. 6) shows a general trend from fresh to salt water along the x-axis. It also suggests a need for re-grouping of sub-communities distinguished in Bridgewater (1985). The ordination at Fig. 6 was used as a basis for recombination of original relevés. For example, the *Sonneratia alba* comm. – *Avicennis marina* sub-comm. and *Rhizophora stylosa* comm. – *Sonneratia alba* sub-comm., grouped closely in the first ordination were combined because of very similar floristics. Other combinations were *Bruguiera exaristata* – *Xylocarpus australasicus* comm. – *Rhizophora stylosa* sub-comm. + *Ceriops tagal* – *Camptostemon schultzii* comm. *Xylocarpus australasicus* sub-comm.; *Bruguiera exaristata* – *Ceriops tagal* comm. (typical) + *Ceriops tagal* – *Camptostemon schultzii* comm. – *Bruguiera exaristata* sub-comm.; *Bruguiera exaristata* – *Ceriops tagal* comm. – *Osbornia octodonta* sub-comm. + *Avicennia marina* – *Ceriops tagal* comm. – *Rhizophora stylosa* sub-comm.; *Avicennia marina* – *Suaeda arbusculoides* comm. – *Aegialitis annulata* sub-comm.

In addition to these recombinations the degree to which communities 15–18 were grouped together and separated from the others forced a re-examination of the raw data. All these communities shared the feature of having the *Avicennia marina* component as a mature plant with a height of 0.5 m–0.75 m. Although there is no evidence as yet that this is a discrete taxonomic entity it was decided to record mature *Avicennia marina* less than or equal to 0.75 m as if it were a separate species.

The recombined phytosociological table is shown as Table 2. As there had been major restructuring of some communities a further analysis seemed likely to clarify the syntaxonomic position. Accordingly the 26 new associations in Table 2 were once again analysed by PAO, as shown in Fig. 7. Here the broad groupings remain intact but there is a clear separation between associations 1–11 and 12–15. Association 16 appears in a totally anomalous position. Associations 12–16 were excluded from a revised data

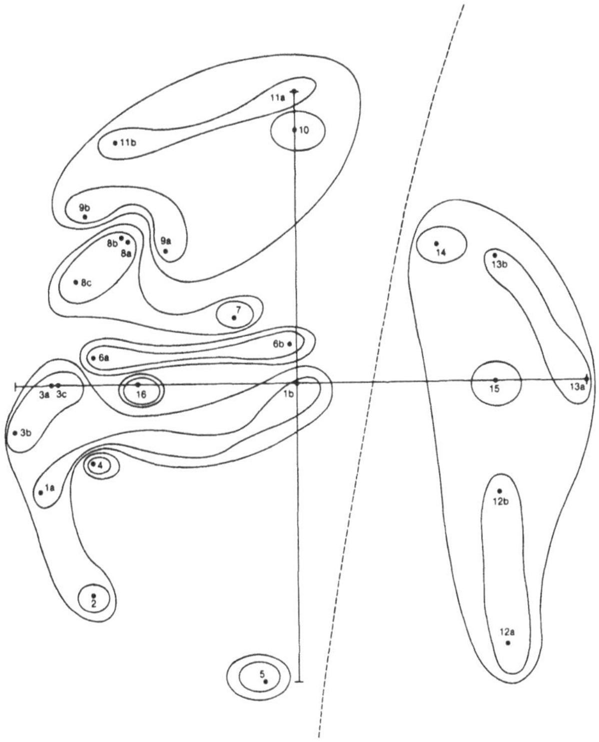

Fig. 6. Ordination of all mangrove communities defined in Table 2.

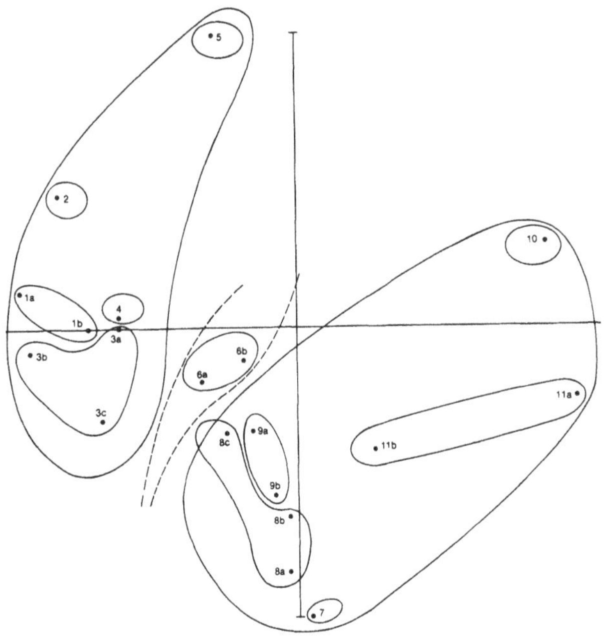

Fig. 7. Ordination of mangrove communities defined in Table 2, with the exception of communities 12–16.

Table 2. Synoptic table formed by re-grouping the relevés as suggested by initial ordinations (Figs. 1–5). Appendix 2 lists the syntaxonomy derived from this synoptic table.

Association Code	1		2	3			4	5	6		7	8			9		10	11		12		13		14	15	16
Sub-Association Code	a	b		a	b	c			a	b		a	b	c	a	b		a	b	a	b	a	b			
Sonneratia alba	5	5																								
Rhizophora stylosa	4	1	1	5	5	5	2					2	2	4	2	2			1							
Avicennia marina	5	5	5		5	2	2		5	5	2	4	4	5	5	5			5							5
Bruguiera exaristata											4	3	5	3												
Xylocarpus australasicus											5	5							1							
Osbornia octodonta									2					5				1								
Ceriops tagal									1	2	2	4	5	5	5	5	5	5	4		2		2			
Camptostemon schultzii						5			5	5	1	3	4	1		5										
Aegiceras corniculatum							5		3	4											1					
Aegialitis annulata	2		1		1		1	5	1	2				2						5	5					
Excoecaria agallocha							1			2		1	1	1				5	5				2	4		
Halosarcia tenuis															1						2				5	
Batis argillicola																					4		5			
Sporobolus virginicus																							3	5		
Halosarcia indica																					4	1	1	4	2	
Suaeda arbusculoides			1							5					2						4	5	3	3		
Avicennia marina (0.75m)																				5	5	5	5		5	
Sarcocornia quinqueflora																										5
Lumnitzera racemosa											4			1					2							
Acrostichum speciosum											3															
Diospyros ferrea											1															
Muellerolimon salicorniaceum															2										2	
Hemichroa diandra																							1		2	
Halosarcia pterygosperma																									2	
Xerochloa imberbis																								2		
Halosarcia halocnemoides																										1
Samolus repens																							1			
Sclerostegia arbuscula																										2
Spartina townsendii																										1
Suaeda australis																										2
Rhizoclonium sp.																										1
Hibiscus tiliaceus																			1							

set which was re-ordinated. That ordination (Fig. 7) confirmed the suggestion that associations 1–5 formed a major group, with 7–11 forming a second group. Association 6 appears to occupy a transitional position.

In syntaxonomic terms, these three major groups correspond well to three of the classes described in Bridgewater (1987);

Avicennietea (associations 1–5)

Avicennio – Ceriopetea (associations 6–11)

Halosarcio –Avicennietea (associations 12–15)

Association 16 (*Avicennio – Sarcocornietum*) was described as part of the *Samolo-Suaedetea* (saltmarsh vegetation) in Bridgewater (1982b). The present ordination analysis confirmed the appropriateness of that view. In essence that association represents the temperate climatic vicariant of the class *Halosarcio-Avicennietea* saltmarsh with an 'emergent' mangal species. The *Camptostemetum*, while possibly transitional between the classes *Avicennietea* and *Avicennio – Ceriopetea*, on present evidence fits the *Aviennio-Ceriopetea* more ap-

Fig. 8. Portion of Australian coastline covered by this paper. Dotted lines indicate coasts where the greatest concentration of mangrove vegetation occurs. The *Avicennion* occurs throughout this coastline.

Fig. 9. Coastline (dotted) where the association *Avicennio – Sarcocornietum* occurs.

Fig. 10. Coastline (dotted) where the *Avicennio – Haloscarcion* occurs.

propriately. The two classes *Aegiceretea* and *Aegilitietea,* although separated by Bridgewater (1987), seem best relegated to orders in the *Avicennietea.*

Conclusions

Use of ordination techniques to clarify syntaxonomic situations is not new, e.g. van Groenewoud (1965), Kortekaas *et al.* (1976), Komárková (1980), van der Maarel (1969). None the less, the iterative approach advocated in this paper could serve as a useful

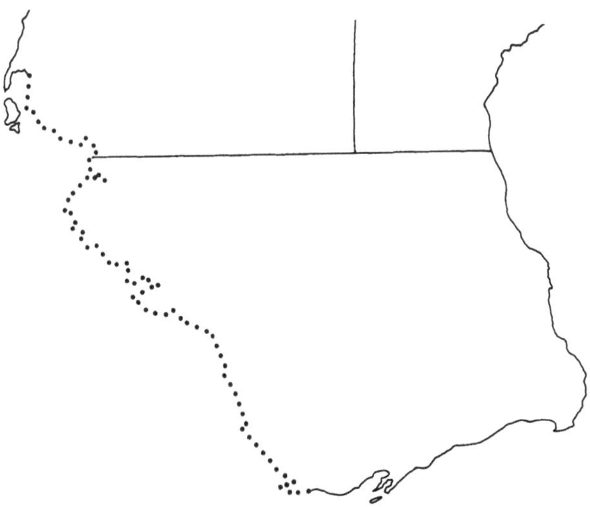

Fig. 11. Coastline (dotted) where the *Rhizophorion, Ceriopion* and *Avicennio – Suaedion* occurs.

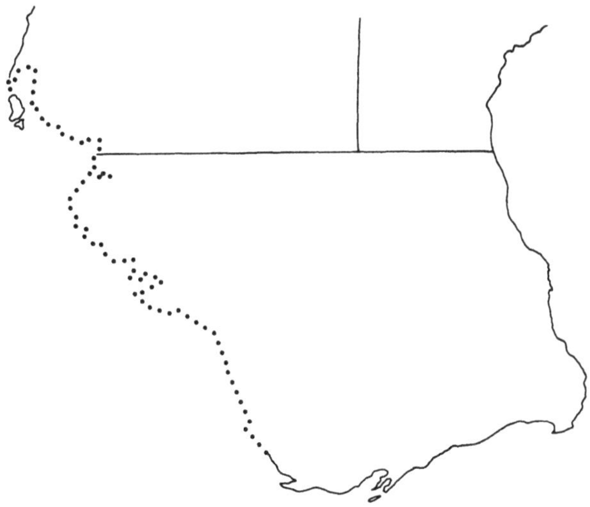

Fig. 12. Coastline (dotted) where the *Aegicerion, Bruguierion, Aegialition* occurs.

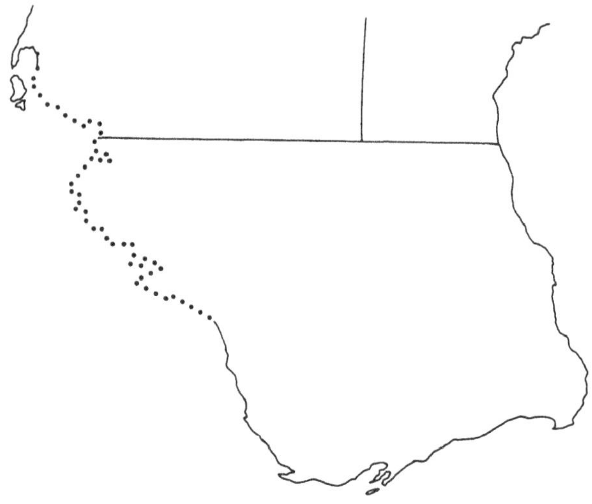

Fig. 13. Coastline (dotted) where the *Camptostemion* occurs.

Fig. 14. Coastline (dotted) where the *Sonneration* and *Avicennio–Aegialition* occurs.

framework for other studies where the data comprise associations from a wide geographic range and influenced by a range of environmental factors.

Peet (1980) also utilised several successive ordinations to analyse complex data sets. This present data set is not complex in the sense of Peet (1980), but it represents considerable biogeographic and ecological variation in species-poor vegetation. The approach of this paper nonetheless follows Peet (1980) in removal of successive groups, and in evaluating the environmental data with each succeeding ordination. Syntaxonomy was, of course, the main focus of the present work – environmental considerations were secondary. Several of Peet's (1980) conclusions are relevant here; in particular his observation that traditional classifications do not allow for ready examination of environmental relationships. He also makes the point that the approach 'has an admitted and inescapable element of subjectivity'. I also freely accept that the approach argued in this paper is subjective, but again cannot do better than quote Peet (1980) '... subjectivity will enter into all useful ecological classification schemes.'

The iterative process demonstrated the value of ordination techniques in testing community hypotheses developed from a data set collected from a wide biogeographic region. The main value of the numerical techniques is thus to help focus the processes of the ecologists intuitive thought, allowing the ecologist to uncritically challenge their assumptions about particular vegetation. As mentioned at the outset, this paper advocates an iterative framework, without specifying any particular classification or ordination techniques as essential. Rather techniques familiar to the investigator, or proven in previous work to be suitable for the vegetation being considered, should be used.

For the mangal ecosystem generally this study suggests:

- the strong influence of littoral position when all communities are considered together;
- the subsidiary importance of fresh/salt water influences
- existence of a saltmarsh/mangal transitional vegetation, clearly separable into two separate classes in the north and south of the continent.

Although this study has been confined to the southern and western coasts of Australia the higher syntaxonomic units should stand for all of Australia, and by extrapolation many other countries in the Asian-Pacific region.

Appendix 1

Classification scheme used in Bridgewater (1985). See also Table 1.

1. *Sonneratia alba* community
   - 1.1 typical sub-community
   - 1.2 *Avicennia marina* sub-community
2. *Rhizophora stylosa* community
   - 2.1 typical sub-community
   - 2.2 *Sonneratia alba* sub-community
   - 2.3 *Avicennia marina* sub-community
   - 2.4 *Camptostemon schultzii* sub-community
3. *Avicennia marina* community
4. *Aegialitis annulata* community
5. *Aegiceras corniculatum community*
6. *Avicennia marina – Camptostemon schultzii* community
   - 6.1 typical sub-community
   - 6.2 *Suaeda arbusculoides* sub-community
7. *Ceriops tagal – Camptostemon schultzii* community
   - 7.1 typical sub-community
   - 7.2 *Bruguiera exaristata* sub-community
   - 7.3 *Xylocarpus australasicus* sub-community
8. *Bruguiera exaristata – Xylocarpus australasicus* community
   - 8.1 *Lumnitzera racemosa* sub-community
   - 8.2 *Rhizophora stylosa* sub-community
9. *Bruguiera exaristata – Ceriops tagal* community
   - 9.1 typical sub-community
   - 9.2 *Osbornia octodonta* sub-community
10. *Avicennia marina – Ceriops tagal* community
    - 10.1 typical sub-community
    - 10.2 *Rhizophora stylosa* sub-community
11. *Excoecaria agallocha* community
    - 11.1 typical sub-community
    - 11.2 *Avicennia marina* sub-community
12. *Ceriops tagal* community
13. *Avicennia marina – Aegialitis annulata* community
14. *Avicennia marina – Sarcocornia quinqueflora* community
15. *Halosarcia halocnemoides – Avicennia marina* community
16. *Avicennia marina – Suaeda arbusculoides* community
    - 16.1 typical sub-community
    - 16.2 *Aegialitis annulata* sub-community
17. *Excoecaria agallocha – Sporobolus virginicus* community
18. *Avicennia marina – Batis argillicola* community
    - 18.1 typical sub-community
    - 18.2 *Aegialitis annulata* sub-community

Appendix 2

Syntaxonomy produced from this present work. Type relevés (published in Bridgewater 1985) are listed in parentheses. See also Table 2 for a synoptic presence table.

Avicennietea
Rhizophorietalia
1. *Sonneration*
1.a *typicum* (696)
1.b *avicennietosum* (832)
(extreme eulittoral, tropical regions)
Avicennion
2. *Avicennietum marinae* (605)
(extreme eulittoral, continent wide)
Rhizophorion
3. *Rhizophoretum stylosae*
3.a *typicum* (250)
3.b *avicennietosum* (206)
3.c *camptostemetosum* (637)
(eulittoral, variable substrate)
Aegiceretalia
Aegicerion
4. *Aegiceretum corniculatae* (243)
(eulittoral, brackish waters)
Aegialitietalia
Aegialition
5. *Aegialitietum annulatae* (235)
(eulittoral, brackish waters)
Avicennio – Ceriopetea
Camptostemetalia
Camptostemion
6. *Camptostemetum*
6.a *typicum* (660)
6.b *suaedetosum* (692)
(eulittoral, sheltered, tropical regions)
Bruguierietalia
Bruguierion
7. *Bruguierio – Xylocarpetum* (808)
8. *Ceriopo – Bruguierietum*
8.a *xylocarpietosum* (810)
8.b *typicum* (675)
8.c *osbornietosum* (816)
(upper eulittoral, substrate varies)
Ceriopietalia
Ceriopion
9. *Avicennio – Ceriopetum*
9.a *typicum* (276)
9.b *camptostemetosum* (676)
10. *Ceriopetum* (259)
11. *Excoecarietum*
11.a *typicum* (688)
11.b *avicennietosum* (679)
(supralittoral, substrate varies)
Halosarcio – Avicennietea
Avicennio – Suaedetalia
Avicennio – Aegialition
12. *Avicennio – Aegialitietum*
12.a *typicum* (652)
12.b *batietosum* (814)
(supra littoral, tropical regions)
Avicennio – Suaedion
13. *Avicennio – Suaedetum*
13.a *typicum* (662)
13.b *batietosum* (663)
14. *Execoecario – Sporoboletum* (690)
(supra littoral, tropical regions)

Avicennio – Halosarcietalia
Avicennio – Halosarcion
15. *Avicennio – Halosarcietum tenuis* (238)
(supra littoral, mainly dry, tropical regions)
Samolo – Suaedetea
Samolo – Suaedetalia
Samolo – Suaedion
16. *Avicennio – Sarcocornietum* (503)
(supra littoral, temperate regions)

References

Anderson, A. J. B. 1971. Ordination methods in ecology. J. Ecol. 59: 713 – 726.

Barkman, J. J., Moravec, J. & Rauschert, S. 1986. Code of phytosociological nomenclature. Vegetatio 67: 145 – 195.

Bridgewater, P. B. 1980. Phytosociological studies in the British heath formation: I. Heaths of the *Ulicetalia minoris* (P. Duvign. 1944) J.M. Géhu 1973 – Phytocoenologia. 8: 191 – 235.

Bridgewater, P. B. 1982a. Mangrove vegetation of the southern and western Australian coastline. In: Clough, B. F. (ed.), Mangrove ecosystems in Australia. pp. 111 – 120 Austr. Nat. University Press, Canberra.

Bridgewater, P. B. 1982b. Phytosociology of coastal salt-marshes in the mediterranean climatic region of Australia. Phytocoenologia 10: 257 – 296.

Bridgewater, P. B. 1985. Variation in the mangal along the west coastline of Australia. Proc. Ecol. Soc. Aust. 13: 243 – 256.

Bridgewater, P. B. 1987. Small and large-scale inventory in mangrove ecosystems. In: Field, C. D. & Dartnall, A. J. (eds), Mangrove ecosystems of Asia and the Pacific: status, exploitation and management. pp 84 – 95. Aust. Inst. of Marine Science, Townsville.

Busby, J. R. & Bridgewater, P. B. 1986. Atlas of distribution of mangrove species in Australia. Aust. Flora and Fauna Series 5. AGPS, Canberra.

Buckley, R. C. 1982. Patterns in north Queensland mangrove vegetation. Aust. J. Ecol. 7: 103 – 106.

Bunt, J. S. & Williams, W. T. 1980. Studies in the analysis of data from Australian tidal forests ('mangroves'). I. Vegetational sequences and their graphic representation. Aust. J. Ecol. 5: 385 – 390.

Chapman, V. J. 1977. Introduction. In: Chapman, V. J. (ed.), Ecosystems of the world I. Wet coastal ecosystems. pp. 1 – 27. Elsevier, Amsterdam.

Dowling, R. & McDonald, T. J. 1987. Mangrove communities of Queensland. In: Clough, B. F. (ed.), Mangrove ecosystems in Australia. pp. 79 – 93. Aust. Nat. University Press, Canberra.

Hutchings, P. & Saenger, P. 1987. Ecology of mangroves. University of Queensland Press, Brisbane.

Komárková, V. 1980. Classification and ordination in the Indian Peaks area, Colorado Rocky Mountains. Vegetatio 42: 149 – 163.

Kortekaas, W., van der Maarel, E. & Beeftink, W. G. 1976. A numerical classification of European *Spartina* communities. Vegetatio 33: 51 – 60.

MacNae, W. 1968. A general account of the flora and fauna of mangrove swamps and forests in the Indo-Pacific region. Adv. Mar. Biol. 6: 73 – 270.

Peet, R. K. 1980. Ordination as a tool for analysing complex data sets. Vegetatio 42: 171 – 174.

Saenger, P., Specht, M. M., Specht, R. L. & Chapman, V. J. 1977. Mangal and coastal salt-marsh communities in Australasia. In: Chapman, V. J. (ed.), Ecosystems of the world. I. Wet coastal ecosystems. pp. 293 – 345. Elsevier, Amsterdam.

Semeniuk, V. 1983. Mangrove distribution in north-western Australia in relationship to regional and local freshwater seepage. Vegetatio 53: 11 – 31.

Semeniuk, V. & Wurm, P. 1987. The mangroves of the Dampier Archipelago, Western Australia. J. Roy. Soc. W.A. 69: 30 – 87.

Specht, R. 1970. Vegetation. In: Lepper, G. W. (ed.), The Australian environment (4th ed.). pp. 44 – 67. CSIRO Melbourne.

Specht, R. 1981. Foliage projective cover and standing biomass. In: Anderson, D. J. & Gillison, A. N. (eds), Vegetation classification in Australia. pp. 10 – 21. Aust. Nat. University Press, Canberra.

Tomlinson, P. B. 1986. The botany of mangroves. Cambridge University Press, New York.

van der Maarel, E. 1969. On the use of ordination models in phytosociology. Vegetatio 19: 21 – 46.

van Groenewoud, H. 1965. Ordination and classification of Swiss and Canadian coniferous forests by various biometric and other methods. Ber. Geobot. Inst. Rübel, Zürich. 36: 28 – 102.

Wells, A. G. 1982. Mangrove vegetation of Northern Australia. In: Clough, B. F. (ed.), Mangrove ecosystems in Australia. pp. 57 – 78. Aust. Nat. University Press, Canberra.

Vegetatio 81: 169–180, 1989

Numerical revision of the *Fagion illyricum* alliance

K. Török, J. Podani & A. Borhidi
Institute of Ecology and Botany, Hungarian Academy of Sciences, H-2163, Vácrátót, Hungary

Accepted 21.10.1988

Keywords: *Aremonio-Fagion*, Balkan Peninsula, Beechwood, Cluster analysis, Consensus classification, Multidimensional scaling, Numerical syntaxonomy, Ordination

Abstract

Cluster analysis and multidimensional scaling are used to revise the syntaxonomy of the *Fagion illyricum* alliance based on presence/absence data. The operational syntaxonomical units are associations and subassociations as suggested in the literature. The results of principal coordinates analysis and nonmetric multidimensional scaling are fairly congruent, whereas some differences occur between the dendrograms obtained by complete linkage and incremental sum of squares clustering. However, a joint evaluation of ordinations and the consensus classification derived from the cluster analysis results facilitates a meaningful revision of the formerly accepted division of the alliance. New low-level syntaxa, and some new names, including *Aremonio-Fagion* to replace the invalid alliance name of *Fagion illyricum*, are proposed and the removal of four associations from the alliance is suggested.

Nomenclature: Ehrendorfer, F. 1973. Liste der Gefässpflanzen Mitteleuropas. G. Fischer, Stuttgart.

Introduction

The *Fagion illyricum* alliance was proposed by Horvat (1938) to include three associations, an oak-hornbeam wood, a beechwood and a maple-ash gorge forest community simultaneously described from the Croatian mountains and colline regions. In this paper numerical methods of classification and ordination are used to revise the within-alliance syntaxonomy of the *Fagion illyricum* established earlier on the basis of traditional phytosociological methods. A new grouping of 36 associations into suballiances is suggested based on analyses of presence/absence data. This paper is not the first attempt for a numerical syntaxonomical revision within the Fagetalia; see Feoli & Lagonegro (1982; see Lausi & Feoli, 1979 for a pioneer attempt with salt marsh communities).

The communities of the *Fagon illyricum* alliance share groups of south European, southeastern European and western Balkan (Illyrian) species which do not occur in the mesophilous forests of central and western Europe. This group comprises ca 30 character-species ('Kennarten') and differential species ('Trennarten', both sensu Braun-Blanquet 1951), of the *Fagon illyricum*. In fact, it was the first instance in phytosociology that a higher syntaxonomic unit was distinguished on the basis of a phytogeographically significant group of character-species. In his monograph of the forest communities of Yugoslavia, Horvat (1950) distinguished as many as 9 associations within the *Fagion illyricum*. Further work in the 1950's and the early 1960's provided more detailed information on these forests (see Borhidi 1963, 1965, 1984, Horvat *et al.* 1974, and references therein), permitting a comparison with other Fagetalia communities and a more elaborated syntaxonomy of the Illyrian beechwoods.

The relationship of the *Fagion illyricum* to other Fagetalia alliances is shown in Soó's (1964) phytogeographic system, which is primarily based on earlier works by Horvat (1938), Oberdorfer (1957), Soó & Borhidi (in Soó 1962), Borhidi (1963) and Vida (1963):

Fagetalia silvaticae Pawlowski 1928
- *Scillo-Fagion* Oberdorfer 1957 – W. Europe
- *Fagion austro-italicum* Soó 1962 – central and southern Italy
- *Fagion medioeuropaeum* Soó (1960) – central Europe (with several ecologically distinct suballiances such as *Carpinion, Cephalanthero–Fagion, Eu-Fagion, Luzulo–Fagion, Abieti–Fagion,* and *Acerion*).
- *Fagion illyricum* Horvat 1938 – SE Alps, SW Hungary, Yugoslavia, Albania, Greece.
- *Fagion dacicum* Soó 1962 – eastern and southern Carpathians, eastern Balkan.

Fagetalia orientalis Soó 1962 – Caucasus, Crimea, northern Asia Minor, SE Balkan.

It is noted that the mesophilous forests of the eastern Balkan were subsequently assigned to a separate alliance (*Fagion moesiacum;* see Horvat *et al.* 1974).

A tentative syntaxonomical division of the *Fagion illyricum* was developed by Borhidi (1963, 1965, 1968; see also Soó 1964) based on the evaluation of 848 relevés most of which were compiled from the literature. Thirty-six community types (associations), divided into four suballiances, are comprised in this system (see Appendix 1). There was a considerable debate among phytosociologists over its validity. Gams (1964, pers. comm.), Soó (1964), and Ozenda (1979) accepted Borhidi's classification, whereas Glavač (1969) and Horvat *et al.* (1974) had several critical comments. Major objections include:
- the oak-hornbeam wood and the submontane beechwood cannot be placed in the same suballiance. The name *Carpinio–Fagion illyricum* (Borhidi 1963) is invalid and should be replaced by *Carpinion illyricum* (Horvat 1963);
- some associations classified into the *Fagion illyricum* should not belong to this alliance; and
- available phytosociological data and evidence are insufficient to construct a 'correct' classification.

Another problem concerns the selection of species groups to be used in the classification. The question is whether ecological species groups of geographically distinct groups of character- and differential species should receive more emphasis in syntaxonomy. Moor (1952, 1960) and Tüxen (1960) are of the opinion that higher syntaxonomic units must be delimited using ecological species rather than geographically 'significant' species. However, it has been suggested over and over again that species with similar ecological requirements but occurring in different geographical regions are equally applicable, and the classifications thus obtained are of more general purpose than those considering ecological differences only. It is implicitly emphasized in numerical approaches to syntaxonomy in which no arbitrary species weights are useful in the analysis.

Material and methods

Ecology of the Fagion illyricum

Communities belonging to this alliance represent a wide ecological range. Oak-hornbeam forests occur on the alluvial terrains of the Drava River at an elevation of 120–150 m, whereas subalpine beechwoods form the timberline in Albania, Macedonia, and eastern Serbia at 1800 m. Within the distributional range of the *Fagion illyricum*, the annual precipitation ranges between 650 and 2000 mm (occasionally up to 4000 mm). The wettest are the Croatian Karst and the central Dinaric Alps. Precipitation decreases and temperature increases towards the east and south, therefore the same vegetation zone develops in higher and higher elevation. These areas belong to the climatic types VI(V), VI(VII) and X (see the bioclimatic map No. 1.1.9 of Bertović *et al.* in Walter & Lieth 1964).

Operational syntaxonomic units

As a starting point in this study the classification of Illyrian beechwoods into 36 associations and geographical variants (Borhidi 1963) is accepted.

Thus, an object in the multivariate analyses, i.e., the operational syntaxonomic unit (OSU; termed on the analogy of operational taxonomic units, OTUs, extensively used in numerical taxonomy, cf. Sneath & Sokal 1973) is any association or geographical variant. Each association is represented by the list of species present, the list being compiled from published data (see Appendix). A rationale for using presence/absence data for the analyses is that due to differences in conditions of sampling and data treatment by various authors, quantitative scores would provide a much less reliable basis. Nevertheless, some unavoidable imbalance is still present in the data, because the number of relevés comprising the OSUs varies greatly (see Appendix, column 4). The total number of species recorded is 644; only vascular plants are considered. The full data matrix is not presented here (but available on request); instead, Appendix 2 provides constancy values for the most important character and differential species.

Multivariate analyses

The 644 by 36 data matrix was analyzed by cluster analysis and multidimensional scaling to reveal an inherent data structure. A 36 by 36 Euclidean distance matrix, an element of which being the square root of the number of species present in only one of the two OSUs compared, served as the starting point in every case. Two cluster analysis procedures were used, complete linkage (CL) and incremental sum of squares agglomeration (SSQ: see Orlóci 1978 for details). Ordination diagrams were obtained by principal coordinates analysis (PCoA; Gower 1966), a metric method, and nonmetric multidimensional scaling (NMDS; Kruskal 1964; Minchin 1987). In the latter case, a random starting configuration of 36 points in two dimensions was improved by the iterations. CL, SSQ and PCoA were performed using the SYN-TAX III package (Podani 1988), while the NMDS ordination was generated by a version of the Brambilla & Salzano (1981) program described in Orlóci & Kenkel (1985). All computations were done on an IBM370 mainframe computer of the Hungarian Academy of Sciences.

Results and discussion

Classification

The dendrogram obtained by SSQ (Fig. 1) provides an answer with respect to the controversy over ecologic vs geographic species groups. The four main clusters recognized on the dendrogram are formed according to geographic separation, i.e., associations from different vegetation zones of the same area rather than the vicarious associations became clustered (see the positions of OSUs 2 and 13, 3 and 14, 4 and 15, 5 and 16, 29 and 32, and 17, 27 and 31). It must be noted, however, that this is not necessarily the case in other high-level syntaxa. In the explanation of this result one must consider that the Balkan Peninsula is a relic-preserving centre with a high tendency to develop diverse floras and that the local diversity of these floras is much higher than in the central European beechwood zone. Therefore, it is not surprising that species shared by vicarious associations are apparently outnumbered by locally distributed species.

Cluster SSQ-1 includes the oak-hornbeam and submontane beechwood communities from Croatia, southern Hungary and Serbia, justifying Borhidi's (1963) view that the oak-hornbeam woods and the submontane beechwoods should not be separated at the level of alliance or suballiance. This is especially

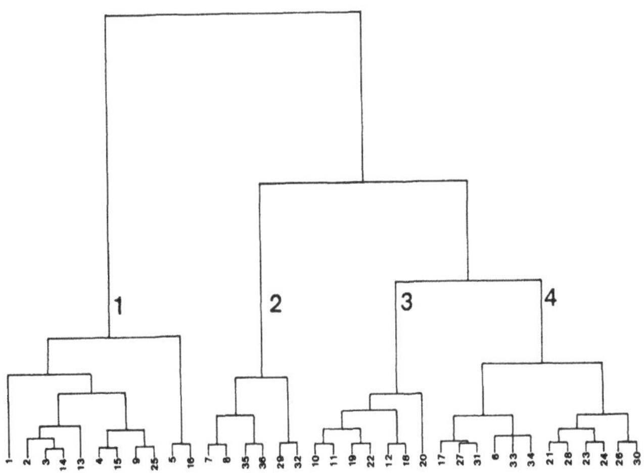

Fig. 1. Incremental sum of squares clustering (SSQ) of associations in *Fagion illyricum*. See text.

true for hilly areas and low mountains where montane and subalpine zones are lacking. Consequently, the suballiance named *Primulo – Fagenion* (Borhidi 1963) should be maintained instead of *Carpinion illyricum* (Horvat 1963).

Cluster SSQ-3 is composed partly of submontane and partly of montane beechwoods which are separated from the oak-hornbeam woods, and the fir-beech forest communities. These communities occur in the submontane region of high mountains, where the montane species descend into the submontane beechwoods primarily through intrazonal gorge forests. Glavač's (1969) observation that they are clearly separated from the oak-hornbeam woods is thus substantiated. From the viewpoint of syntaxonomy, these communities are the most typical representatives of the *Fagion illyricum*, and it seems straightforward to include them in a new suballiance, *Lamio orvalae – Fagenion* with the *Lamio orvalae – Fagetum* as a typical association.

The above proposal necessitates two other nomenclatural changes: a) *Enneaphyllo-Fagetum* (Košir 1962) is an invalid name and for associations 11 and 12 a new name, *Epimedio alpino – Fagetum* is to be used, and b) the name *Vicio oroboidi – Fagetum* should be applied exclusively to the colline beechwoods of SW Hungary.

Cluster SSQ-2 includes some azonal rocky beechwoods (35, 36) and some zonal forests with a transitional syntaxonomic position (7, 8, 29, 32). All are rich in xero – thermophilous species for edaphic or geographic/climatic reasons, and all are strongly influenced by the species from the Quercetalia pubescentis zone.

The dendrogram of Fig. 2 is in a closer agreement with the starting hypothetical syntaxonomic division of the alliance (Borhidi 1963). Cluster CL-1 contains oak-hornbeam woods and submontane forests, whereas CL-4 comprises the montane and subalpine beechwoods. Cluster CL-2 contains an oak-hornbeam and a submontane beechwood community from central and eastern Serbia; their position indicates a transitional situation towards the *Fagion moesiacum* from the eastern Balkan. Group CL-3 is more heterogeneous syntaxonomically; the OSUs of this cluster are similar to one another in that the species of the *Quercus petraea – Q. cerris* zone predominate. The results of complete linkage sorting suggest the following conclusions:

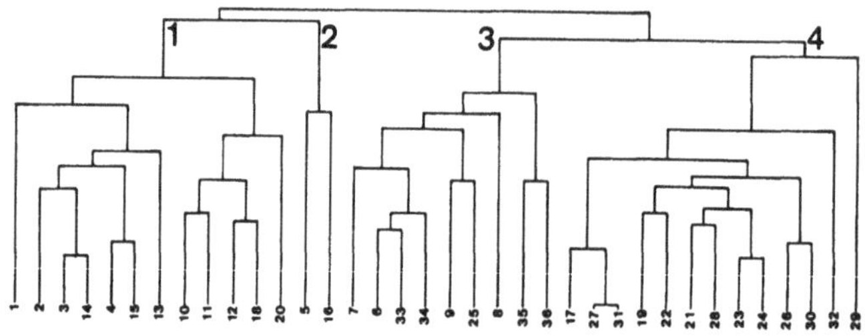

Fig. 2. Complete linkage clustering (CL) of associations and subassociations in *Fagion illyricum*. See text.

– The west Illyrian rocky beechwoods (*Ostryo-Fagion*) and the Turkish hazel-mixed relic beechwoods of eastern Serbia are separated from the zonal beechwood and oak-hornbeam communities. It seems reasonable to treat these associations in a separate group.

– The oak-hornbeam woods of the south Illyrian and central Balkan distribution differ considerably from the remaining *Carpinio – Fagion illyricum* associations. Therefore, the oak-hornbeam woods of Montenegro, Macedonia and Albania (OSUs 6, 7, and 8) are replaced from the *Asperulo taurinae – Carpinetum* association into a new syntaxon, the *Aremonio-Carpinetum*, which has three geographic variants. The nomenclatural type of this association is relevé No. 5 of the *Querceto – Carpinetum montenegrinum* (in Blečić 1958).

Ordination

The first three axes of the PCoA ordination of 36 OSUs explain 19, 10 and 7% of the total variance, respectively. The first axis (Fig. 3) arranges OSUs according to altitude and humidity: the associations from wet, high-mountain areas are grouped on the negative pole, whereas the forests from drier, hilly regions are positioned at the opposite end. The submontane beechwoods are in a transitional situation. The interpretation of axis 2 is not straightforward, yet there is a slightly obscured NW – SE geographic gradient along this component. Axis 3 (not illustrated) produces a clear distinction between three associations from E Serbia (16, 29, 32), with high positive

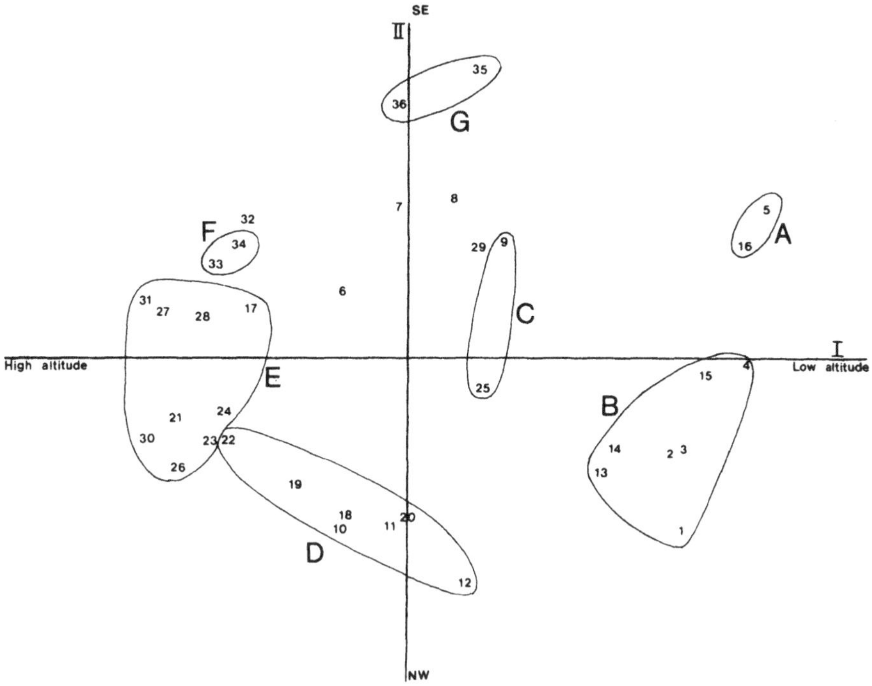

Fig. 3. Principal co-ordinates ordination of the *Fagion illyricum* alliance. Consensus groups used in the revision are outlined (see text for explanation).

loadings, and the other OSUs. These three associations are poor in Illyrian elements while rich in Dacic species, so they represent an obvious transition towards the beechwoods of Transylvania.

The two-dimensional NMDS solution (Fig. 4) is in close agreement with the PCoA ordination along axes 1–2. Its rotation with −90° would result in a configuration that is essentially similar to the PCoA ordination.

Comparative evaluation of results

Whereas the two scaling techniques produced similar configurations in two dimensions, the CL and SSQ dendrograms do not imply the same partition of OSUs at the four-cluster level. Agreements and disagreements between the two classifications are best illustrated by a cross-classification table (cf. Kortekaas *et al.* 1976), see Fig. 5. OSUs occurring in the same cell of this table were classified together by both SSQ and CL at the four-cluster level, and therefore the non-empty cells of the table represent

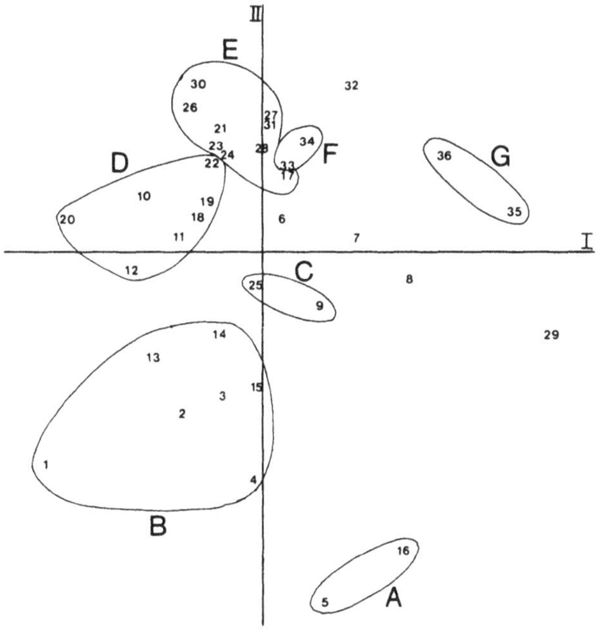

Fig. 4. Nonmetric multidimensional scaling of the *Fagion illyricum* alliance. Consensus groups are superimposed onto the scattergram.

	CL-1	CL-2	CL-3	CL-4
SSQ-1	(1,2,3,4, 13,14,15) B	(5,16) A	(9,25) C	-
SSQ-2	-	-	7,8 (35,36) G	(29,32) H
SSQ-3	(10,11,12, 18,20) D	-	-	(19,22)
SSQ-4	–	–	6 (33,34) F	(17,21,23, 24,26,28, 28,30,31.) E

Fig. 5. Cross-classification table illustrating the strict consensus partition of 36 OSUs derived from the four-cluster partitions implied by SSQ and CL. Outlined groups are used in the revision (see text).

groups of the strict consensus partition (cf. Podani in press) defined for the two alternative classifications. The consensus groups may be identified in the ordination scattergrams, allowing the simultaneous interpretation of two classifications and the ordinations.

Consensus group A (Figs 3–5) includes two associations from Serbia, group B comprises submontane beechwoods and oak-hornbeam woods from Croatia and southern Hungary, and group C contains two intrazonal associations, a rocky and a gorge forest, from the Mecsek Mountains, Hungary. These groups form cluster SSQ-1 and they fall onto the positive side of the PCoA ordination along axis 1. Their inclusion into the same alliance seems straightforward according to our recent knowledge, even though the CL dendrogram is somewhat contradictory.

Group D includes the inframontane beechwoods and gorge forests from Croatia and Slovenia. Their treatment as a separate suballiance was already suggested above. Group E corresponds to the *Lonicero–Fagenion* suballiance. Two communities from Slovenia, the *Lamio orvalae–Fagetum slovenicum* (OSU #19) and the *Arunco–Fagetum* (OSU #22) represent a somewhat transitional situation between the two suballiances.

Two OSUs (#33 and 34) are assigned to group F which corresponds to the *Ostryo–Fagenion* suballiance of rocky beechwoods from the western Balkan. According to the two ordinations and SSQ, its separation from the *Lonicero–Fagenion* is less clear-cut, however.

Group G (35 and 36) contains the *Corylus colurna–Fagus moesiaca* mixed rocky beechwoods of SE Serbia. It is situated relatively far from the other OSUs in the ordinations, suggesting that its more adequate syntaxonomical position is in the Quercetalia pubescentis rather than in the Fagetalia, similarly to the Pannonian karstic beechwoods (*Orno–Fagetum*). This group may be treated as a separate alliance.

Consensus group H (29 and 32) includes two montane beechwoods from SE Serbia. It is not recognized in the first two dimensions of the ordinations, but both OSUs involved have high positive loadings on the 3rd PCoA axis. It was already mentioned that these are close to the Transylvanian Fagetalia communities and may belong to the *Fagion moesiacae*.

There are three remaining OSUs (6, 7, 8) representing southern Illyrian communities, whose class membership is doubtful. Their inclusion in a separate association, as proposed above, is perhaps the best temporary solution as long as no more data are obtained.

A new syntaxonomy of the *Fagion illyricum*

As several authors note (see Gauch 1982: pp. 33–34), community ecology is partially an art even though multivariate methods themselves offer an objective way of analyzing data. The interpretation of numerical syntaxonomic analyses is in a similar situation, especially if different objective methods suggest contradictory solutions for the same syntaxonomic problem. In the present study, there was a considerable agreement among the results, and 'subjective' expectations were rarely contrasting with the ordinations and numerical classifications. The ordination results and the consensus groups created by synthetizing two four-cluster classifications facilitate the construction of a new syntaxonomic system of the *Fagion illyricum* (Appendix 3). The table also suggests a number of nomenclatural corrections of illegitimate or invalid names (according to the Phytosociological Code, Barkman *et al.* 1986) applied by the former authors. The application of the new name *Aremonio–Fagion* for the alliance *Fagion illyricum* is also justified. The name is based on the East-Submediterranean 'Trennart' *Aremonia agrimonioides* occurring with high frequency values in nearly all *Fagion illyricum* communities. The geographic epithetons used under the binomials of the associations indicate geographic races.

The classification suggested here is subject to future refinement. Intensive research is needed to create an improved data base for the Illyrian beechwoods, so that differences in sampling conditions will be less influential on the results. The next step towards a better understanding of the syntaxonomy of European beechwoods is perhaps an extended numerical analysis of all Fagetalia associations.

References

Aichinger, E. 1933. Vegetationskunde der Karawanken. Pflanzensoziologie 2. G. Fischer, Jena.

Barkman, J. J., Moravec, J. & Rauschert, S. 1986. Code of phytosociological nomenclature. Vegetatio 67: 145–195.

Blečić, V. 1958. Šumska vegetacija stena i točila doline Reke Pive. (Végétation des forêts et celle des rochers et des éboulis dans la vallée de la rivière Piva (Montenegro).) Bull. Mus. Hist. Nat. Pays Serbe. series B. 11.

Borhidi, A. 1963. Die Zönologie des *Fagion illyricum* Verbandes. I. Allgemeiner Teil. Acta Bot. Acad. Sci. Hung. 9: 259–297.

Borhidi, A. 1965. Die Zönologie des Verbandes *Fagion illyricum*. II. Systematischer Teil. Acta Bot. Acad. Sci. Hung. 11: 53–102.

Borhidi, A. 1968. Die geobotanischen Verhältnisse der Eichen-Hainbuchenwälder Südosteuropas. Feddes Repert. 78: 109–130.

Borhidi, A. 1984. The forests of Zselic. Transdanubian Stud., Ser. Nat. Hist. 4: 1–145. Janus Pannonius Museum, Pécs, Hungary (in Hungarian with English summary).

Brambilla, C. & Salzano, G. 1981. A nonmetric multidimensional scaling method for non-linear dimension reduction. Theory and computer program. Inst. Appl. Calc. 'Mauro Picone', CNR, Rome. Series III, Nr 121.

Braun-Blanquet, J. 1951. Pflanzensoziologie. 2nd ed. Vienna.

Feoli, E. & Lagonegro, M. 1982. Syntaxonomical analysis of beech woods in the Apennines (Italy) using the program package IAHOPA. Vegetatio 50: 129–173.

Gauch, H. G. 1982. Multivariate analysis in community ecology. Cambridge Univ. Press, Cambridge, Mass.

Glavač, V. 1969. Über Eichen-Hainbuchenwälder Kroatiens. Feddes Repert. 79: 115–138.

Gower, J. C. 1966. Some distance properties of latent root and vector methods used in multivariate analysis. Biometrika 53: 325–338.

Horvat, I. 1938. Biljnosociološka istrazivanija šuma u Hrvatskoj (Pflanzensoziologische Walduntersuchungen in Kroatien). Glasn. sum. pok. 6: 127–279.

Horvat, I. 1949. Nauka o biljnim zajednicama. Zagreb, Nakl, Zav. Hrvatska.

Horvat, I. 1950. Šumske zajednice Jugoslavije (Les associations forestières Yugoslavie). Nakl. Zav. Hrvatska, Zagreb.

Horvat, I. 1963. Leitende Gesichtspunkte für eine pflanzensoziologische Gliederung Europas. Ber. Symp. Vegetationskart. 1959, Stolzenau-Weser. pp. 61–94.

Horvat, I., Glavač, V. & Ellenberg, H. 1974. Vegetation Südosteuropas. G. Fischer, Jena.

Jovanović, B. 1955a. Šumske fitocenoze i staništa Suve Planine (Waldphytozönosen und Standorte der Suva-Planina). Beograd.

Jovanović, B. 1955b. Šumske fitocenoze Rtanja (Rtanj's Waldphytozönosen). Glasn. šum. fak. Beograd 10: 99–127.

Kortekaas, W. M., van der Maarel, E. & Beeftink, W. G. 1976. A numerical classification of European *Spartina* communities. Vegetatio 33: 51–60.

Košir, Z. 1962. Übersicht über die Buchenwälder Sloweniens. Internat. Koll. über die Systematik der europäischen Buchenwälder. 1962. Stolzenau-Weser.

Kruskal, J. B. 1964. Nonmetric multidimensional scaling: a numerical method. Psychometrika 29: 115–129.

Lausi, D. & Feoli, E. 1979. Hierarchical classification of European salt marsh vegetation based on numerical methods. Vegetatio 39: 171–184.

Minchin, P. R. 1987. An evaluation of the relative robustness of techniques for ecological ordination. Vegetatio 69: 89–107.

Moore, M. 1952. Die *Fagion*-Gesellschaften im Schweizer Jura. Beitr. Geobot. Landesauf. Schweiz 31: 1–201.

Moor, M. 1960. Zur Systematik der Querco-Fagetea. Mitt. flor.-soz. Arbeitsgem. N.F. 8: 263–293.

Oberdorfer, E. 1957. Süddeutsche Pflanzengesellschaften. Pflanzensoziologie 10. Fischer, Jena.

Orlóci, L. 1978. Multivariate analysis in vegetation research. 2nd ed. Junk, The Hague.

Orlóci, L. & Kenkel, N. C. 1985. Introduction to data analysis with applications from population and community ecology. Internat. Co-operative Publ. House, Fairland, Maryland, USA.

Ozenda, P. 1979. Sur la correspondence entre les Hêtraies medioeuropéennes et les Hêtraies atlantiques et subméditerranéennes. Docum. Phytosoc. 4: 768–782.

Podani, J. 1988. SYN-TAX III. A package of programs for data analysis in community ecology and systematics. Coenoses 3: 111–119.

Podani, J. in press. A method for generating consensus partitions and its application to community classification. Coenoses

Sneath, P. H. A. & Sokal, R. R. 1973. Numerical taxonomy. 2nd ed. Freeman, San Francisco.

Soó, R. 1962. Systematische Übersicht der pannonischen Pflanzengesellschaften V. Die Gebirgswälder I. Acta Bot. Acad. Sci. Hung. 8: 335–366.

Soó, R. 1964. Die regionale *Fagion*-Verbände und Gesellschaften Südosteuropas. Studia Bot. Hung. 4: 1–104.

Tüxen, R. 1960. Zur Systematik der west- und mitteleuropäischen Buchenwälder. Bull. Inst. Agron. Stat. Rech. de Gemblou 2: 45–58.

Vida, G. 1963. Die zonalen Buchenwälder des ostkarpatischen Florenbezirkes (Transsilvanicum) auf Grund von Untersuchungen im Paring-Gebirge. Acta Bot. Acad. Sci. Hung. 9: 177–196.

Walter, H. & Lieth, H. 1964. Klimadiagramm-Weltatlas. G. Fischer, Jena.

Appendix 1 Origin of the 36 units described within the *Fagion illyricum* alliance (Borhidi 1963).

OSU number	Syntaxonomical unit	Localities	Nr of relevés	Authors
	PRIMULO – FAGION OR CARPINO – FAGION ILLYRICUM SUBALLIANCE: *Oak-hornbeam woods*			
1.	Epimedio – Carpinetum	Croatia, Bosnia, Herzegovina	57	Horvat (1938, 1949); Wraber (1962)
2.	Fraxino angustifoliae – Carpinetum	SW Hungary	35	Soó & Borhidi in Soó (1962), Borhidi & Járai-Komlódi (1959)
3.	Helleboro dumetorum – Carpinetum	SW Hungary	50	Soó & Borhidi in Soó (1962)
4.	Asperulo taurinae – Carpinetum – mecsekense	S Hungary and NE Serbia	72	Horvat (1958); Soó & Borhidi in Soó (1962); Borhidi ined., Janković & Mišić (1960)
5.	– serbicum	Middle and E Serbia	31	Rudski & Ćernjavski (1949); Borisavljević, Jovanović-Dunjić & Mišić (1955), Jovanović (1955a), Rajevski & Borisavljević (1956)
6.	– montenegrinum	Piva valley, Montenegro	7	Blečić (1958)
7.	– macedonicum	W Macedonia	5	Em (1952)
8.	– albanicum	Albania	6	Markgraf (1927, 1932)
9.	Tilio argenteae – Fraxinetum	Mecsek hills, S Hungary	5	Horvat (1961)
	Submontane beechwoods			
10.	Hacquetio – Fagetum	Slovenia	28	Košir (1962)
11.	Vicio oroboidi – Fagetum – subpannonicum	E Slovenia	37	Košir (1962)
12.	– croaticum	E Slovenia	18	Horvat (1938)
13.	– saladiense	SW Hungary	40	Soó & Borhidi in Soó (1962)
14.	– somogyicum	SW Hungary	50	Soó & Borhidi in Soó (1962)
15.	Hellebro odoro – Fagetum – mecsekense	Mecsek hills, S Hungary and NE Yougoslavia	69	Horvat (1959); Soó & Borhidi in Soó (1962); Janković & Mišić (1960)
16.	– serbicum	Middle and E Serbia	66	Rudski & Černjavski (1949); Jovanović & Dunjić (1951); Jovanović (1955a); Mišić (1956); Grebenščikov (1950); Knapp (1944a)
17.	Erythronio-Fagetum	Montenegro	9	Blečić (1958)
25.	Scutellario – Aceretum	Mecsek hills, S Hungary and E Serbia	5	Borhidi ined.; Knapp (1944a)
	LONICERO – FAGION SUBALLIANCE: *Montane beechwoods*			
18.	Lamio orvalae – Fagetum – croaticum	Croatia	10	Horvat (1938)
19.	– slovenicum	Slovenia	23	Košir (1962)
20.	Anemoni trifoliae – Fagetum – subcarinthiacum	Lower Carinthia	18	Knapp (1944b)
21.	– austroalpinum	SE Austria	10	Aichinger (1933)
22.	Arunco – Fagetum	Slovenia	26	Košir (1962)
23.	Isopyro – Fagetum	Slovenia	26	Košir (1962)
24.	Chrysanthemo macrophyllo – Aceretum	Croatia	1	Horvat (1938)

Appendix 1. Continued.

OSU number	Syntaxonomical unit	Localities	Nr of relevés	Authors
	Mixed fir-beechwoods			
26.	Calamintho grandiflorae – Abietetum – dinaricum	Croatia, Dinarian Alps	15	Horvat (1938)
27.	– montenegrinum	Piva valley, Montenegro	9	Blečić (1958)
28.	– albanicum	Albania	5	Markgraf (1927)
29.	Doronico columnae (Abieti) – Fagetum	SE Serbia	53	Jovanović (1955a, b); Grebenščikov (1950)
	Subalpin or elfin beechwoods			
30.	Homogyni – Fagetum	Croatia	15	Horvat (1938)
31.	Pancicio – Fagetum	Montenegro	8	Blečić (1958)
32.	Hyperico umbellato – Fagetum	Suva Planina, SE Serbia	12	Jovanović (1955a)
	OSTRYO – FAGION SUBALLIANCE			
33.	Seslerio autumnali – Fagetum	Piva valley, Montenegro	5	Blečić (1958)
34.	Colurno – Ostryetum carpinifoliae	Piva valley, Montenegro	10	Blečić (1958)
	FAGO – COLURNION SUBALLIANCE			
35.	Fago – Hyrcano – Colurnetum	Suva Planina, SE Serbia	7	Jovanović (1955a)
36.	Colurno – Fagetum	Rtanj hills, SE Serbia	5	Jovanović (1955b)

Appendix 2 Constancy values (+: up to 10%, 1: 11–20%, etc.) for character- and differential species in the 36 OSUs analyzed.

Syntaxa	Primulo-Fagenion														Lamio-Fagenion							Lonicero-Fagenion									O-F		F.m		F-C	
OSU No.	1	2	3	4	5	6	7	8	9	13	14	15	16	25	10	11	12	18	19	20	22	17	21	23	24	26	27	28	30	31	33	34	29	32	35	36
Acer tataricum	4	+	1	1	5								+																							
Asperula taurina				3	2	4	4	5	3			+	+	1																			+		1	
Tilia tomentosa	+	1	7	5	3		5	3	5		8	6	2	1								1														
Ficaria verna	3	3	4	+	2	8	5	1	1	1	4	+	+	5			1																+			
Knautia drymeia	5	7	4	2	+			3	3	5	2	1		3			2			+													2	3	1	
Luzula forsteri	+	1	3	1	2		1	1		1	4	+	1	1																			+			
Polygonatum latifolium		3	+	+	2		1				2	+	+																							
Polytrichum setiferum	+	+	+	1	+			3	1	+	3	+	2	5														1					3			
Quercus robur	1	7	2	+	2		1			1		+	+				+			+																
Ruscus aculeatus	+	5	6	4	3				3		4	3		2																						
Rosa arvensis	7	1	4	2	5				2		2	+	2	1	6		9	1		+																1
Cyclamen purpurascens	4	5	+							3	+				8	6	8	8	3	2	9		+	1		2										
Euphorbia dulcis	6	1	3	+						+	6					5	7	2	+	3						+			2							
Festuca drymeia			+	2	1						1	4	1	2			+	1															2	1	1	
Galanthus nivalis	2	1	5	1		8			3		2	+	+	3	+		5	5	1			5		6												
Primula vulgaris	8	9	4	3	1		9	1	1	4	3	2	+	7	6	1	2					2							1							
Ruscus hypoglossum	+		1	4	1				3	+	6	5	+	2	6	4	6	2	6		4			3												
Tamus communis	4	9	4	2	3	4	1	3	1	2	4	1	2	3	4	4	6	2			1					+									2	
Vicia oroboides	2		2							3	4				3	4	9	1	2	1	1			1		+			+							
Dentaria enneaphyllos		+	+	+					5	1	1	3	2	5	8	+	7	9	4	5	9	4	6	7	1	8	7	5	9	7						
Aremonia agrimonioides	1		2	1	2	9	7	5		+	4	1	4	1	6	2	2	4	+		1	5	2		+	9	8	3	8	8	7	8	7	3	2	7
Anemone nemorosa	7	5	8		2	9				2	8		+		8	8	7	6	8	2	9	8	8	8	2	9	8	9	9	8	7	9	1			
Galium odoratum	4	8	8	5	2		3		5	9	9	8	6	7	7	9	9	9	9	+	2	6	4	9	3	9	9	7	9	6			6	5		
Oxalis acetosella	1	2	3	+						3	3	+	+		3	4	2	3	5	5	1	1	9	3		9	8	1	6	2			+			
Paris quadrifolia	2	3	3	1		5				+	3	+	+	7	7	6	6	9	8	2	5	4	2	8	1	9	3		8	5			2	2		
Acer campestre	7	2	5	2	8	9	6	6	7	4	5	2	3	1	2	1	8	2	3	1	3	4			+						1	5			1	7
Fraxinus ornus	3		5	6	4	4	2	6	1		4	5	2	9	4	1	3			3		1									9	9			9	9
Quercus petraea	9	1	9	9	6	5	9	5	3		4	2	4	2	2	7	2		1	+												9				5
Cornus mas	+		2	3	5	5	5	5	5		+	+	1	1			2															1			7	5
Quercus cerris	1	1	4	1	6	8	3	5	6	2	2	+	+	+	1							3										9			1	1
Helleborus odorus	+			6	7	8		9	7			4	2	2			+					2						2					1	2	8	3
Lathyrus venetus	+	6	8	2	7		7	7	1	2	8	1	4	1																			1	+	1	3
Potentilla micrantha	1		2	2	5		5	3	3			+	2	1			1	2				2											+	+	1	+
Aposeris foetida	8														9	8	8		+	1			6			+			2							
Epimedium alpinum	6				+								+		9	8	1									+										
Hacquetia epipactis	3														9	1	8	3		8	+					+										
Lamium orvala	6	+								1					4	3	7	5	7	1	+	1							4							
Phyteuma spicatum										+					3	3		+	+										6							
Helleborus niger ssp. macr.															7		1	1			6			+												
Abies alba															5	4	+	4		3					+	9	9	5	4	2			1			
Dentaria polyphylla	+														+			1	9				6	7		6			3							
Dentaria latifolia															+		1	4			1			5	1	4			1							
Picea abies	+														3	4	+	2	1	5			2		+	6	2		1	1			+	2		
Pronanthes purpurea	+												+		2		3	2	3	8	9	8	2			6	8	1	8		7		4	4		
Senecio fuchsii	1												+		2	1	3	7	4	1	6			2	3	5			4	3			2	1		
Gentiana asclepiadea	4														7	1	2	2	+	5	3	1			+	2						5	3	4		
Lonicera alpigena													+		2		2	2	1	1	8	3		5		5	7		6	7	1	8	2	9		
Acer visianii																											3			7	+					
Calamintha grandiflora							2	1							2	3						5				4	6	2	1	6						
Hordelymus europaeus												+	+									1		1		2	3	1	+				1	+		
Lonicera nigra																										2	9						1	5		
Polygonatum verticillatum													+									4	2	1		4			8				5	4		
Veronica latifolia	+												+							7		2	2			6	4		4	5			1			
Polystichum lonchitis															+						7	1				+	4		4	5			1	5		
Rosa pendulina																				+	8	1		1		2	7		5	6		3	3	9	1	1
Rhamnus fallax								2														2				2	6	1	1	6	1	6				
Saxifraga lasiophylla								2					+									3				+	4	1	4	3			2	7	2	2
Rubus saxatilis																						2					9		1	1		7	2	7		
Vaccinium myrtillus	+															+			5			1	2			1	5		2	3		6	1	5		
Sesleria autumnalis						8																1									9	2				
Corylus colurna							1						+																			6				7
Cotinus coggygria							1															1										7			1	2
Vincetoxicum officinale	1	+							2				+																		3	4	+	+	9	9
Cotoneaster tomentosa							+																				3					8	1	5	5	2
Hypericum umbellatum																																	3	8		
Moehringia muscosa																+			1														3	4		
Primula columnae						9		9					+									4											1	2	8	9

Appendix 2. Continued.

Sorbus graeca																																	6	9	4	2	
Doronicum columnae					+		7						+									1									5		5	5	2	5	
Geranium macrorrhizum					+								+														1			1		8	5	5	9	9	
Euphorbia polychroma					+						+	+																							4	3	
Origanum vulgare	+							1	1				+																						5	9	
Syringa vulgaris													+													+									4	9	
Teucrium chamaedrys					+				1																										4	3	
Helleborus atrorubens	3				+												2	1																			
Fraxinus angustifolia		5			+									1																							
Dianthus barbatus	+	3																																			
Helleborus dumetorum	+		4												+									1													
Doronicum orientale				2					5																												
Trifolium pignatti					2																																
Acer obtusatum						4	5	5									1									2		1			7						
Crocus neapolitanus	6					9									+	1	2	1																			
Helleborus cyclophyllus							9																														
Anemone appennina								7																													
Scrophularia vernalis				+					7			+					1	2	+					2													
Melandrium rubrum	+											+		7			2								+												
Daphne laureola												+	+		4		3	4																			
Polygala chamaebuxus																				4																	
Melampyrum velebiticum																					5																
Dentaria pentaphyllos	+																			+			8														
Saxifraga rotundifolia																				+			8														
Chrysanthemum macrophyllum													+				+								1								+				
Galium rotundifolium	+																									5	6	1									
Geranium nodosum																										3											
Myrrhis odorata																								+			3										
Podanthum trichocalycinum																												3									
Adenostyles glabra																										+			2								
Pancicia serbica																														7							
Ranunculus platanifolius																													6	6			+				
Anthericum ramosum																															7						
Rhamnus rupestris																																5					
Senecio rupester																																	2				
Daphne oleoides																																		2			
Acer hyrcanum							+	+					+																				+		9		
Galium purpureum																																			4		
Acer monspessulanum							+	+					+																			+					5

Appendix 3.

Suggested syntaxonomic system of the *Fagion illyricum* and the related associations. OSU numbers as used in this paper are in brackets.

Order: *Fagetalia silvaticae* Pawlowski 1928

Alliance: *Aremonio-Fagion* Borhidi nom. nov. hoc loco (*Fagion illyricum* I. Horvat 1938)

Suballiance: *Primulo vulgaris-Fagenion* Borhidi 1963 (*Carpino-Fagenion illyricum* Borhidi 1963, *Carpinion illyricum* I. Horvat 1963)

Associations: *Epimedion-Carpinetum* (I. Horvat 1938) Borhidi 1963 (*Querceto-Carpinetum croaticum* I. Horvat 1938); 57 relevés (1)

Fraxino pannonicae-Carpinetum Soó & Borhidi in Soó 1962; 35 relevés (2)

Helleboro dumetorum-Carpinetum Soó & Borhidi in Soó 1962; 50 relevés (3)

Asperulo taurinae-Carpinetum Soó & Borhidi in Soó 1962

mecsekense (A.O. Horvát 1958) Soó & Borhidi in Soó 1962 (*Querceto-Carpinetum mecsekense* A.O. Horvát 1958); 72 relevés (4)

serbicum (Rudski 1949) Soó & Borhidi in Soó 1962 (*Querceto-Carpinetum serbicum* Rudski 1949); 31 relevés (5)

Aremonio-Carpinetum Borhidi nom. novum hoc loco (*Asperulo taurinae-Carpinetum* Soó & Borhidi 62 p.p.)

montenegrinum (Blečić 1958) Borhidi hoc loco (*Querceto-Carpinetum montenegrinum* Blečić 1958); 7 relevés (6)

macedonicum (Em 1950) Borhidi hoc loco (*Querceto-Carpinetum macedonicum* Em 1950); 5 relevés (7)

albanicum (Soó 1962) Borhidi hoc loco (*Asperulo taurinae-Carpinetum albanicum* Soó 1962); 6 relevés (8)

Tilio tomentosae-Fraxinetum (A.O. Horvát 1961) Soó & Borhidi in Soó 1962 (*Tilio-Fraxinetum mecsekense* A.O. Horvát 1961); 5 relevés (9)

Vicio oroboidi-Fagetum Pócs & Borhidi 1960 em. Borhidi hoc loco

saladiense Borhidi & Pócs 1960; 40 relevés (13)

somogyicum Borhidi & Pócs 1960; 50 relevés (14)

Helleboro odoro-Fagetum Soó & Borhidi in Soó 1962

mecsekense (A.O. Horvát 1959) Soó & Borhidi in Soó 1962 (*Fagetum mecsekense* A.O.Horvát 1959); 69 relevés (15)

serbicum (Rudski 1949) Borhidi 1963 (*Fagetum serbicum montanum* Rudski 1949); 67 relevés (16)

Scutellario-Aceretum Soó & Borhidi in Soó 1962; 5 relevés (25)

Suballiance: *Lamio orvalae-Fagenion* Borhidi suball. nova

Associations: *Hacquetio-Fagetum* Košir 1962; 28 relevés (10)

Epimedio-Fagetum Borhidi nom. novum hoc loco (*Querco-Fagetum* Košir p.p., *Enneaphyllo-Fagetum* Košir p.p.)

slovenicum (Košir 1962) Borhidi hoc loco (*Querco-Fagetum* Košir 1962 non Tx. 1937); 37 relevés (11)

croaticum (I. Horvat 1938) Borhidi hoc loco (*Fagetum croaticum boreale lathyretosum* I. Horvat 1938); 18 relevés (12)

Lamio orvalae-Fagetum I. Horvat 1938 em. Borhidi 1963

croaticum I. Horvat 1938; 10 relevés (18)

slovenicum (Košir 1962) Borhidi 1965 (*Enneaphyllo-Fagetum* Košir 1962); 23 relevés (19)

Anemoni trifoliae-Fagetum Tregubov 1957

subcarinthiacum (Knapp 1944) Borhidi 1965 (*Fagetum subcarinthiacum* Knapp 1944); 18 relevés (20)

Arunco-Fagetum Košir 1962; 26 relevés (22)

Suballiance: *Lonicero-Fagenion* Borhidi 1963

Associations: *Erythronio-Fagetum* (Blečić 1958) Borhidi 1963 (*Fagetum silvaticae montenegrinum montanum* Blečić 1958); 9 relevés (17)

Anemoni trifoliae-Fagetum Tregubov 1957

austro-alpinum (Aichinger 1933) Borhidi 1963 (*Fagetum austro-alpinum* Aichinger 1933); 10 relevés (21)

Isopyro-Fagetum Košir 1962; 26 relevés (23)

Chrysanthemo macrophyllo-Fagetum (I. Horvat 1938) Borhidi 1963 (*Acereto-Fraxinetum croaticum* I. Horvat 1938); 1 relevé (24)

Calamintho grandiflorae-Abietetum Borhidi nom. nov. (*Calamintho grandiflorae-Abieti-Fagetum* Borhidi 1963 nom. invalid.

croaticum (I. Horvat 1938) Borhidi 1963 (*Fagetum croaticum montanum abietetosum* I. Horvat 1938; *Abieti-Fagetum dinaricum* Tregubov 1957); 15 relevés (26)

montenegrinum (Blečić 1958) Borhidi 1963 (*Fagetum montenegrinum abietetosum* Blečić 1958); 9 relevés (27)

albanicum Borhidi 1963; 5 relevés (28)

Homogyno silvestris-Fagetum (I. Horvat 1938) Borhidi 1963 (*Fagetum croaticum australe subalpinum* I. Horvat 1938); 15 relevés (30)

Pancicio-Fagetum (Blečić 1958) Borhidi 1963 (*Fagetum montenegrinum subalpinum* Blečić 1958); 8 relevés (31)

Suballiance: *Ostryo-Fagenion* Borhidi 1963

Associations: *Seslerio autumnali-Fagetum* (I. Horvat 1950) Wraber 1960 (*Fagetum croaticum seslerietosum* I. Horvat 1950); 8 relevés (33)

Corylo colurnae-Ostryetum Blečić 1958; 10 relevés (34)

Alliance: *Fagion moesiacae* Horvat, Glavač & Ellenberg 1974

Associations: *Doronico columnae-Abietetum* (Jovanović 1955) Borhidi nom. novum (*Abieti-Fagetum serbicum* Jovanović 1955); 53 relevés (29)

Hyperico umbellatae-Fagetum (Jovanović 1955) Borhidi 1963 (*Piceo-Fagetum serbicum* Jovanović 1955); 12 relevés (32)

Order: *Quercetalia pubescentis-petraeae* Br.-Bl. 1931

Alliance: *Fago-Corylion colurnae* Borhidi 1963 stat. novus

Associations: *Aceri hyrcano-Fagetum moesiacae* (Jovanović 1955) Borhidi nom. nov. (*Fago-Hyrcano-Colurnetum* Jovanović 1955); 7 relevés (35)

Corylo colurnae-Fagetum (Jovanović 1955) Borhidi 1963 (*Fagetum serbicum colurnetosum* Jovanović 1955); 5 relevés (36)

Vegetatio **81**: 181–187, 1989.

Numerical syntaxonomy of the tall-forb and tall-grass communities in the Tatra Mountains

Ján Šeffer[1], Eva Šefferová[2] & Zuzana Dúbravcová[3]
[1]*Institute of Experimental Biology and Ecology, S.A.S., Sienkiewiczova 1, CS-814 34 Bratislava, Czechoslovakia;* [2]*Institute of Geography, S.A.S., Obrancov mieru 49, CS-800 00 Bratislava, Czechoslovakia;* [3]*Department of Geobotany, Comenius University, Mlynská dolina B-2, CS-842 15 Bratislava, Czechoslovakia*

Accepted 10.1.1989

Keywords: Alpine vegetation, *Calamagrostietalia villosae*, Czechoslovakia, Numerical classification, Ordination, Poland

Abstract

The tall-forb and tall-grass subalpine and alpine vegetation belonging to the *Mulgedio-Aconitetea* from the Tatra Mountains was subjected to reclassification with Complete Linkage Clustering. The communities are classified into 3 alliances: *Adenostylion*, *Calamagrostion villosae* and *Trisetion fusci*. It was concluded that the orders *Adenostyletalia* and *Calamagrostietalia villosae* should be fused into one unit. Synecological relations of the communities was clarified using a series of Detrended Correspondence Analyses.

Abbreviations: CLC = Complete Linkage Clustering; DCA = Detrended Correspondence Analysis

Nomenclature: of taxa follows Ehrendorfer, F. (ed.) 1973. Liste der Gefässpflanzen Mitteleuropas. 2. Aufl. Fischer, Stuttgart; of syntaxa Mucina, L. & Maglocký, Š. (eds) 1985. A list of vegetation units of Slovakia. Doc. Phytosociol. 9: 175–220.

Introduction

Tall-forb and tall-grass vegetation belonging to the *Mulgedio-Aconitetea* in the subalpine and alpine belts of the Tatra Mountains has been studied both by Polish and Czechoslovakian phytosociologists, starting with Szafer *et al.* (1923) and Pawłowski *et al.* (1928). Syntaxonomical studies include Krajina (1933), Hadač (1956), Lišková (1960), Unar (1961), Komárková (1964), Hadač *et al.* (1969), Dúbravcová *et al.* (1976), Králik (1979), Šeffer (1984), Šefferová (1984), Unar *et al.* (1984, 1985), Šeffer & Šefferová (in press).

These studies were mostly limited to local areas, and the large-scale variation and synsystematic relations among the communities remained hidden. The only synthetical attempt was to compare the tall-forb and tall-grass communities in the western Tatra Mts. (Dúbravcová *et al.* 1976).

The aim of the present paper is to reclassify the communities of the *Mulgedio-Aconitetea* and to analyse their synsystematical and synecological relations with regard to their occurrence along major environmental gradients.

Two systems of classification exist for these communities. Originally, all alliances were included within one order, either the *Calamagrostietalia villosae* (Pawłowski *et al.* 1928; Hadač 1956), or the *Adenostyletalia* (Braun-Blanquet 1950; Krajina 1933).

According to the second conception only the *Calamagrostion villosae* belongs to the *Calamagrostietalia villosae*, whereas the *Adenostylion* and *Trisetion fusci* belong to the *Adenostyletalia* (Klika & Hadač 1944; Holub *et al.* 1967; Moravec *et al.* 1983; Unar *et al.* 1984, 1985).

Methods

A synthetic table with constancy values 1–5 was composed on the basis of local phytosociological tables of communities of the orders *Adenostyletalia* and *Calamagrostietalia villosae* from various areas of the western and eastern Tatra Mountains (Appendix 1). This table contained 40 columns (based on 322 relevés) and 442 taxa. It was analysed by Complete Linkage Clustering, CLC, based on Wishart's Similarity Ratio (Podani 1984), and by Detrended Correspondence Analysis, DCA (Hill 1979; Hill & Gauch 1980).

Results and discussion

The first ordination revealed two basic complex gradients (Fig. 1), which represent a gradient between lime-rich and lime-poor substrates (axis 1) and a moisture gradient (axis 2).

The *Delphinion elati* communities from the calcareous Belianske Tatry Mountains were separated along axis 1. The calciphilous species such as *Aconitum variegatum*, *Delphinium elatum*, *Sesleria tatrae* and *Corthusa mathiolii*, all abun-

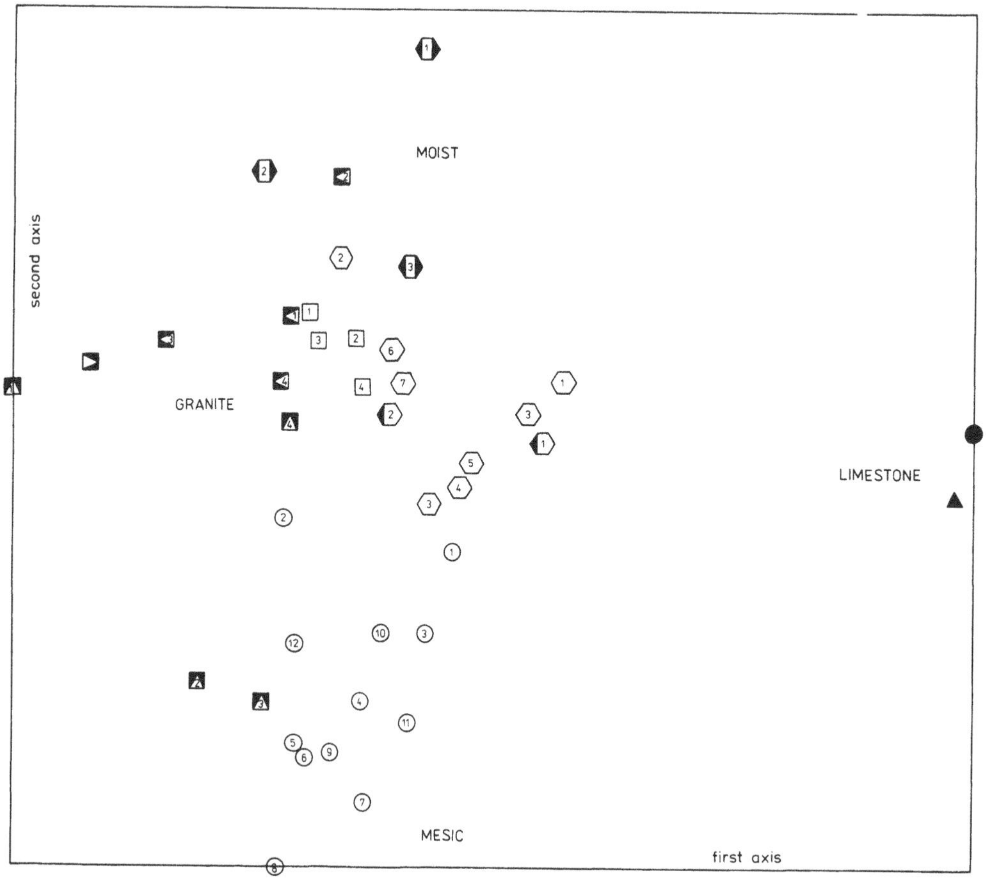

Fig. 1. DCA ordination of 40 community types (see Appendix) including *Delphinion elati* communities.

dant in the *Delphinion elati* characterize the positive end of axis 1, whereas the acidophilous species *Eriophorum vaginatum*, *Salix helvetica*, *Nardus stricta* and *Oreochloa distycha* show the highest scores on the negative end of this axis.

The moisture gradient underlies a coenocline from the mesophilous *Calamagrostion villosae*, to the tall-forb *Adenostylion* and the *Trisetion fusci* at the edges of mountain rills.

The *Aconitetum firmae* has the highest score on the positive end of axis 2. This type is transitional between the *Mulgedio-Aconitetea* and the *Montio-Cardaminetea*. Hygrophilous species such as *Stellaria nemorum*, *Philonotis fontana*, *Cardamine opizii* and *Geum rivale* occur in the latter community.

The scatter diagram of the 40 types DCA (Fig. 1) shows, that the structure was influenced by two outliers at the positive end of axis 1. These are the two *Delphinion elati* types. Therefore these types were removed and the remaining 38 types were reordinated (Fig. 2). The soil-moisture gradient now became the most important one as the communities were arranged between mesic to moist along axis 1 of this ordination. An altitudinal gradient was revealed along axis 2; it links the *Salix helvetica* communities at high altitudes with the *Aconito firmi-Deschampsietum alpicolae* at lower altitudes.

The taxa with an optimum in the upper part of the forest belt are situated at the positive end of axis 2: *Carum carvi*, *Ranunculus lanuginosus*, *Rumex alpinus*, *Ranunculus acris* and *Achillea sudetica*. The species with an optimum in the subalpine and alpine belts are found at the negative end: *Salix helvetica*, *Salix herbacea*, *Pedicularis oederi*, *Saxifraga androsacea* and *Swertia perennis* ssp. *alpestris*.

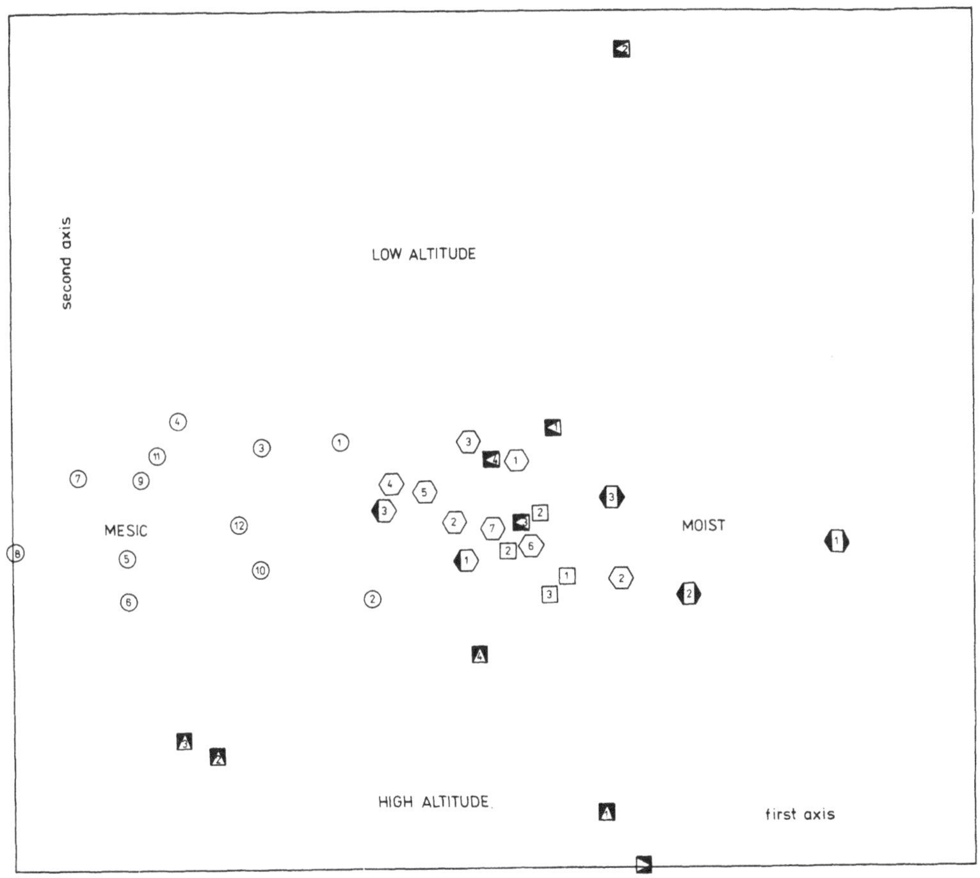

Fig. 2. DCA ordination of the same communities except the two deviating *Delphinion elati* types.

The synthetic table made on the basis of the CLC clustering (Fig. 3) shows three strong groups, which we can interpret as three alliances – the *Calamagrostion villosae*, the *Adenostylion* and the *Trisetion fusci*. Each of them has numerous diagnostic species (Table 1). The analysis also showed a strong group of species common to all the alliances discussed (Table 1). These are considered as the diagnostic taxa of one order only, the *Calamagrostietalia villosae*.

Results of the CLC (not shown in detail) also showed, that the *Delphinion elati* should be eliminated from the *Calamagrostietalia villosae*. It strongly differs from other communities of this order in both species composition and habitat conditions. Another change concerns the classification of the *Aconitetum firmi*, which should be assigned to the *Trisetion fusci*. The resulting syntaxonomy of the Tatra tall-forb and tall-grass communities is presented in Appendix 2.

Acknowledgements

We thank J. Hrabovcová and P. Králik for unpublished data, G. Schlosser for assistance with computations, I. Jarolímek for valuable comments, and L. Mucina for constructive criticism.

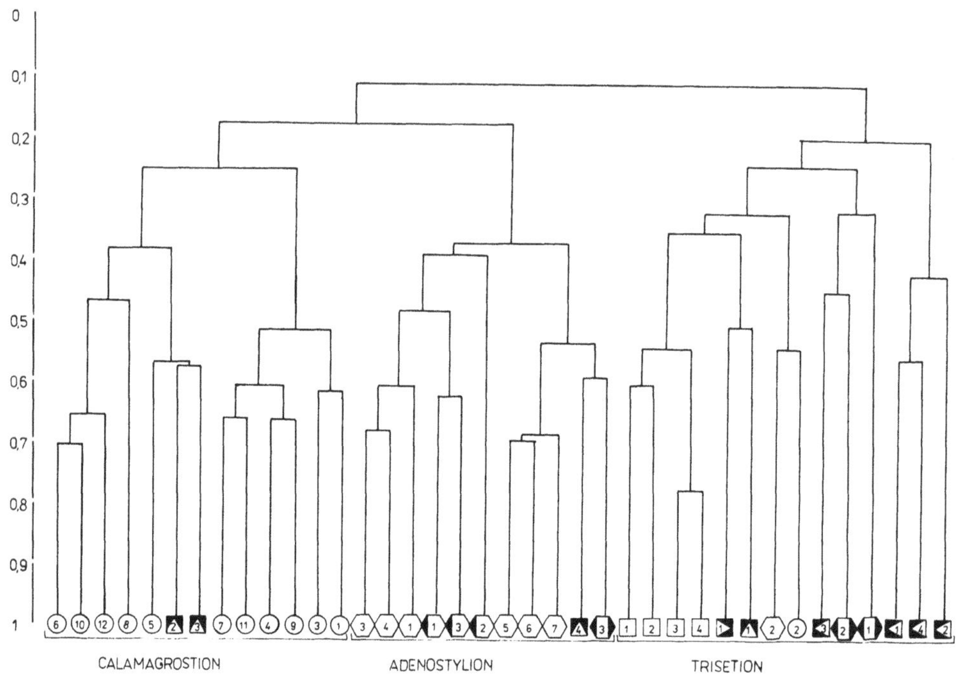

Fig. 3. Complete linkage clustering after omitting the *Delphinion elati*.

Table 1. Part of the synthetic table of the *Calamagrostietalia villosae* with the distribution of four diagnostic species groups.

Number of table	11 33 1 6028512714931	11122211132 56302179835	234322331 2323 66478980424235
Diagnostic species of the *Calamagrostietalia villosae:*			
Calamagrostis villosa	5555544555555	45534443354	2 1 433445 221
Adenostyles alliariae	123 1 121223	55555155523	1 1 352 54 11
Festuca picta	3551521342355	34412345544	32555535555543
Rumex arifolius	121111 231 25	55555344413	2 2322 532234
Alchemilla sp.	1 2 1 33	3131 133133	45454554433353
Aconitum firmum	122 13 2 114	2253 325355	5333555253535
Veratrum lobelianum	3442 22443555	54554255543	1 31555453 12
Gentiana punctata	4432433221445	54454154552	1 2135455314
Geranium sylvaticum	121 2 333143	3225 142 21	412223 1 311
Phleum alpinum	1111 21124	123 123321	23523224531325
Hypericum maculatum	23122 443334	434 4 2	212323 22 112
Solidago x alpestris	3333221544555	231 3 2222	1 3231 231 1
Phyteuma spicatum	121 333344	312 4111	1 1124 3 1
Diagnostic species of the *Calamagrostion villosae:*			
Vaccinium myrtillus	4435523453545	2 12 21	2 2 1 3
Campanula tatrae	2352 11432424		133 2 1 1
Sempervivum montanum	1121 3412 2		1
Hieracium sp.	31124 422 13		1 2
Carex sempervirens	3224 2 535523	1 1 1	422 315 3 3
Poa chaixii	12 11 2 32	1 1	3 24 1
Gentiana asclepiadea	11132 113324	1	1332 3
Pulsatilla alba	1132 2142 422	1 2113	1 3 2
Crepis conyzifolia	1 12 532332		31 112 2
Luzula luzuloides	2425311455434	121 2 1	113112 14 11
Hypochaeris uniflora	2121 5235 2	1	1 2
Bartsia alpina	1 3 2213	1	1 1
Diagnostic species of the *Adenostylion:*			
Cicerbita alpina	1 1	33533 2 1	1
Doronicum austriacum	121 2 2	422 1221421	2211 54 2
Milium effusum	22 11 2	23552244421	1 1 2 21 2
Silene dioica	11 2 11 1112	5221 54412	1 1 2 31111
Ranunculus platanifolius	111 1 132354	54442 3331	11 113 2 1
Athyrium distentifolium	1 1 1211 12	32555522	31 11 21 1
Diagnostic species of the *Trisetion fusci:*			
Trisetum ciliare	1 3 13	1 1 231	22 525111441
Rhodiola rosae	11 3 11 133	11 125332	52 25554122 1
Deschampsia caespitosa	1 1 1 1	1 1 2 423	55552555523345
Swertia x alpestris	1 2	2	2 1435 21
Caltha x laeta		32	23223355112
Carex atterima	1 1 11	1 1 1	2111425111
Taraxacum alpinum	1 1 23	11 12213	541355331 15 3
Cerastium fontanum		1 2	32531 1 111
Chrysosplenium alternifolium	1	2 112	2 1 11 1 15
Chaerophyllum hirsutum	2	121 1 1 12	212 1 1 3242

Appendix 1

Festuco pictae-Calamagrostietum villosae:

1. Pawlowski et al. (1928)
2. Krajina (1933)
3. Hadač (1956)
4. Komárková (1964)
5. Unar et al. (1984)
6. Dúbravcová et al. (1976)
7. Dúbravcová et al. (1976)
8. Dúbravcová et al. (1976)
9. Králik (1976)
10. Dúbravcová unpubl.
11. Šefferová (1984)
12. Šefferová (1984)

Ranunculo platanifolii-Adenostyletum alliariae:

13. Pawlowski et al. (1928)
14. Krajina (1928)
15. Hadač (1956)
16. Dúbravcová et al. (1976)
17. Králik (1979)
18. Šefferová (1984)
19. Šefferová (1984)

Acetoso alpestris-Athyrietum alpestris:

20. Hadač (1956) + Unar et al. (1984)
21. Dúbravcová et al. (1976)
22. Šefferová (1984)

Aconitetum firmi:

23. Krajina (1933)
23. Komárková (1964)
25. Šeffer (1984)

Rhodiolo-Deschampsietum caespitosae:

26. Dúbravcová et al. (1976)
27. Dúbravcová unpubl.
28. Šeffer (1984)
29. Šeffer (1984)

Salicetum helveticae:

30. Krajina (1933)
31. Dúbravcová et al. (1976)
32. Dúbravcová unpubl.
33. Šeffer (1984)

Aconito firmi-Deschampsietum alpicolae:

34. Krajina (1933)
35. Lišková (1960)
36. Dúbravcová et al. (1976)
37. Šeffer (1984)

Trisetetum fusci:

38. Krajina (1933)

Doronicetum austriaci:

39. Hadač (1969) + Unar et al. (1984)

Senecioni-Adenostyletum alliariae:

40. Hadač et al. (1969)

Appendix 2

Syntaxonomical revision of the vegetation units discussed.

Mulgedio-Aconitetea Hadač et Klika in Klika et Hadač 1944

Calamagrostietalia villosae Pawłowski *et al.* 1928
Calamagrostion villosae Pawłowski *et al.* 1928
Festuco pictae-Calamagrostietum villosae Pawłowski *et al.* 1928
Adenostylion Br.-Bl. 1926
Ranunculo platanifolii-Adenostyletum alliariae (Kraj. 1933) Dúbravcová et Hadač in Mucina et Maglocký 1985 Acetoso alpestris-Athyrietum alpestris (Hadač 1956) Hadač in Mucina et Maglocký 1985
Trisetion fusci Kraj. 1933
Rhodiolo-Deschampsietum caespitosae Kraj. 1933
Aconito firmi-Deschampsietum alpicolae (Kraj. 1933)
Hadač in Mucina et Maglocký 1985
Salicetum helveticae (Kraj. 1933) Dúbravcová et Hrabovcová in Mucina et Maglocký 1985
Trisetetum fusci Kraj. 1933
Aconitetum firmi Kraj. 1933

Seslerietalia tatrae Hadač 1962
Delphinion elati Hadač 1962
Doronicetum austriaci Hadač *et al.* 1969
Senecioni-Adenostyletum alliariae Hadač *et al.* 1969

References

Braun-Blanquet, J. 1950. Übersicht der Pflanzengesellschaften Rätiens (V). Vegetatio 2-3: 214–237.

Dúbravcová, Z., Foltinová, J., Paclová, L. & Turečková, J. 1976. Vegetácia subalpinskeho a alpinskeho stupňa Západných Tatier. Manuscript, Katedra geobotaniky PvF UK, Bratislava.

Hadač, E. 1956. Rastlinné spoločenstvá Temnosmrečinovej doliny vo Vysokých Tatrách. Biol. Práce 2/1: 5–78.

Hadač, E. *et al.* 1969. Die Pflanzengesellschaften des Tales 'Dolina Siedmich Prameňov' in der Belaer Tatra. Vegetácia ČSSR, Ser. B 1: 1–343.

Hill, M.O. 1979. DEORANA A Fortran program for detrended correspondence analysis and reciprocal averaging. Ecology and Systematics, Cornell University, Ithaca, 52 pp.

Hill, M.O. & Gauch Jr., H.G. 1980. Detrended Correspondence Analysis: an improved ordination technique. Vegetatio 42: 47–58.

Holub, J., Hejný, S., Morevec, J. & Neuhäust, R. 1967. Übersicht der höheren Vegetationseinheiten der Tschechoslowakei. Rozpr. Čs. Adad. Věd, 75 pp.

Klika, J. & Hadač, E. 1944. Rostlinná společenstva středni Evropy. Přiroda 36: 249–259, 281–295.

Komárková, V. 1964. Alpinská vegetace Roháčů. Thesis, Katedra botaniky PvF KU, Praha.

Krajina, V. 1933. Die Pflanzengesellschaften des Mlynica-Tales in den Vysoké Tatry. Beih. Bot. Cbl., Dresden 50: 1–224.

Králik, P. 1979. Rastlinnė spoločenstvá Spálenej doliny. Thesis, Katedra geobotaniky PvF UK, Bratislava.

Lišková, M. 1960. Rostlinná společenstva Tomanovej doliny v Západných Tatrách. Thesis, Katedra biologie rostlin PvF UUEP, Brno.

Moravec, J. *et al.* 1983. Rostlinná společenstva Českė socialistickė republiky a jejich ohroženi. Severočes. Přir., Litoměřice 1: 1–110.

Pawłowski, B., Sokołowski, N. & Wallisch, K. 1928. Die Pflanzenassoziationen des Tatra-Gebirges. Teil VII. Die Pflanzenassoziationen und die Flora des Morskie Oko-Tales. Bull. Int. Acad. Polon. Sci. Lettr., Cracovie, Suppl. 2: 13–18.

Podani, J. 1984. SYN-TAX II. Computer programs for data anlysis in ecology and systematics. Abstr. Bot. 8: 73–94.

Szafer, W., Pawłowski, B. & Kulczyński, S. 1923. Pflanzenassoziationen des Tatra-Gebirges. Teil III. Die Pflanzenassoziationen des Kościeliska-Tales. Bull. Int. Acad. Polon. Sci. Lettr., Cracovie, Suppl. 2: 13–78.

Šeffer, J. 1984. Spoločenstvá niv a horských lúk vo Vysokých Tatrách I. Thesis, Katedra geobotaniky PvF UK, Bratislava.

Šefferová, E. 1984. Spoločenstvá niv a horských lúk VO Vysokých Tatrách II. Thesis, Katedra geobotaniky PvF UK, Bratislava.

Šeffer, J. & Šefferová, E. in press. Spoločenstvá s Adenostyles alliariae vo Vysokých Tatrách – numerický pristup. Biologia.

Unar, J. 1961. Rostlinná společenstva Javorové doliny v Západnich Tatrách. Thesis, Katedra biologie rostlin PvF UJEP, Brno.

Unar, J., Unarová, M. & Šmarda, J. 1984. Vegetačni poměry Tomanovy doliny a Žlebu spod Diery v Západnich Tatrách. Část 2. Charakteristika přirodnich poměrů a rostlinných společenstev. Folia Fac. Sci. Natur. Univ. Purky. 84/14: 5–78.

Unar, J., Unarová, M. & Šmarda, J. 1985. Vegetačni poměry Tomanovy doliny a Žlebu spod Diery v Západnich Tatrách. Část 1. Fytocenologické tabulky. Folia Fac. Sci. Natur. Univ. Purkyn. 80/10: 5–101.

Vegetatio **81**: 189–198, 1989

Partitioning an elevation gradient of vegetation from southeastern Ethiopia by probabilistic methods

Zerihun Woldu[1], Enrico Feoli[2] & Lisanework Nigatu[3]
[1]*Department of Biology, Addis Ababa University, P.O. Box 1176, Addis Ababa, Ethiopia;* [2]*Department of Biology, University of Trieste, Via A. Valerio 30-32, I-342100 Trieste, Italy;* [3]*Jimma Agricultural College, Jimma, Ethiopia*

Accepted 29.11.1988

Keywords: Environmental fuzzy sets, Gradient, Probabilistic similarity index, Stratocoenon

Abstract

Vegetation and environment have been analyzed along an altitudinal gradient in Harena Forest, Bale Mountains National Park, southeastern Ethiopia. Vegetation data include numbers of each tree and shrub species and cover-abundance values of each herbaceous species. Environmental data comprise edaphic factors, altitude and topography. The two vegetation layers data were analysed separately.

Probabilistic similarity coefficients were computed between the relevés, and these values were used in subsequent computations for classification and ordination. Two sets of stratocoena, comprising 6 types each, derived on the basis of separate analyses of tree-shrub and herb layers of the forest were recognised. A combination of the two sets of stratocoena produced a total of 11 vegetation types. Environmental fuzzy set analysis was applied to determine the strength of the relationship of the relevés to the environmental factors. Autocorrelation analysis was applied to the eigenvectors of probabilistic similarity matrices and environmental data. Altitude appears to be more important than the other environmental factors in controlling the zonation of the forest. Other important environmental influences on the vegetation include pH, organic matter content and texture of the soil. It is suggested that the whole forest be included in the National Park to create suitable conditions for adequate protection.

Abbreviations: EFS – Environmental fuzzy sets, PROSIM – Probability similarity index, PCA – Principal components analysis

Nomenclature: follows Cufodontis, G. 1953–1972. Enumeratio Plantarum Ethiopiae. Bull. Jard. Bot. Belg., and Hedberg, I. & Edwards, S. (in press). Flora of Ethiopia. Vol. III. Addis Ababa University, Addis Ababa.

Introduction

The study of vegetation is particularly urgent in countries such as Ethiopia, since the natural environment has been severely altered and only a few relict patches of natural vegetation remain. The long history of human influence in Ethiopia has changed the country into an anthropogenic landscape. Bare hills and small patches of forest and overgrazed grasslands are characteristic features of the landscape, especially that of the Central Plateau. Some relict vegetation on the Central Plateau of Ethiopia, however, permits a plausible reconstruction of the original dry evergreen montane forest.

The Harena State Forest in southeastern Ethiopia is one patch that has escaped heavy exploitation, mainly due to its inaccessibility. A large part of the forest is situated in the Bale Mountains National Park and lies at altitudes between 1510 to 3300 m. This forest is the easternmost moist forest in Ethiopia and forms an important catchment basin for some rivers flowing to the neighbouring lowlands.

The parent material of the region is of volcanic origin (Murphy 1958; Weinert & Masurek 1984). The climate is humid subtropical (Walter & Lieth 1960; Daniel Gamachu 1977). Rainfall is bimodal (April – June and October – December) with an average annual of 1441 mm.

Logging and grazing are taking place in the southern part of the forest which is outside the National Park. A village is being formed on the northern edge of the forest. There are also old abandoned farms around the forest suggesting that human interference is not only recent. Barley cultivation and honey gathering are the two major occupations of the people in the forest.

Little has been published about the vegetation of this area. Friis (in press) has given a floristic characterisation of a subjectively estimated zonation on the southern slope of the Bale Mountains, including the Harena Forest.

The aims of the present work are: (1) To quantitatively characterise this remnant vegetation and to find the relationships between the vegetation and some environmental factors; (2) To suggest management plans and stimulate protective action for this unique national heritage.

In this work, a method of analysis based on probability was applied to distinguish and define vegetation types (Goodall 1964, 1966, 1968). The method has recently been refined for analysing hierarchical classifications of phytosociological data previously defined using the Braun-Blanquet (1964) approach (Goodall *et al.* 1987; Goodall & Feoli 1988). A relatively new method of ordination, environmental fuzzy set analysis (EFSA) (Zimmerman 1986; Roberts 1986; Feoli & Zuccarello 1986) has been applied to elucidate the relationships between vegetation and environmental factors.

Data and data analysis

Vegetation data

There are several methods of sampling vegetation and ecological systems (e.g. Mueller-Dombois & Ellenberg 1974; Orlóci 1978). Typification and classification of vegetation is usually based on preferential sampling according to the Braun-Blanquet approach (Orlóci 1978; Westhoff & van der Maarel 1978). In the present work systematic sampling (Orlóci 1978) along the complex altitudinal gradient was performed with no intention of delimiting preconceived vegetation types.

A total of 150 quadrat relevés (400 m^2 each) was sampled at regular intervals along a transect of 75 km in an altitudinal range of 1470 to 3300. 200 species of vascular plants were recorded, 89 of which were trees and shrubs and 111 herbs including lianas and epiphytes. Trees and shrubs were counted in each relevé while cover and abundance of the herbaceous species were estimated following the Braun-Blanquet (1964) scale transformed into the 1 – 9 ordinal scale according to van der Maarel (1979). The data were arranged in two matrices corresponding to the tree-shrub and the herb layer.

Environmental data

Altitude, slope and aspect of each relevé were recorded. Aspect was codified according to LoBue (pers. comm.) such as N=0, E=2, S=4, W=2.5, which represents a scale of total solar energy. Soil samples were collected in 40 quadrats evenly distributed along the gradient. Chemical and physical analyses of soil were done for depths of 1 to 10 cm (top layer) and 40 to 50 cm (lower layer); pH, organic matter, sand and clay, and basic cations (K, Na, Ca, Mg), were determined by standard methods (Jou 1978).

Data analysis and definition of vegetation types

The classification of quadrats was based on the probabilistic similarity index (PROSIM) of Goodall (1964, 1966) modified by Goodall & Feoli (1988) in order to combine binary and quantitative values of species occurrence in the same data set. The program PROSIM considers relevés (OTU's) two by two, and for each pair computes the probability that the two values of each attribute would be as alike as they are, if the two OTU's simply constitute a random sample of attribute values from the homogeneous set. The probabilities for the various attributes are combined, and the overall probability that each pair would be as alike as they are by random assortment of attribute values is obtained. Average linkage clustering (Lagonegro & Feoli 1984; Wildi & Orlóci 1983) was applied to the probabilistic resemblance matrices.

The affinity between each individual and the clusters was determined using the program AFFIN (Goodall 1968; Goodall *et al.* 1987; Goodall & Feoli 1988). AFFIN computes the probability that a particular OTU could have attribute values as similar as those of a defined cluster of OTU's if its attribute values were simply a random assortment of those in the data set.

After the probabilistic definition of the clusters, the hierarchical nested model of divergence (Feoli *et al.* 1984) was applied to identify the discriminating species. Finally, the classifications for the two vegetation strata were compared by means of a contingency table.

Gradient analysis

The relevés were ordered according to the eigenvectors of the similarity matrices of PROSIM. Since these matrices are non-centered, the second and third eigenvectors were used for analysing the ordination pattern (Feoli 1977) by the ellipses of equal concentration (Lagonegro & Feoli 1985). Autocorrelation analysis (Ganis 1985; Feoli & Ganis 1986) was applied to test the correlation between environmental factors and the vegetation represented by the eigenvectors of the similarity matrices of PROSIM. Ordination of relevés with respect to environmental factors was done by PCA and by EFS.

PCA was applied to the 40 relevés which had data on physical and chemical properties of soil samples and to the EFS. Biplot scattergrams (Gabriel 1971; ter Braak 1983) of the relevés and environmental data were obtained by superimposing the PCA projections of the respective data sets. Fuzzy sets were defined for environmental factors following Feoli & Zuccarello (in prep.). The steps include:

a) Fuzzy sets are defined on the basis of vegetation data. They constitute matrix F with k rows (fuzzy sets) and n columns (relevés).
b) Mean values for each environmental factor are computed for each cluster. They constitute a matrix E with p-rows (environmental factors) and k columns (fuzzy sets).
c) Matrix E, normalised by rows is multiplied by matrix F, normalised by columns. Each value in the new matrix G (with p rows and n columns) is a cosine, i.e. a measure of similarity.˙ The column vectors of matrix G represent the EFS. The EFS are axes which can be used for a direct ordination of the relevés with respect to each environmental factor. These axes retain the information summarised by the classification of the vegetation data. The ellipses of equal concentration were used to show and test the degree of fuzziness of each type for the EFS.

Results and discussion

Vegetation types

Six clusters appeared at a high level of the hierarchy in the classification of both layers. These can be defined as stratocoena (Barkman 1978). The computation of affinity between each cluster and the relevés other than in the cluster by the probabilistic method reinforced the finding that the six clusters of both layers represented an optimal partitioning. A few relevés from some neighbouring clusters, however, had to be reallocated.

The centroids of the six clusters of stratocoena are shown in Tables 1 and 2. On the basis of the discriminating species we considered the 6 clusters representing the vegetation types of the two layers. Figure 1 shows the hierarchical relationships between clusters and the proportional exention of the types along the altitudinal gradient. The altitudinal ranges of the tree-shrub layer communities are

Table 1. Centroids of the stratocoena for the tree-shrub layer. Only the most important species characteristic of the respective stratocoena are listed (a full species list is available upon request). Values are 10×

Species	Stratocoenon					
	A	B	C	D	E	F
Teclea simplicifolia	57	23	3			
Syzygium guineense var.	32					
Combretum aculeatum	30					
Scutia myrtina	30					
Euclea schimperi	15					
Combretum collinum	12					
Rhus natalensis	12					
Combretum molle	7					
Dicrostachys cinerea	3					
Coffea arabica	28	591	9			
Diospyros abyssinica		35				
Strychnos mitis		20				
Celtis gomphophylla		6				
Macaranga capensis		6	6			
Margaritaria discoidea		5				
Suregada procera		5				
Mimusops kummel		4				
Oxyanthus speciosus		24	37	29	33	
Podocarpus gracilior	10	15	15			
Turraea holstii		4	15	2		
Apodytes dimidiata		1	10			
Prunus africana		1	9	10	1	
Adhatoda schimperi			8			
Alangium chinense			3			
Linociera giordani			3	1		
Lepidotrichilia volkensii		2	18	110	117	
Teclea nobilis		9	17	82	4	
Vepris daineillii	12	2	18	35		
Dracaena afromontana				33	15	
Aningera adolfi-friderici			4	8		
Erythrina brucei				8		
Arundinaria alpina				5	159	5
Hagenia abyssinica				4	31	18
Ilex mitis					25	
Nuxia congesta					9	
Erica arborea						89
Discopodium penninervium						68
Schefflera volkensii					18	44
Rapanea simensis					23	36

1470–1510 m (A), 1510–1600 m (B), 1600–1800 m (C), 1800–2350 m (D), 2350–2800 m (E), and 2800–3300 m (F). A and B lie outside the present southern border of the National Park. In the tree-shrub layer two major points of discontinuity can be distinguished, one between types B and C at an alti-

Table 2. Centroids of the stratocoena for the herb layer. Only the most important species characteristic of the respective stratocoena are listed (a full list is available upon request). Values are 10×

Species	Stratocoenon					
	a	b	c	d	e	f
Panicum monticolum	9					
Desmodium repandum	7	4				
Jasminum floribundum	6	2				
Geophila repens	4					
Tectaria gemmifera	4					
Caesalpinia volkensii	3	1				
Pentas lanceolata		13				
Isoglossa punctata	5	8	2			
Didymodoxa caffra	1	7				
Peperomia retusa		5	2			
Scadoxus nutans		3				
Acanthus eminens	2	5	8	9	2	3
Mimulopsis solmsii			7			
Bassella alba				10	3	
Galium spurium			1	4		
Anthoxanthum aethiopicum				2	14	
Impatiens ethiopica				7	14	5
Canarina eminii					13	6
Girardinia diversifolia					12	
Selaginella kraussiana			2	3	11	1
Asystacia gangetica	6	7			10	
Asplenium protensum			2		9	5
Cerastium afromontanum			5	1	9	
Uebelina kiwuensis					9	5
Solanum benderianum					8	
Asplenium aethiopicum	1		2	2	7	1
Lycopodium dacrydioides					3	
Helichrysum citrispinum						13
Lobelia scebelii				5	8	13
Agrocharis melanantha						12
Kniphofia foliosa					3	9
Salvia merjamie						9
Pseudocarum eminii				2		9
Pimpinella oreophila				1		9
Hebenestreitia dentata						8
Umbellicus botryoides					3	9
Trifolium semipilosum						8
Satureja simensis						7
Senecio ochrocarpoides					2	7
Zehneria scabra					2	7
Scabiosa columbaria						5
Galium thunbergianum						5
Lotus discolor						5
Alchemilla haumanii						4
Galium simense						4
Trifolium burchellianum						4
Cardamine obliqua						3

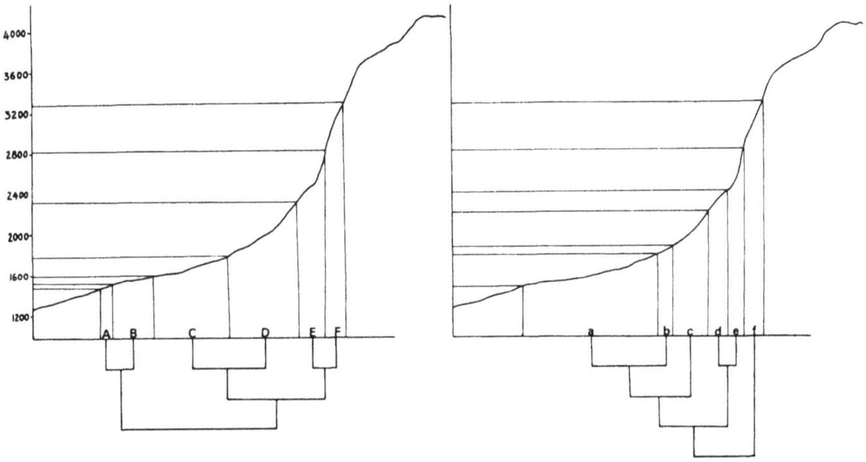

Fig. 1. Dendrograms of the hierarchical relationships between the tree-shrub-layer stratocoena (A–F) and the herb-layer stratocoena (a–f).

tude of 1600 m and the other between types D and E at 2350 m. Factors such as temperature and precipitation that have not been measured may be responsible for these discontinuities. The first discontinuity marks the transition between the thermophilous and mesophilous forest and the second one marks the transition between the mesophilous and the afroalpine vegetation zone, where shrubs are dominating and the herb cover becomes greater. These discontinuities agree with the major discontinuities found by Friis (in press).

The altitudinal ranges of the herb communities are 1470–1820 m (a), 1820–1900 m (b), 1900–2200 m (c) 2200–2400 m (d), 2400–2800 m (e), and 2800–3300 m(f). The variation between the types in this layer is more gradual. There is only one point of discontinuity between the high-altitude herb-layer type (occurring under open tree canopy) and the low-altitude ones, which occur under the more closed forest canopy.

Table 3. Contingency table comparing the classification of the vegetation in the tree-shrub and herb layers. The capital and lowcase letters refer to the stratocoena of the respective layers.

	a	b	c	d	e	f
A	6	-	-	-	-	-
B	32	-	-	-	-	-
C	15	17	-	-	-	-
D	-	6	17	13	-	-
E	-	-	-	4	12	-
F	-	-	-	-	2	26

The stratocoena can be combined into 11 vegetation types, such as Aa, Ba, Ca, Cb, Db, Dc, Dd, Ed, Ee, Fe, and Ff (Table 3). The relevés in the clusters of the two layers do not show a perfect matching, indicating that one herbaceous layer type can occur under different tree-shrub layer types. The combinations of the stratocoena types, however, are relatively few (only 11 of 36 possible). The degree of matching between the tree-shrub layer and the herb layer increases with increasing altitude.

Gradient analysis

The ellipses of equal concentration (Figs. 2 & 3) show that the centroids are significantly different, which supports the assumption that an effective definition of types has been achieved. The ordination scattergrams show that altitude is the overriding environmental factor. This can be reflected by a line joining the centers of the types. No other environmental factor changes regularly along the line connecting the centroids of the types.

The biplot scattergram of PCA of environmental data of the 40 relevés is presented in Fig. 4. The 11 vegetation types are highly dispersed in the biplot

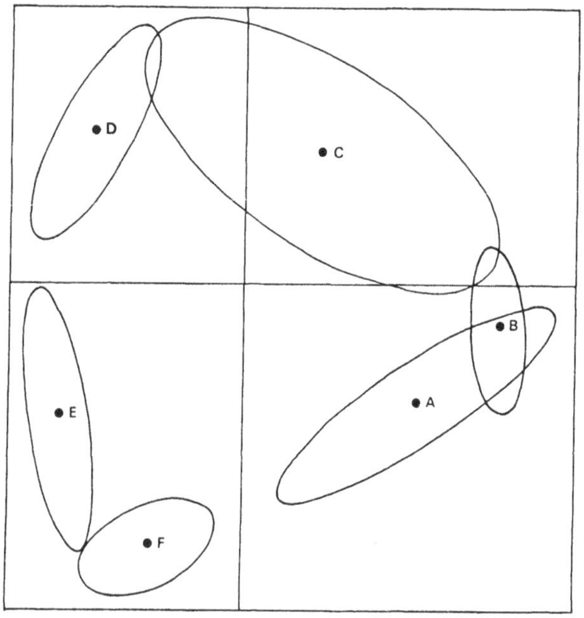

Fig. 2. Ordination of relevés of the tree-shrub layer based on the second and third eigenvectors of PROSIM. The ellipses of equal concentration are drawn around the centroids of the vegetation types at the 5% probability level.

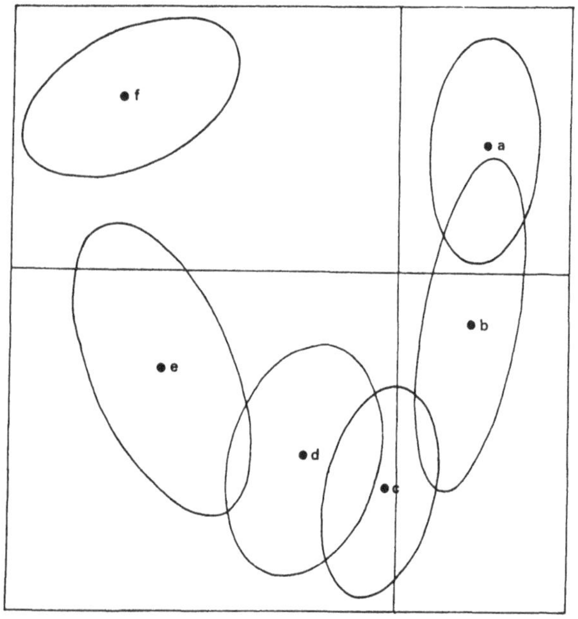

Fig. 3. Ordination of relevés of the herb layer based on the second and third eigenvectors of PROSIM. The ellipses of equal concentration are drawn around the centroids of the vegetation types at the 5% probability level.

scattergram suggesting that many of the factors considered are involved in controlling the variation. This suggestion is supported by the relatively low cumulative percentage (Table 4) of variance accounted for by the first three eigenvalues. There are positive correlations of pH, Ca and Mg of the top horizon and clay of the lower horizon with the first component, while altitude, slope and, to a lesser degree, organic matter are negatively correlated (Table 4). The autocorrelation analysis shows (Table 5) that not all the environmental factors are correlated with the variation in vegetation ($Z=1.92$ with 0.05 probability).

The ellipses of equal concentration based on PCA$^-$ EFS are presented in Fig. 5. Both ordinations show that the sequence of relevés along the lines joining the centroids of the types follows the altitudinal gradient. The dispersion of the types and the intersections between the ellipses in the herb layer show that the fuzziness is higher in this layer than in the tree-shrub layer. This may suggest that many of the factors considered have their share in influencing the variation in the herb layer (Table 5). The efficiency of the ordination based on EFS is higher than that based on environmental data alone (Table 6),

Table 4. Correlation coefficients between environmental factors and the first three principal components.

No.	Factor		1st	2nd	3rd
1	pH	10 cm	0.91	0.26	0.01
2		50 cm	0.61	0.36	0.28
3	OM	10 cm	−0.54	0.51	0.03
4		50 cm	−0.69	0.44	0.21
5	Sand	10 cm	0.01	0.73	−0.40
6		50 cm	−0.58	0.57	−0.25
7	Clay	10 cm	0.68	−0.59	−0.10
8		50 cm	0.86	−0.46	0.03
9	K	10 cm	0.42	0.51	0.49
10		50 cm	0.01	0.20	0.56
11	Na	10 cm	0.58	0.20	0.33
12		50 cm	0.23	0.39	−0.30
13	Ca	10 cm	0.87	0.29	−0.07
14		50 cm	0.54	0.33	0.39
15	Mg	10 cm	0.79	0.35	−0.07
16		50 cm	0.54	0.37	0.35
17	Slope		−0.78	−0.12	0.19
18	Aspect		0.09	0.45	−0.63
19	Altitude		−0.86	0.29	0.18
	Cum.		39%	56%	66%

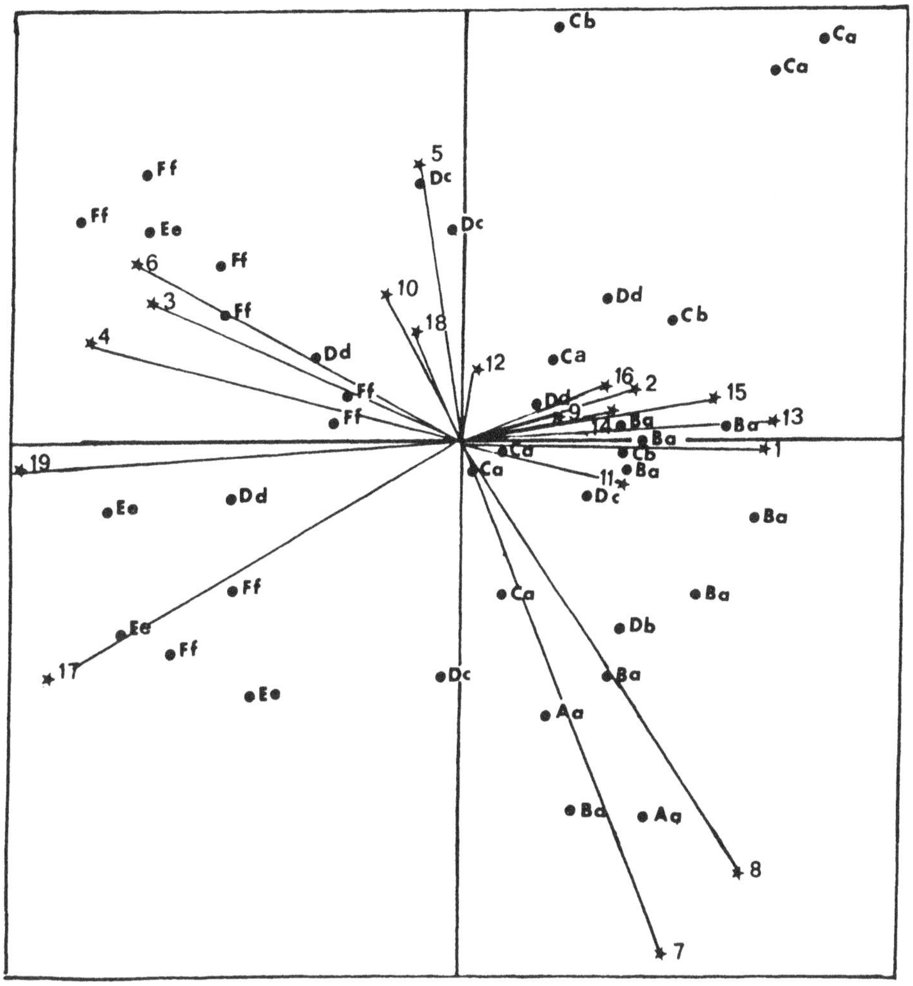

Fig. 4. Biplot scattergram of the relevés and environmental data. The numbers denoting the centroids (1–19) refer to the environmental data (see Table 6).

both in terms of the cumulative percentage of eigenvalues and correlation of EFS with the first three components. This high efficiency is the outcome of incorporating the information on the classification of the vegetation data with EFS which constrained the environmental data.

Conclusions and recommendations

(1)Six stratocoena were identified for both the tree-shrub and herb layers in Harena Forest by probabilistic methods. Combinations of the six types for each of the two layers resulted in 11 distinct vegetation types; the herb stratocoenon types can be used to distinguish particular ecological conditions within the tree-shrub stratocoena.

(2)The vegetation types are ordered along the altitudinal gradient which is also positively correlated with other environmental factors such as organic matter and slope and negatively correlated with pH, clay and calcium.

(3)Altitude is the most important factor in the tree-shrub layer while the herb layer is controlled by several other environmental factors with relatively equal magnitude.

(4)At a time when the need for more land and wood for fuel and construction is acute, an under-

Table 5. *Z* values of the autocorrelation analysis of the environmental factors based on the second, third, and fourth eigenvectors of PROSIM matrix for the tree-shrub and herb layers.

			Tree-shrub	Herb
1	pH	10 cm	1.39	5.31
2		50 cm	0.29	2.18
3	OM	10 cm	2.83	2.45
4		50 cm	1.93	4.49
5	Sand	10 cm	1.41	4.53
6		50 cm	3.54	4.50
7	Clay	10 cm	3.65	5.90
8		50 cm	4.03	7.09
9	K	10 cm	0.47	0.98
10		50 cm	0.12	0.87
11	Na	10 cm	1.32	0.63
12		50 cm	−2.60	0.84
13	Ca	10 cm	1.21	4.27
14		50 cm	−0.31	1.45
15	Mg	10 cm	1.54	3.83
16		50 cm	0.35	1.98
17	Slope		0.47	4.19
18	Aspect		−0.09	0.89
19	Altitude		4.49	8.53

standing of the conflicting desire for sustainable utilization and exploitation and their resolution deserves a prominent place in the agenda of environmental protection. We hope that this work will stimulate further research and also contribute to a

Table 6. Correlation coefficients of EFS with the first three principal components.

	Tree-shrub			Herb		
1	0.98	0.04	0.18	−0.93	0.27	0.19
2	0.77	−0.15	−0.58	−0.36	0.53	0.64
3	−0.96	−0.12	−0.22	0.98	0.13	0.08
4	−0.99	−0.01	−0.06	0.99	−0.05	0.02
5	−0.76	−0.37	0.24	0.30	0.83	−0.36
6	−0.96	0.05	0.12	0.97	0.00	0.17
7	0.94	0.30	−0.09	−0.95	0.23	0.19
8	0.98	0.12	−0.13	−0.97	0.14	0.18
9	−0.03	−0.90	0.42	−0.55	−0.64	−0.26
10	−0.80	−0.46	0.26	0.93	−0.27	0.10
11	0.92	0.29	0.11	−0.95	0.24	0.00
12	0.05	0.78	0.58	−0.28	0.62	−0.69
13	0.97	−0.13	0.09	−0.99	−0.10	−0.03
14	0.86	−0.45	0.21	−0.84	−0.50	−0.14
15	0.96	−0.13	0.23	−0.99	0.00	−0.09
16	0.79	−0.51	0.25	−0.67	−0.62	−0.23
17	−0.97	0.14	−0.15	0.99	0.02	0.07
18	0.00	0.84	0.40	−0.30	0.90	−0.11
19	−0.97	0.18	−0.10	0.99	0.09	−0.04
Cum.	70%	88%	96%	0.69%	0.87%	0.95%

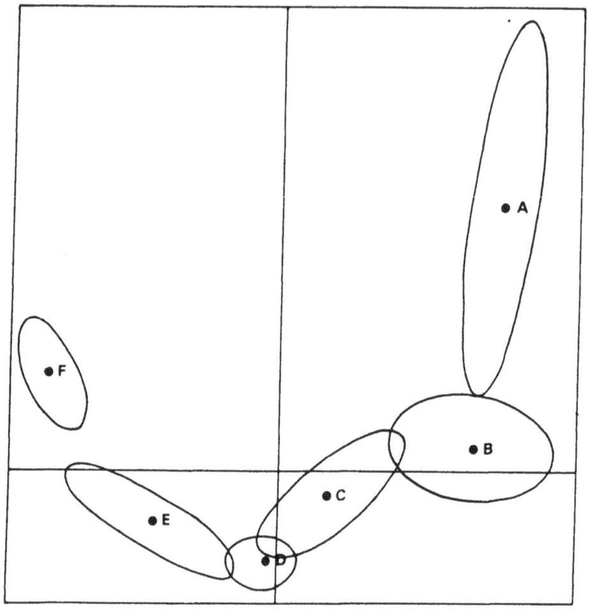

A

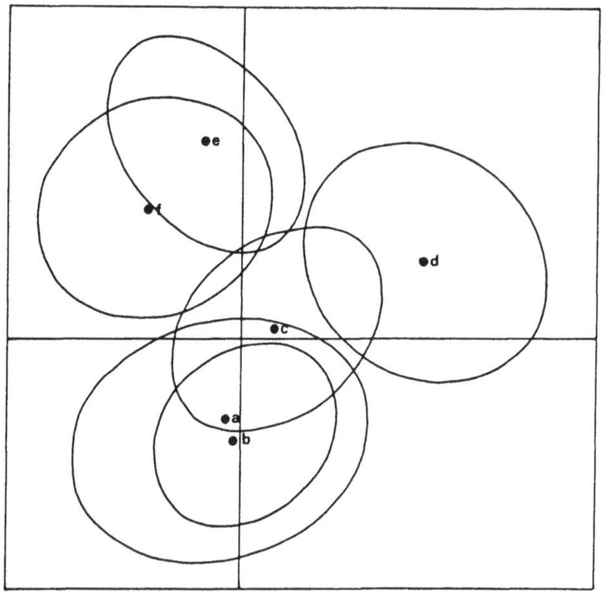

B

Fig. 5 A. Ordination of relevés of the herb layer based on the first and second components of PCA EFS. The ellipses of equal concentration are drawn around the centroids of the vegetation types at the 5% probability level. B. Ordination of relevés of the herb layer based on the first and second components of PCA EFS. The ellipses of equal concentration are drawn around the centroids of the vegetation types at the 5% probability level.

better understanding of the forest and help conserve the yet unexplored gene pools of several useful plants in the forest (e.g. *Coffea arabica, Aningeria adolfi-fredericii, Podocarpus gracilior*). The continually improving access to the forest has increased the risk of its eventual destruction as has happened to the forests on the central plateau in the distant past. We would like to bring to the attention of the authorities that activities such as logging, grazing and cultivation are incompatible with protecting all parts of the forest. Type A and B lie south of the 1680 m a.s.l. border of the National Park where logging is being intensified to feed the saw mill nearby. The whole forest should therefore be included in the National Park for adequate protection of the different vegetation types as well as the fauna in the forest.

Acknowledgements

The first author is thankful to the Third World Academy of Science (TWAS) for a fellowship grant in 1987 which made his stay and work in the Department of Biology of the University of Trieste possible. All expenses incurred in data collection and soil analysis were covered by the Swedish Agency for Research Cooperation with Developing Countries (SAREC).

The authors are grateful to Prof. Eddy van der Maarel, Prof. D. W. Goodall and Dr. Laco Mucina for their critical reading of the manuscript, to Dr. Paula Ganis for her invaluable assistance in the computations and Dr. I. Friis for his comments on the species list. Mrs. Dagmar Mucina kindly retyped the text.

References

Barkman, J. J. 1978. Synusial approaches to classification. In: Whittaker, R. H. (ed.), Classification of plant communities, pp. 113–165. Dr. W. Junk, The Hague.

Braun-Blanquet, J. 1964. Pflanzensoziologie. Grundzüge der Vegetationskunde. 3. Aufl. Springer-Verlag, Wien.

Daniel Gamachu 1977. Aspects of climate and water budget in Ethiopia. Addis Ababa Univ. Press, Addis Ababa.

Feoli, E. 1977. On the resolving power of principal components analysis in plant community ordination. Vegetatio 33: 119–125.

Feoli, E. & Ganis, P. 1986. Autocorrelation for measuring predictivity in community ecology: An example with structural and chorological data from mixed forest in NE Italy. Coenoses 1: 53–56.

Feoli, E., Lagonegro, M. & Orlóci, L. 1984. Information analysis of vegetation data. Dr. W. Junk, The Hague.

Feoli, E. & Zuccarello, V. 1986. Ordination based on classification: Yet another solution? Abstr. Bot. (Budapest) 10: 203–219.

Friis, I. (in press). Zonation of the vegetation on the south slope of the Bale Mountains. Proc. Sect. Int. Symp. Flora Ethiopia.

Gabriel, K. R. 1971. The biplot graphic display of matrices with application to principal components analysis. Biometrika 58: 453–467.

Ganis, P. 1985. Spatial autocorrelation analysis, a computer program and examples of application with different data from grassland vegetation under natural reforestation process in the karst near Trieste. Stud. Geobot. 5: 127–142.

Goodall, D. W. 1964. A probabilistic similarity index. Nature 203: 1098.

Goodall, D. W. 1966. A new similarity index based on probability. Biometrics 22: 883–907.

Goodall, D. W. 1968. Affinity between an individual and a cluster in numerical taxonomy. Biometr. Praximetr. 9: 52–55.

Goodall, D. W. & Feoli, E. 1988. Application of probabilistic methods in the analysis of phytosociological data. Coenoses 3: 1–10.

Goodall, D. W., Ganis, P. & Feoli, E. 1987. Probabilistic methods in classification: A manual for seven computer programs. Gruppo Elabor. Automat. Dati – Ecol. Quant., Trieste 7: 1–50.

Jou, A. S. R. (ed.) 1978. Selected methods of soil analysis. Int. Inst. Trop. Agricult., Ibadan, Nigeria.

Lagonegro, M. & Feoli, E. 1984. THREE-Package for classification and ordination of multivariate data. Libreria Goliardica, Trieste.

Lagonegro, M. & Feoli, E. 1985. The use of ellipses of equal concentration to analyse ordination vegetation patterns. Stud. Geobot. 5: 143–165.

Mueller-Dombois, D. & Ellenberg, H. 1974. Aims and methods of vegetation ecology. J. Wiley & Sons, N.Y.

Murphy, H. P. 1958. The fertility status of some soils in Ethiopia. College of Agriculture, Jimma, Ethiopia.

Orlóci, L. 1978. Multivariate analysis in vegetation research. 2nd ed. Dr. W. Junk, The Hague.

Roberts, D. W. 1986. Ordination on basis of fuzzy set theory. Vegetatio 66: 123–131.

ter Braak, C. J. F. 1983. Principal components biplots and alpha and beta diversity. Ecology 64: 454–462.

van der Maarel, E. 1979. Transformation of cover-abundance values in phytosociology and its effects on community similarity. Vegetatio 39: 97–114.

Walter, H. & Lieth, H. 1960. Klimadiagramm-Weltatlas. Gustav Fischer, Stuttgart.

Westhoff, V. & van der Maarel, E. 1978. The Braun-Blanquet approach. In: Whittaker, R. H. (ed.), Classification of vegetation, pp. 617–726. Dr. W. Junk, The Hague.

Weinert, E. & Masurek, A. 1984. Notes on the vegetation and soils of Bale province of Ethiopia. Feddes. Repert. 95: 373–380.
Wildi, O. & Orlóci, L. 1983. Management and multivariate analysis of vegetation data. 2nd ed. Eidgenössische Anstalt für das forstliche Versuchswesen, Birmensdorf (ZH), Switzerland.
Zimmerman, H. J. 1986. Fuzzy sets. The theory and its application. Kluwer, Boston.

Vegetatio **81**: 199–208, 1989.

Numerical phytosociology of the subalpine belt of the Kazbegi region, Caucasus, USSR

P. Pyšek[1] & M. Šrůtek[2]
[1] *Institute of Applied Ecology and Ecotechnology, University of Agriculture, CS-281 63 Kostelec nad Černými lesy, Czechoslovakia* [2] *Department of Plant Ecology, Institute of Botany, Czechoslovak Academy of Science, Dukelská 145, CS-379 82 Třeboň, Czechoslovakia*

Accepted 30.1.1989

Keywords: Alpine vegetation, Glacial valley, Grazing, Numerical classification, Syntaxonomy, Transect sampling

Abstract

A vegetational transect across a glacial valley was investigated. Structure and species composition, as well as influence of disturbance by biotic factors (grazing of sheep and cattle) and abiotic factors (topography, wind, water) are discussed.

According to species composition and community structure, 4 main vegetation types were distinguished by comparing field data from phytosociological relevés with results of numerical treatments (3 methods of numerical classification and DCA ordination): 1) *Chamaesciadium acaule-Polygonum viviparum*, 2) *Festuca cf. woronowii-Carex tristis*, 3) *Veratrum lobelianum-Taraxacum stevenii* and 4) *Rhododendron caucasicum-Vaccinium myrtillus* communities.

Species diversity (evenness, and number of species) was low in the ungrazed dwarf-shrub alpine communities, which were dominated by *Rhododendron caucasicum*, in comparison with the grazed meadow. The highest values of these characteristics were found in the transitions among the types.

Nomenclature: Cherepanov, S.K. 1981. Plantae Vasculares USSR, Nauka, Leningrad

Introduction

Although the vegetation of the Caucasus Mountains has been extensively treated in the Russian literature, detailed phytosociological investigations are less frequent (e.g. Bedoshvili 1985a, b; Korotkov 1986, Rabotnov 1987).

Our study had the following objectives:

1. To give a detailed description of the vegetation along a representative transect across a glacial valley situated in the subalpine belt of the Kazbegi region.
2. To identify the main types of vegetation of a selected locality by subjecting our field data to numerical treatments.
3. To identify the pattern of variation along the transect, and to try to explain the factors accounting for the variability.

The vegetation in the study area is strongly affected by long-term and extensive grazing, as in most high-mountain grasslands (e.g., Dolukhanov 1978; Plesník 1978; Seibert 1983; Tukel 1984). Comparable grazing effects are reported from other climatic and geographical zones as well (e.g. Looman, 1983).

Rangelands occur extensively in the Caucasus and increase in extent from North to South and from West to East gradients, along with increasing continentality (Dolukhanov 1978). The most favourable grazing areas are subalpine grasslands and man-made meadows created by cutting and burning of *Rhododendron* stands (Gadziev *et al.* 1986).

Grazing comprises both biological factors (destruction of biomass, fertilization by faeces) and physical ones (trampling, digging). Thus it is an important disturbance factor affecting the composition and structure of plant communities (Sousa 1984; Collins & Barber 1985; Pickett & White 1985).

Much attention has been focussed on the hypothesis that species diversity in grasslands will be maximal at intermediate levels of disturbance (e.g., Grime 1979; Collins & Barber 1985). Grazing appears to increase diversity in some cases, and to reduce it in others, depending on the intensity of grazing as well as on the relationship between the palatability and competitive ability of plants (Harper 1969; Dirzo 1985).

This problem is also touched upon in this paper.

Study area

The study area is situated at the foot of Mount Kazbek (5047 m a.s.l.; 42°48′ N and 44°39′ E) at the eastern border of the Central Caucasus. The Central Caucasus is a part of the Great Caucasus which is 1500 km long and covers an area of 145 000 km^2. The volcanic activity of Mt Kazbek determined the geological features of the area in the Late Pliocene and Early Pleistocene. The surface is built up of andezite-dacite lava sheets covering primary shale-sandstones (Ketskhoveli *et al.* 1975). Erosion by rivers reshaped the glacial valleys into deep canyons whose precipitous slopes are 30-40 m high (e.g. the Chkheri River valley, adjacent to the locality under investigation). The glacial zone surrounding Mt Kazbek covers an area of 81 km^2. The largest glacier (the Gergeti) is located above the study valley. The area is drained to the Caspian Sea by the Terek River.

The climate is influenced by atmospheric circulation from the Middle Asian deserts and generally dry, with most of the precipitation during the summer. (Gulisashvili *et al.* 1975). The area is characterized by the climate diagram in Fig. 1 (Walter 1984, see also Zanina 1961).

In the Kazbegi region the following altitudinal zonation of vegetation can be distinguished (Gulisashvili *et al.* 1975; Ketskhoveli *et al.* 1975):

– 1700–2200 m: montane forest dominated by *Betula litwinowii* and *Pinus hamata*. Generally, the timberline in the Caucasus lies between 2200 and 2750 m (Dolukhanov 1978).

– 2200–2500 m: subalpine belt characterized by *Rhododendron* and *Betula* krummholz, subalpine herb communities and pastures.

– 2500–3000 m: alpine belt with grasslands and pastures.

– 3000–3900 m: subnival and nival belts with sparse tussocks and patches of vascular plants and scattered mosses and lichens.

The locality studied is a glacial valley situated along a path leading from Kazbegi village to the peak. It has a characteristic shape (Fig. 2), and ranges between 2370 and 2500 m.

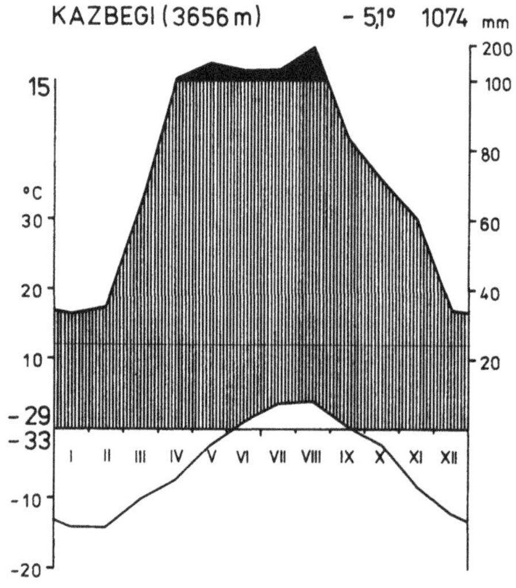

Fig. 1. Climate diagram (after Walter 1984) of the Mount Kazbek station.

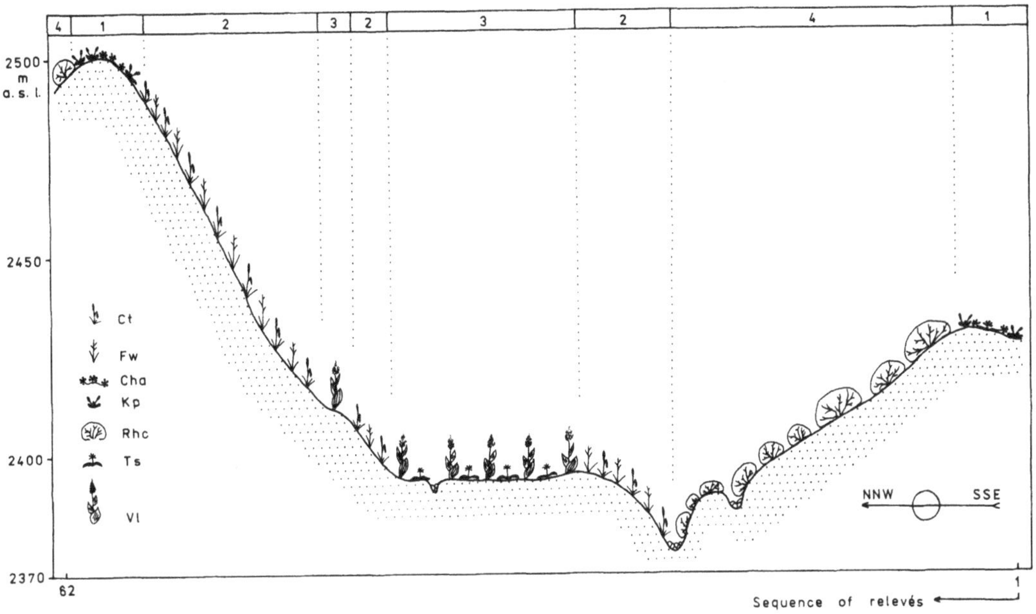

Fig. 2. The vegetational transect across the glacial valley; 1 – *Chamaesciadium acaule-Polygonum viviparum*-community, 2 – *Festuca cf. woronowii-Carex tristis*-community, 3 – *Veratrum lobelianum-Taraxacum stevenii*-community, 4 – *Rhododendron caucasicum-Vaccinium myrtillus*-community (symbols: Ct – *Carex tristis*, Fw – *Festuca cf. woronowii*, Cha – *Chamaesciadium acaule*, Kp – *Kobresia persica*, Rhc, – *Rhododendron caucasicum*, Ts – *Taraxacum stevenii*, Vl – *Veratrum lobelianum*).

The distribution of plant communities on the site is related to slope orientation and inclination: NNW slopes (15-25°) are occupied by dense *Rhododendron* stands, while the opposite flanks (SSE, 20-40°) as well as the valley bottom are dominated mostly by grasses. Marginal ridges of the valley are exposed to strong wind and water erosion. At the valley bottom, there are several periodically dry stream beds that contribute to the habitat diversity. Grazing and migration of herds of livestock upwards to the pastures and back to lower altitudes occurs regularly during the whole growing season.

Material and methods

The vegetation was analyzed along a 690 m line transect, running across the valley in NNW-SSE direction. In 25 m^2 sample plots, the cover values of species were estimated at 5% intervals; species with <5% cover were scored *r*, + or 1 Braun-Blanquet symbols and arbitrarily given the cover values 0.02, 0.1 and 2.5% (van der Maarel 1979). The distance between individual plots differed according to the structure of vegetation: 10 m in homogeneous grasslands of the SSE slope and 5 m on the NNW flank and at the valley bottom with more heterogeneous vegetation.

Numerical classification included (1) average linkage clustering, (2) complete linkage clustering, (3) TWINSPAN (Orlóci 1978; Hill 1979). Complete linkage was used for a detailed description and interpretation (cf. Mucina 1982). In addition, DCA, Detrended Correspondence Analysis (Hill & Gauch 1980) was applied.

Various transformations of the cover data were tried (cf. Orlóci 1978). For the DCA ordination, we used (a) original cover values, (b) ordinal transformed values (van der Maarel 1979), and (c) presence-absence.

Original cover values were used in the TWINSPAN procedure, and transformed values $x = \log(x + 1)$ (cf. Jensén 1978) were used for the methods of agglomerative hierarchical clustering. Standardized Euclidean distance (chord distance of Orlóci 1978) was used as resemblance measure.

According to Lepš (Kovář & Lepš 1986), for each group defined by complete linkage, the following 'synthetic' characteristics are indicated (see Table 1):

1. *s* – belonging to the characteristic species combination, including species with a constancy of at least 90%
2. *c* – characteristic species defined as follows: constancy at least 25% and *V* statistics higher than 0.4, where *V* is computed from the following 2 × 2 table:

		relevé belonging to the group	other
species	present	*a*	*b*
	absent	*c*	*d*

and

$$V = \frac{(ad - bc)}{\sqrt{(a + b)\cdot(a + c)\cdot(b + d)\cdot(c + d)}}$$

3. *d* – dominant species, with mean cover greater than 1
4. *D* – characteristic dominant, species with the common *t* statistics greater than 4 (where *t* indicates the mean importance value of the species in the group and outside the group). These statistics are not used as a test of significance, but rather for comparative purposes (Kovář & Lepš 1986).

It should be pointed out that the characteristics used by Kovář & Lepš are different from the terms 'characteristic species combination' and 'characteristic species' as defined in Braun-Blanquet syntaxonomy (Westhoff & van der Maarel 1978).

The species diversity *H* was computed using the Shannon formula (see, e.g. Peet 1974):

$$H = -\sum_{i=1}^{S} p_i \ln p_i \qquad (1)$$

where p_i is relative importance of the species *i*, and *S* is the number of species in the relevé.

Evenness *J′* was expressed according to Pielou's formula (Pielou 1966):

$$J' = \frac{H'}{\ln S} \qquad (2)$$

where *S* is the number of species in the relevé.

The phytosociological data are summarized in Table 1.

Results and discussion

Description of communities

According to species composition and community structure, four main vegetation types were distinguished in the field (Fig. 2). The distribution of these community types along the transect accords well with topography and disturbance intensity. Because of lack of comparative descriptions these communities cannot be described in a more definitive manner.

1. Marginal ridges of the valley harbour specific vegetation reflecting extreme effects of wind, water erosion, frost action (soil polygons), and intensive grazing. The *Chamaesciadium acaule-Polygonum viviparum*-community with *Alchemilla retinervis agg.*, *Festuca cf. woronowii* and *Kobresia presica* is found in these habitats. *Alchemilla caucasica, Androsace lehmanniana, Polygonum viviparum* and *Thalictrum alpinum* are considered differential species. The species composition of particular ridges is influenced by the steepness and size of the ridge as well as by contact communities. Evergreen shrubs (*Rhododendron caucasicum*, *Vaccinium vitis-idaea*, *Daphne glomerata*) occur on the marginal ridges close to stones, scattered on the ground. Presumably, these serve as a shelter which protects the seedlings from grazing (Livingstone quoted by Harper 1977) and provides a more favourable microclimate. On the opposite margins of the valley the stands have some symmorphological

Table 1. Mean cover (a) and constancy classes (b) of 40 important species in the 4 vegetation types distinguished along the transect. (Full table available on request.) Numbers of species correspond to those in the ordination diagram, see Fig. 5. Species important for the classification of the vegetation types, are marked with special symbols (c, d): *s* – species belonging to the characteristic species combination, *c* – characteristic species, *d* – dominant species, *D* – characteristic dominant. Three phytosociological relevés not belonging to any group (nrs 18, 19 and 26) have been excluded from the table.

Community	(1) *Chamaesciadium -Polygonum*				(2) *Festuca -Carex*				(3) *Veratrum -Taraxacum*				(4) *Rhododendron -Vaccinium*			
Relevé numbers	1–3, 55–61				21–25, 36–39 41–54				27–35, 40				4–17, 20, 62			
Mean cover (%)	91.5				95.2				92.0				84.4			
Average number of species	28.3				35.5				30.8				23.9			
	a	b	c	d	a	b	c	d	a	b	c	d	a	b	c	d
1 *Bromopsis variegata*	<1	V, s			2.2	V		D	<1	III			·	·		
2 *Campanula* cf. *saxifraga*	4.5	V, s		D	<1	III			<1	II			<1	I		
3 *Carex tristis*	5.2	V, s			8.2	V		d, D	·	·			<1	I		
4 *Chamaesciadium acaule*	25.0	V, s		d, D	1.3	IV			2.8	V, s			<1	IV		
5 *Festuca* cf. *woronowii*	19.0	V, s		d	30.1	V, s		d, D	1.5	III			<1	I		
6 *Kobresia persica*	17.8	V, s		d, D	5.3	V			<1	II			<1	I		
7 *Minuartia oreina*	<1	V, s			<1	V			<1	II			<1	I		
8 *Taraxacum* sp.	1.5	V, s			1.4	V			<1	V, s			<1	IV		
9 *Thalictrum alpinum*	1.1	V, s, c		D	<1	I			·	·			·	·		
10 *Polygonum viviparum*	1.3	III		D	·	·			<1	I			<1	II		
11 *Alchemilla retinervis agg.*	10.7	III		d	7.3	V, s			12.0	V, s		d	<1	V		
12 *Betonica grandiflora*	3.3	III			1.5	V, s, c			·	·			·	·		
13 *Campanula* cf. *stevenii*	<1	III			2.1	V, s		D	<1	V, s			<1	III		
14 *Cirsium obvallatum*	2.0	III			1.6	V, s			<1	IV			<1	IV		
15 *Euphrasia amblyodonta*	<1	II			<1	V, s, c			<1	III			·	·		
16 *Leontodon* cf. *crispus*	<1	I			1.7	V, s		D	<1	V, s			<1	I		
17 *Nardus stricta*	<1	I			16.7	V, s		d	29.5	V, s		d	6.7	V, s		
18 *Plantago caucasica*	3.3	IV			4.6	V, s			4.5	V, s			<1	I		
19 *Polygonum carneum*	<1	IV			1.2	V, s, c		D	<1	II			<1	I		
20 *Potentilla crantzii*	<1	III			2.2	V, s		D	<1	II			<1	I		
21 *Ranunculus* cf. *caucasicus*	1.0	IV			4.5	V, s		D	2.0	V			<1	III		
22 *Trifolium canescens*	1.0	III			5.0	V, s		D	1.0	III			<1	I		
23 *T. repens*	3.5	IV			5.2	V, s			19.5	V, s		D	<1	I		
24 *Sedum* cf. *rupestre*	<1	II			<1	IV	c		<1	I			·	·		
25 *Polygala alpicola*	<1	I			<1	III		D	·	·			·	·		
26 *Gnaphalium supinum*	<1	I			<1	II			<1	V, s, c		D	<1	II		
27 *Poa alpina*	<1	II			<1	III			<1	V, s			<1	V, s		
28 *Sibbaldia parviflora*	<1	I			<1	III			1.5	V, s			<1	III		
29 *Veratum lobelianum*	·	·			<1	I			22.3	V		d, D	<1	I		
30 *Phleum alpinum*	·	·			<1	I			<1	V		D	<1	IV		
31 *Taraxacum stevenii*	·	·			<1	I			8.0	V		D	·	·		
32 *Rhododendron caucasium*	1.0	I			·	·			·	·			67.2	V, s, c, d, D		
33 *Anthoxanthum alpinum*	<1	I			<1	IV			<1	III			<1	V, s		
34 *Vaccinium myrtillus*	<1	I			·	·			<1	I			11.7	V, s, c, d		
35 *Deschampsia flexuosa*	<1	I			<1	I			<1	I			<1	V	c	D
36 *Dolichorrhiza caucasica*	·	·			·	·			<1	I			<1	IV	c	
37 *Festuca* sp.	·	·			·	·			·	·			<1	IV	c	
38 *Geranium sylvaticum*	·	·			·	·			<1	I			<1	IV	c	D
39 *Oxalis acetosella*	·	·			·	·			·	·			1.9	V	c	D
40 *Briza marcowiczii*	·	·			·	·			<1	I			<1	III		D

features in common. Most striking is the prostrate character of the dominant species, which is also found in African montane grasslands (Belsky 1987, 1988).

The average values of species diversity (H'), evenness (J') and total number of species (S) are 2.94, 0.61 and 71, respectively.

2. Grassland communities prevail along the transect. They can be characterized by the following constant grassland species: *Bromopsis variegata, Campanula cf. stevenii, Festuca cf. woronowii, Kobresia persica, Minuartia oreina, Plantago caucasica, Potentilla crantzii, Trifolium canescens* and *T. repens* (Table 1).

The SSE slope is covered by the *Festuca cf. woronowii-Carex tristis*-community, with a cover of 90-100% and composed of rigid tussocks of *Festuca cf. woronowii* and *Nardus stricta*. These tough grasses are less palatable than broad-leaved forbs (Grant *et al.* 1985). Due to the precipitous slope (20°-40°), grazing by sheep is much more frequent than by cattle. As reported by Grant *et al.* (1985), the sheep diet consists of a greater portion of dicots and contains less *Nardus stricta* than the fodder of cattle. The distribution of *Nardus stricta* along the transect seems to confirm this statement. The rarer occurrence of the species at the valley bottom might be related to the more intensive grazing by cattle (Welch 1986).

The *Festuca-Carex*-community differs from other grassland types by the presence of *Betonica grandiflora, Polygala alpicola, Polygonum carneum* and *Primula algida*. The occurrence of this community, linked with south-facing slopes, was recorded also on the banks of seasonal streams in the valley bottom (Fig. 2).

The average values of H', J' and S are 3.19, 0.62 and 85, respectively.

3. *The Veratrum lobelianum-Taraxacum stevenii*-community prevails in the valley bottom. The upper layer, reaching a height of 1.5-2.0 m, is exclusively formed by *Veratrum lobelianum*. The development of this forb is probably due to the fact that it is not grazed by cattle. Broad ground-leaves of *Veratrum lobelianum* protect the nearby area from thorough grazing, which favours the occurrence of *Arenaria rotundifolia, Anthriscus ruprechtii* and *Gentiana aquatica*. Scattered tussocks of *Nardus stricta*, the other matrix species, build the lower layer with the open space occupied by species with typical horizontal spreading (cf. Gomez Sal *et al.* 1986; Belsky 1987, 1988), such as *Taraxacum stevenii* and *Trifolium repens*. The community differs from other grasslands by the presence of *Carex cf. melananthiformis, Phleum alpinum* and *Rumex alpestris*. On the SSE slope, *Veratrum lobelianum* always occurred at places with a horizontal ground surface. Its penetration was presumably limited to places with deeper soil, enriched with nutrients, necessary for the successful establishment of larger plants. Expansion of *Veratrum lobelianum* and spreading of *Rumex alpinus, Cirsium obvalatum* or *Urtica dioica* due to intensified human-induced changes of pastures, were reported by Kuliev *et al.* (1986).

The average values of H', J' and S are 2.69, 0.54 and 74, respectively.

4. The *Rhododendron caucasicum-Vaccinium myrtillus*-community is strongly dominated by the two species appearing in its name (Table 1). Stands occupying the NNW slope are differentiated by: *Dolichorrhiza caucasica, Empetrum caucasicum, Geranium sylvaticum, Oxalis acetosella, Solidago virgaurea* and others. These stands are homogeneous in structure and relatively poor in species. This is due to the high cover of the dominant shrubs.

The average values of H', J' and S are 1.50, 0.33 and 67, respectively.

Variation in species diversity along the transect

The H' index generally shows the highest values in the *Festuca-Carex*-community on the SSE slope, where the grazing pressure (mostly by sheep) is less heavy than on the valley bottom. Trampling of grass by resting cattle adds moreover to the disturbance of the latter habitat. In the *Veratrum-Taraxacum*-community, a low species number together with the low number of dominant species lead to a decreased species diversity.

Transition at sites between the valley bottom and a small ridge situated above the periodical stream bed are slightly richer than the surrounding sites (limes divergens environment, see van der Maarel 1976). The borders between the *Veratrum-Taraxacum* and *Festuca-Carex*-communities, which grow on the bank, show a local increase of species richness and of H' value. The same holds for marginal parts of the *Rhododendron* stand of the NNW slope. The topographic diversity pattern supports the conclusion of Collins & Barber (1985) that species diversity increases with increased habitat diversity. The species diversity on the marginal ridges with the *Chamaesciadium-Polygonum* community is higher than that of the *Rhododendron-Vaccinium*-community, but lower than that of the *Festuca-Carex*-community.

Numerical methods

Complete linkage gives results which correspond most closely with the situation observed in the field. As Ward's clustering method (Wishart 1978) it produces relatively homogeneous and well-defined clusters. This is why these methods are considered a good numerical approximation to Braun-Blanquet classification of plant communities as well (Westhoff & van der Maarel 1978; Mucina 1982). Four groups of relevés are formed (Fig. 3) which correspond to the vegetation types distinguished in the field: 1. *Chamaesciadium-Polygonum*, 2. *Festuca-Carex*, 3. *Veratrum-Taraxacum*, 4. *Rhododendron-Vaccinium*. The delimitation between the grassland clusters 1-3 and the shrubland cluster 4 (Fig. 3) reflects the differences between opposite ridges of the valley, as mentioned in the preceding section.

The results of TWINSPAN are close to those obtained by the average linkage method. Both methods did not delineate the four groups as clearly ad Complete linkage did.

The results of the DCA ordination are presented in Fig. 4 and 5. Regardless of the data transformation, the coenocline produced by the first axis reflects the general topography. The best

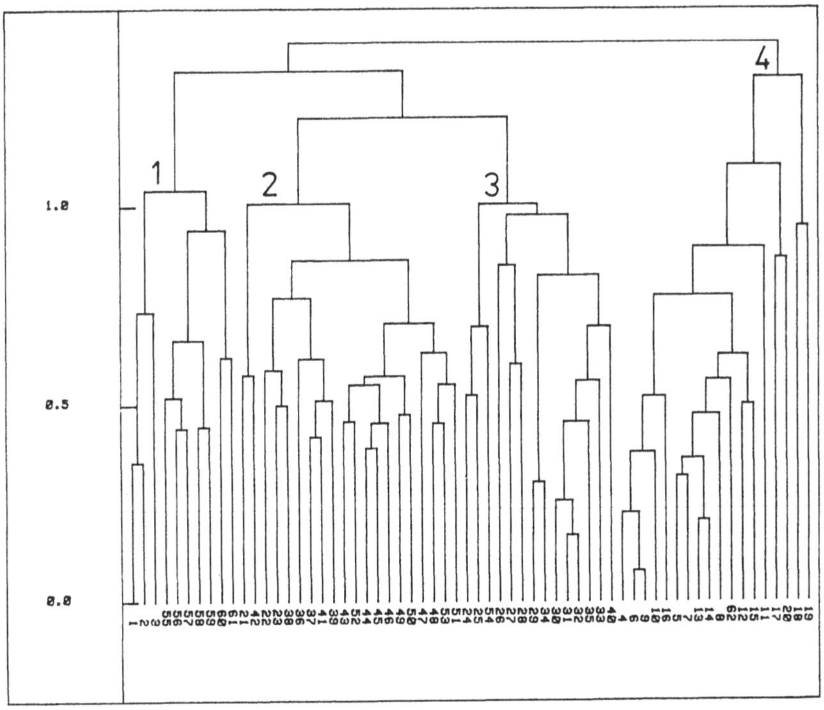

Fig. 3. Complete linkage clustering (transformed data) with chord distance as disimilarity measure. The numbers 1-4 correspond to the 4 vegetation types (see Fig. 2); the numbers below correspond to the phytosociological relevés 1 to 62.

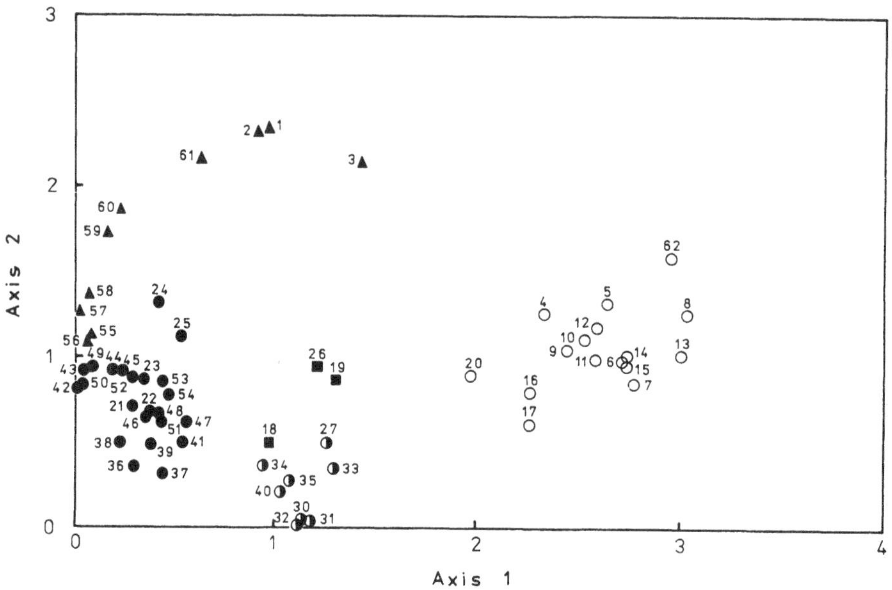

Fig. 4. DCA diagram of the 62 relevés (data subjected to ordinal transformation). Symbols: ▲ (1) *Chamaesciadium-Polygonum* ● (2) *Festuca-Carex* ◑ (3) *Veratrum-Taraxacum* ○ (4) *Rhododendron-Vaccinium* ■ transitional relevés.

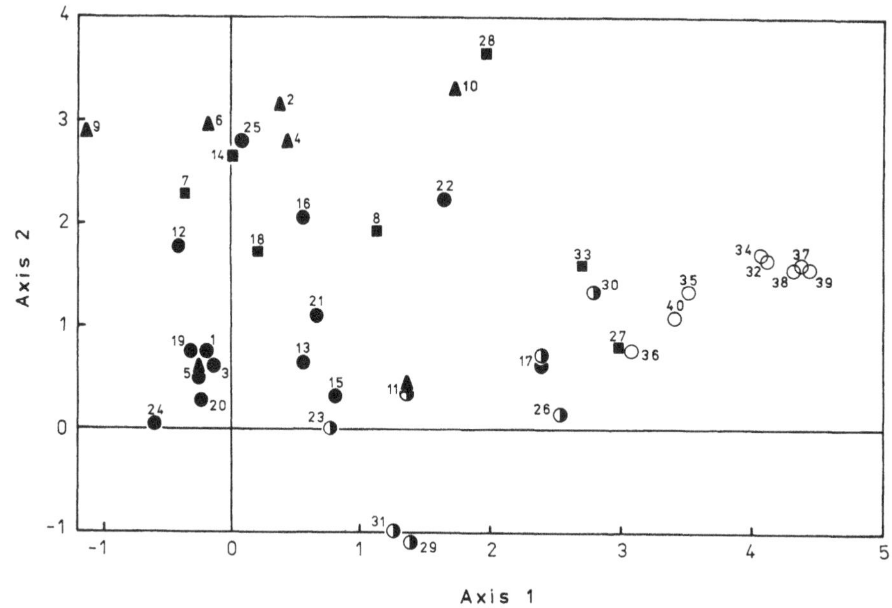

Fig. 5. DCA diagram of 40 selected species (See Table 1). (original data). Symbols indicate prevailing occurrence in vegetation types 1 to 4 (see Fig. 4).

results were obtained using transformed data. The results confirm the heterogeneity of community type 1 occupying especially the marginal ridges.

The DCA ordination of the species (Fig. 5) was based on original values; we have taken into account only the characteristic species (Kovář & Lepš 1986) of the communities 1 to 4 in the ordination display. This delimitation was based on the complete linkage clustering method. The computed characteristics are comparable with diagnostic values in the sense of the Braun-Blanquet approach (Westhoff & van der Maarel 1978). Especially the *c* species are suitable for the qualitative differentiation of particular communities, because they are comparable with the category of differential species as understood by the Central European phytosociologists. As in Fig. 4, the first axis of the species ordination (Fig. 5) reflects the main line of variation. The species of the SSE slope appear in the left part of the ordination space, while those of the NNW slope are concentrated in the right part.

The position of the species along axis 2 appears to correspond with altitude. This gradient becomes obvious when Fig. 5 is compared with Fig. 2.

Acknowledgements

Our thanks are due to T. Herben and J. Lepš for data processing, J. Holub, J. Chmelař, J. Chrtek, J. Kirschner, A. Plocek, F. Procházka, J. Soják and J. Štěpánek for identification of critical species; J. Jeník and K. Prach for critically reading the manuscript and Miss L. Klečková for the drawings and the tables.

References

Bedoshvili, D.O. 1985a. Groups of associated species in the plant communities of the alpine meadows in the Kazbegi region (Central Caucasus). Bot. Zh. 70: 1523–1528 (in Russian).

Bedoshvili, D.O. 1985b. An attempt at classifying the alpine meadow vegetation of the Caucasus using groups of associated species (as exemplified by the Kazbegi region). Bot. Zh. 70: 1605–1613 (in Russian).

Belsky, A.J. 1987. Revegetation of natural and human-caused disturbances in the Serengeti National Park, Tanzania. Vegetatio 70: 51–60.

Belsky, A.J. 1988. Regional influences on small-scale vegetational heterogeneity within grasslands in the Seregeti National Park, Tanzania. Vegetatio 74: 3–10.

Collins, S.L. & Barber, S.C. 1985. Effect of disturbance on diversity in mixed-grass prairie. Vegetatio 64: 87–94.

Dirzo, R. 1985. The role of grazing animals. In: White, J. (ed.), Studies on plant demography. pp. 343–355. Academic Press, London.

Dolukhanov, A.G. 1978. The timberline and the subalpine belt in the Caucasus Mountains, USSR. Arct. alp. Res. 10: 409–422.

Gadziev, V.D., Yevstratova, O.I. & Efendiev, P.M. 1986. *Rhododendron* stands at high altitudes of Great Caucasus, Azerbaijan. In Kamelin, R.V. (ed.) Vegetation of high altitudes, pp. 109–113. Nauka, Moscow (in Russian).

Gomez Sal, A., De Miguel, J.M., Casado, M.A. & Pineda, F.D. 1986. Successional changes in the morphology and ecological responses of a grazed pasture ecosystem in Central Spain. Vegetatio 67: 33–44.

Grant, S.A., Suckling, D.E., Smith, H.K., Torvell, L., Forbes, T.D.A. & Hodgson, J. 1985. Comparative studies of diet selection by sheep and cattle: The hill grasslands. J. Ecol. 73: 987–1004.

Grime, J.P. 1979. Plant strategies and vegetation processes. J. Wiley and Sons, Chichester.

Gulisashvili, V.Z., Vakhatadze, L.B. & Prilipkov, L.I. 1975. Vegetation of the Caucasus Mountains. Nauka, Moscow. (in Russian).

Harper, J.L. 1969. The role of predation in vegetational diversity. Brookhaven Symp. 22: 48–62.

Harper, J.L. 1977. Population biology of plants. Academic Press, London.

Hill, M.O. 1979. TWINSPAN – A FORTRAN program for arranging multivariate data in an ordered two-way table by classification of the individuals and attributes. Ecology and Systematics, Cornell University, Ithaca, New York.

Hill, M.O. & Gauch, H.G. 1980. Detrended correspondence analysis: An improved ordination technique. Vegetatio 42: 47–58.

Jensén, S. 1978. Influences of transformation of cover on classification and ordination of lake vegetation. Vegetatio 37: 19–31.

Ketskhoveli, N.N., Kharadze, A.L., Ivanishvili, M.I. & Gagnidze, M.I. 1975. Botanical description of the Georgian Military Road (Tbilisi-Kazbegi-Ordjonikidze). Acad. Sci. Georg. SSR, Botanical Institute, Tbilisi.

Korotkov, K.O. 1986. Vegetation of the Guton Mountain, East Caucasus. In: Kamelin, R.V. (ed), Vegetation of high altitudes, pp. 127–130. Nauka, Moscow (in Russian).

Kovář, P. & Lepš, J. 1986. Ruderal communities of the railway station Česká Třebová (Eastern Bohemia, Czechoslovakia) – remarks on the application of classical and numerical methods of classification. Preslia 58: 141–163.

Kuliev, V.Sh., Vagabov, Z.V. & Magomedi, B.R. 1986. Control of *Veratrum lobelianum* in high mountain pastures and hay meadows of the Caucasus, USSR. In: Kamelin R.V. (ed.), Vegetation of high altitudes, pp. 225–227. Nauka, Moscow (in Russian).

Looman, J. 1983. Grassland as natural or semi-natural vegetation. In: Holzner, W., Werger, M.J.A. & Ikusima, I. (eds), Man's impact on vegetation. pp. 173–184. Junk, The Hague.

Mucina, L. 1982. Numerical classification and ordination of ruderal plant communities (Sisymbrietalia, Onopordetalia) in the western part of Slovakia. Vegetatio 48: 267–275.

Mueller-Dombois, D. & Ellenberg, H. 1974. Aims and methods of vegetation ecology. J. Wiley and Sons, New York.

Orlóci, L. 1978. Multivariate analysis in vegetation science. 2nd ed. Junk, The Hague.

Peet, R.K. 1974. The measurement of species diversity. Ann. Rev. Ecol. Syst. 5: 285–307.

Pickett, S.T.A. 1980. Non-equilibrium coexistence of plants. Bull. Torrey Bot. Club 107: 238–248.

Pickett, S.T.A. & White, P.S. (eds), 1985. The ecology of natural disturbance and patch dynamics. Academic Press, London.

Pielou, E.C. 1966. The measurement of diversity in different types of biological collections. J. Theor. Biol. 13: 131–144.

Plesník, P. 1978. Man's influence on the timberline in the West Carpathian Mountains, Czechoslovakia. Arct. Alp. Res. 10: 491–504.

Rabotnov, T.A. (ed.) 1987. Biogeocoenoses of alpine deserts. Nauka, Moscow (in Russian).

Seibert, P. 1983. Human impact on landscape and vegetation in the Central High Andes. In Holzner, W., Werger, M.J.A. & Ikusima, I. (eds), Man's impact on vegetation. Junk, The Hague.

Sousa, W.P. 1984. The role of disturbance in natural communities. Ann. Rev. Ecol. Syst. 15: 353–392.

Takhtajan, A.L. (ed.) 1954–1980. Flora of Armenia. Vol. 1-7. Acad. Sci. Armen. SSR, Jerevan (in Russian).

Tukel, T. 1984. Comparison of grazed and protected mountain steppe rangeland in Ulukisla, Turkey. J. Range Manage. 37: 133–135.

van der Maarel, E. 1976. On the establishment of plant community boundaries. Ber. Deutsch. Bot. Ges. 89: 415–443.

van der Maarel, E. 1979. Transformation of cover-abundance values in phytosociology and its effects on community similarity. Vegetatio 39: 97–114.

Walter, H. 1984. Vegetation und Klimazonen. Eugen Ulmer, Stuttgart.

Welch, D. 1986. Studies in the grazing of heather moorland in north-east Scotland. V. Trends in *Nardus stricta* and other unpalatable graminoids. J. Appl. Ecol. 23: 1047–1058.

Westhoff, V. & van der Maarel, E. 1978. The Braun-Blanquet approach. In: Whittaker, R.H. (ed.), Classification of plant communities, pp. 287–399. Junk, The Hague.

Wishart, D. 1978. CLUSTAN IC. User manual. Inter-University Research Council Series. Report 47. London.

Zanina, A.A. 1961. Climate of the USSR. Caucasus. Gimiz, Leningrad. (in Russian).

Vegetatio **81**: 209–215, 1989

Phytogeographical analysis of a treeline community in northern Yukon (NW-Canada)

Pier Luigi Nimis
Dipartimento di Biologia, Università degli Studi di Trieste, I 34100 Trieste, Italy

Accepted 8.9.1988

Keywords: Beringia, Phytogeography, Vegetation, Yukon Territory

Abstract

The phytogeography of a treeline *Salix pulchra* community from the northern Yukon is studied by means of classification and ordination on a matrix of species and Operational Geographic Units. The basic data were obtained from the global distributions of the species. Results suggest that this community is an example of pre-glacial treeline vegetation which survived in ice-free areas within ancient Beringia.

Abbreviations: OGV = Operational Geographical Unit.

Nomenclature: Hultén (1968).

Introduction

An alternative title for this paper could be 'Eric Hultén revisited: a quantitative approach to the phytogeographical study of plant communities'. In his 'History of Boreal and Arctic Biota' Hultén (1937) presented the application of an original approach, called 'Method of equiformal progressive areas', to the phytogeographical study of the Boreal and Arctic zones. The method is based on the analysis of concentric ranges of species, assuming that such species had spread after glaciation from the same area. It is surprising that, although Hultén's work is generally appreciated as a milestone in the development of modern phytogeography, this method is rarely referred to in the subsequent literature. Probably there are two main reasons for this: 1) the method can be employed only in areas where distribution maps are available for the whole flora 2) the method contains strong elements of subjectivity, of which Hultén himself was fully aware.

In recent years, such monumental works as those of Meusel *et al.* (1965) and of Hultén & Fries (1986) have increased our knowledge of plant distributions in the northern Hemisphere. At least in North- and Central Europe distribution maps are now available for the majority of plant species. Furthermore, the development of computerized database analysis allow us to modify Hultén's approach in such a way that its elements of subjectivity are reduced to a minimum.

In this paper I present an example of the phytogeographical analysis of a single plant community, based on a quantitative methodology which strongly resembles Hultén's method of equiformal progressive areas. The main difference is that the selection of the species whose ranges are jointly mapped is not subjective, but is the result of a numerical classification of the ranges.

Data and methods

The plant community studied in this paper is a *Salix pulchra* scrub vegetation recorded at the treeline in northern Yukon. The study area is located at North Fork Pass, along the Dempster Highway, 64°32′N, 138°16′W, at an elevation of 1400–1500 m, on siliceous substrate. The area shows signs of extensive glaciation, but in its immediate surroundings there are mountain ridges which always have been ice-free (Oswald & Senyk 1977).

From the phytosociological table (Lausi & Nimis unpubl.) only the 30 species which were present in more than 20% of the relevés have been retained for the analysis. The range of each species (from Hultén 1968) has been indicated on a map which has been subdivided into squares used in the analysis as Operational Geographic Units (OGU's, see Crovello 1981). The size of the OGU's is based on a compromise between degree of detail and computing load.

The data have been organized in a matrix of species and OGU's, which was submitted to classification and ordination. Classification of both species and OGU's was obtained by Complete Linkage Clustering, with the correlation coefficient as a resemblance measure; the ordination is a reciprocal ordering based on the contingency table of species and groups of OGU's, obtained with the package of programs of Wildi & Orlóci (1980).

Results

Figure 1 shows the dendrogram of species, the dendrogram of OGU's groups and the frequency distribution of each species in each group of OGU's. Five main species groups are recognized as follows:

Group 1, with *Calamagrostis canadensis, Stellaria longifolia, Galium boreale,* the joint distribution of which is shown in Fig. 2a: they have discontinuous boreal-circumpolar ranges with extensions in the

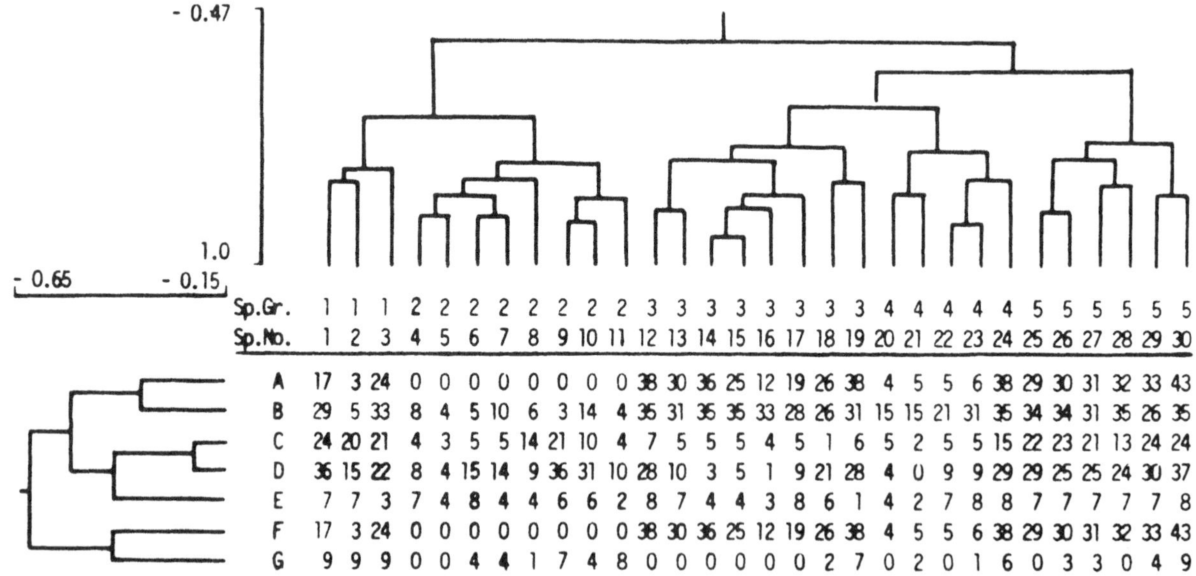

Sp.Gr.	1	1	1	2	2	2	2	2	2	2	2	3	3	3	3	3	3	3	3	4	4	4	4	4	5	5	5	5	5	5
Sp.No.	1	2	3	4	5	6	7	8	9	10	11	12	13	14	15	16	17	18	19	20	21	22	23	24	25	26	27	28	29	30
A	17	3	24	0	0	0	0	0	0	0	0	38	30	36	25	12	19	26	38	4	5	5	6	38	29	30	31	32	33	43
B	29	5	33	8	4	5	10	6	3	14	4	35	31	35	35	33	28	26	31	15	15	21	31	35	34	34	31	35	26	35
C	24	20	21	4	3	5	5	14	21	10	4	7	5	5	5	4	5	1	6	5	2	5	5	15	22	23	21	13	24	24
D	36	15	22	8	4	15	14	9	36	31	10	28	10	3	5	1	9	21	28	4	0	9	9	29	29	25	25	24	30	37
E	7	7	3	7	4	8	4	4	6	6	2	8	7	4	4	3	8	6	1	4	2	7	8	8	7	7	7	7	7	8
F	17	3	24	0	0	0	0	0	0	0	0	38	30	36	25	12	19	26	38	4	5	5	6	38	29	30	31	32	33	43
G	9	9	9	0	0	4	4	1	7	4	8	0	0	0	0	0	0	2	7	0	2	0	1	6	0	3	3	0	4	9

Fig. 1. Dendrogram of OGU groups and species, with frequency values per species per group. The groups are indicated by letters, see text, the species are indicated by numbers: *Calamagrostis canadensis* (1), *Stellaria longifolia* (2), *Galium boreale* (3), *Delphinium glaucum* (4), *Carex microchaeta* (5), *Pedicularis sudetica* ssp. *interior* (6), *Senecio lugens* (7), *Mertensia paniculata* (8), *Betula glandulosa* (9), *Veronica wormskjoldii* (10), *Anemone parviflora* (11), *Trisetum spicatum* ssp. *molle* (12), *Polemonium acutiflorum* (13), *Valeriana capitata* (14), *Salix pulchra* (15), *Petasites hyperboreus* (16), *Poa arctica* (17), *Salix reticulata* (18), *Polygonum bistorta* ssp. *plumosum* (19), *Artemisia arctica* (20), *Festuca altaica* (21), *Gentiana algida* (22), *Gentiana glauca* (23), *Luzula parviflora* (24), *Vaccinium uliginosum* (25), *Empetrum nigrum* (26), *Andromeda polifolia* (27), *Equisetum arvense* (28), *Rubus arcticus* (29), *Rubus chamaemorus* (30).

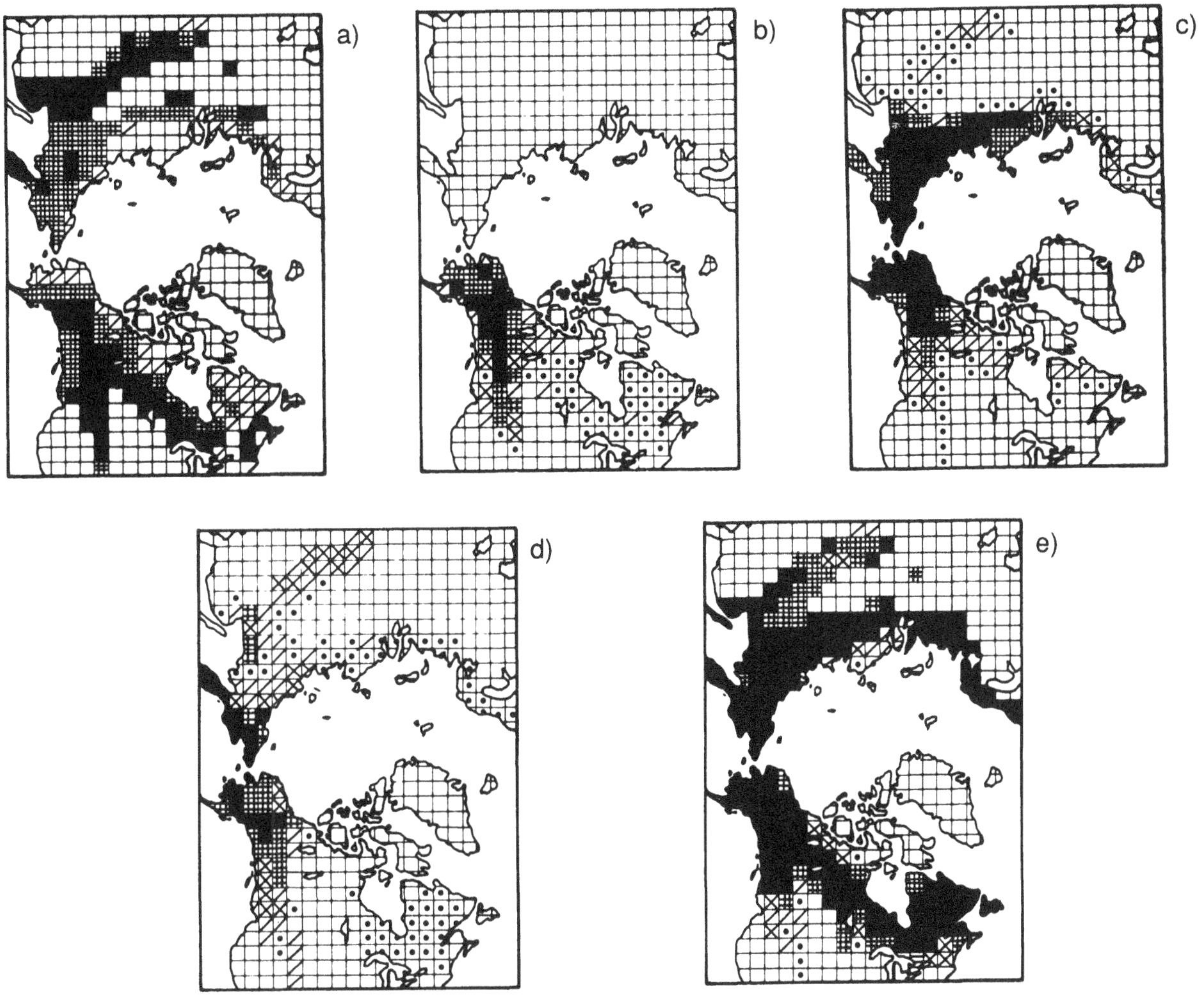

Fig. 2(a–e). Joint distribution of the species of the five species groups obtained by classification: a) species of group 1; b) species of group 2; c) species of group 3; d) species of group 4; e) species of group 5. Different shadings refer to the percent of species occurring in the OGU's, on a 5-class scale with intervals of 20%.

mountains of Central Asia and along the American Cordilleras.

Group 2, with *Delphinium glaucum, Carex microchaeta, Pedicularis sudetica* spp. *interior, Senecio lugens, Mertensia paniculata, Betula glandulosa, Veronica wormskjoldii, Anemone parviflora,* the joint distribution of which is shown in Fig. 2b. They have a northwestern American distribution with continental-montane affinities.

Group 3, with *Trisetum spicatum* spp. *molle, Polemonium acutiflorum, Valeriana capitata, Salix pulchra, Petasites hyperboreus, Poa arctica, Salix reticulata,* and *Polygonum historta* ssp. *plumosum.* Their ranges (Fig. 2c) are clearly centered on the whole of Beringia, with a pronounced extension towards eastern Asia along the Arctic coast; only a few reach northern Europe and eastern North America.

Group 4, with *Artemisia arctica, Festuca altaica, Gentiana algida, Gentiana glauca,* and *Luzula parviflora.* These species (Fig. 2d) and those of group 3 have a similar amphi-beringian distribution, but the plants of group 4 appear to be more southern-oceanic and less bound to an arctic climate; most of them reach their northern distributional limit in the study area.

Group 5, with *Vaccinium uliginosum, Empetrum nigrum, Andromeda polifolia, Equisetum arvense, Rubus arcticus* and *Rubus chamaemorus.* They have more or less continuous circumpolar ranges (Fig. 2e); most of them reach both northern Europe and the coast of northeastern America, and extend to the mountains of Central Asia.

The classification of OGU's (Figs 1 and 3) produced 7 main groups:

Group A:
: Arctic coast of Eurasia (except northeastern Siberia).

Group B:
: northeastern Siberia and western Alaska. The area corresponds well with the limits of Beringia.

Group C:
: Arctic North America and the central part of the Cordilleras. This area corresponds with several ranges of arctic-alpine plants in North America.

Group D:
: Boreal North America, including interior Alaska.

Group E:
: western slopes of the northern Cordilleras, with oceanic climate.

Group F:
: eastern slopes of the Cordilleras, with a continental para-steppic climate.

Group G:
: Central Asiatic mountains and Boreal Central Asia with continental climates, often at the border with the steppe-zone.

A synthetic representation of the relations between species and groups of OGU's is provided by Fig. 4, which presents their joint arrangement according to the first two canonical variates of the Reciprocal Ordering, based on the data in Fig. 1. The following pattern can be recognized: Beringia (Group B) is located at the negative extreme of the first axis, the other regions are disposed along the first axis in order of decreasing distance from Beringia. The second axis separates regions in Eurasia (positive scores) from those in North America (negative scores), Beringia holding a central position. The species groups obtained by classification are still recogniza-

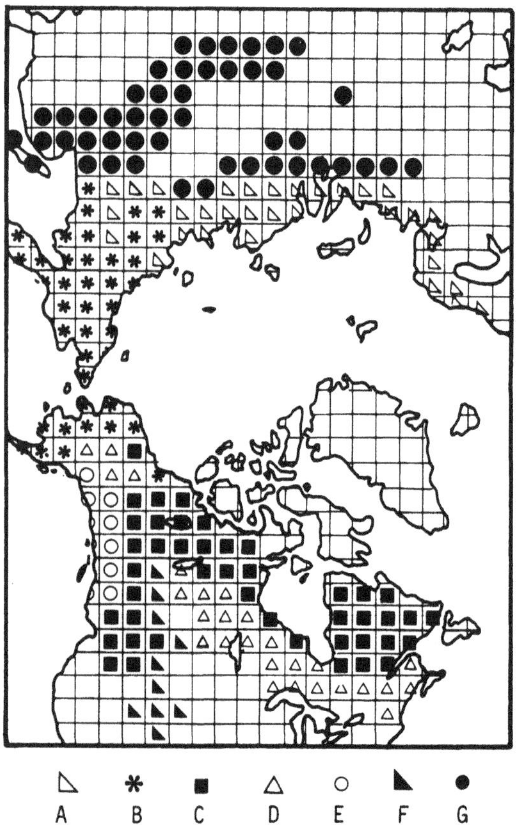

Fig. 3. Geographic distribution of OGU groups. Symbols refer to the different groups, indicated by letters, as in the text.

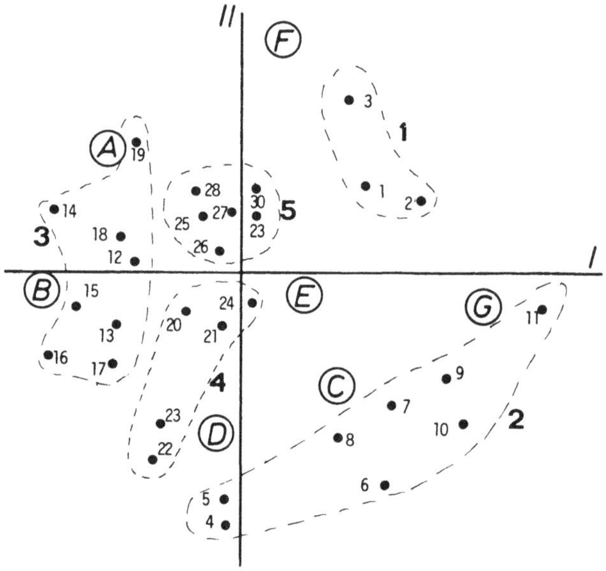

Fig. 4. Reciprocal ordering of species (numbers, as in Fig. 1), and OGU groups (letters, as in the text), based on the data in Fig. 1.

ble in the ordination: most species (groups 3, 4, 5, altogether 19 species) are highly correlated with Beringia, whereas those of groups 1 and 2 have higher correlation with other phytogeographical regions; in particular, the North American-continental radiants from Beringia (group 2) are highly correlated with the subdivisions of North America with a more continental, dry climate (groups C and G). No species is highly correlated with the oceanic western slopes of the Cordilleras (group E).

Discussion

The results underline the great importance of Beringian species in this community. The species of groups 2, 3 and 4 form more than 70% of the total. The central importance of Beringia is also evident from its position in the ordination of Fig. 4. Furthermore, the classification of OGU's produced a meaningful subdivision of the boreal and arctic zones: the groups of OGU's correspond with well-known phytogeographical regions of the northern Hemisphere. It is remarkable that this result could be obtained on the basis of only a small number of plant ranges, referring to a single plant community, recorded on a single mountain.

The great phytogeographical importance of Beringia as a refugial area for several boreal species has been already stressed by Hultén (1937) and by many subsequent authors (e.g. Hopkins 1967; Young 1971; Gjaerevoll 1980; Murray 1981; Vitt *et al.* 1987). Hultén distinguished between species which radiated west- and eastwards from Beringia after the retreat of ice ('plastic' species) and those with present ranges limited to parts of ancient Beringia ('rigid' species). Within the group of 'rigid' Beringian species we can distinguish two further groups: a) truly amphiberingian species occurring at both sides of the Bering Strait, b) species present only at one side of the strait (endemic of eastern Siberia or of Alaska-Yukon). The species of group 2 belong to the latter element: they correspond well with the 'rigid American radiants of northern Beringia' of Hultén (1937). The main problem here is to explain their absence in Siberia. The group is probably heterogeneous in its history: some of the species (e.g. *Pedicularis sudetica*) are evidently old elements, with a very large pre-glacial range (isolated stations occur in Europe), the absence of which in northeastern Siberia is difficult to explain; others, such as *Carex microchaeta*, probably derive from an amphiberingian population which was split after the opening of the Bering Strait and underwent differentiation in postglacial times (a closely related taxon, *Carex melanostoma*, exists in northeastern Siberia). Others, such as *Anemone parviflora*, probably survived glaciation along the North American Arctic coast, which was much less severely affected by glaciation than its Siberian counterpart (Hopkins 1972); finally, species such as *Veronica wormskjoldii* and *Mertensia paniculata* are old North American endemics which survived glaciation partly in the Alaska-Yukon area (*Mertensia paniculata* var. *alaskana* and var. *eastwoodae*), partly south of the North American ice-sheet (var. *paniculata*): they could not cross the Bering Land Bridge since they are bound to permafrost-free, rather dry Brunisols, whereas there is evidence that the bridge was mostly covered by tundra or taiga vegetation on Cryosols (Lausi & Nimis 1985).

The 'rigid' amphiberingian species of groups 3 and 4 form a more homogeneous phytogeographical group. The northern tendency of the former and the southern tendency of the latter are evident from the respective maps. Following Lausi & Nimis (1985) the endemic element of the Yukon flora can be divided into two widely different chorological groups: a) a group of southern species of para-steppic character, b) a group of northern species of arctic-montane character. On this basis Lausi & Nimis (1985) assumed the existence of two main refugial centres in the Yukon Territory: the periglacial steppes in the South (Nimis 1982; Murray *et al.* 1983) and the unglaciated mountains in the North. As far as the southern species of group 4 are concerned, only *Festuca altaica* and *Artemisia arctica* are often bound to a steppe or tundra-steppe vegetation; most of the species of this group have isolated stations in areas located south of the ice-sheets both in Asia and in North America; in the case of *Festuca altaica* the southern populations in North America underwent differentiation and are sometimes considered as a distinct species (*Festuca scabrella* Torr.); apparently, group 4 includes old boreal-montane species with a main refugial centre in Beringia and others located south of the continental ice-sheets; the most 'plastic' species of the group is *Luzula parviflora,* with a very narrow circumpolar extension to Scandinavia.

The 'northern' taxa of group 3 represent a transition between the 'rigid' amphiberingian-montane plants of group 4 and the circumboreal (-polar) plants of group 5. They may have survived glaciation in different parts of Beringia, with later, incomplete radiation east- and westwards to Boreal-arctic Asia and North America. A few of them have attained an almost complete circumboreal (-polar) range, like the plants of group 5. Some of the species included in group 5 are not homogeneous throughout their ranges. Within *Rubus arcticus* three subspecies can be recognized, ssp. *arcticus,* with a mainly Eurasiatic range, ssp. *acaulis,* limited to boreal North America, and ssp. *stellatus,* ranging from southern Alaska to eastern Siberia: it should be noted that the ranges of the three subspecies overlap only in areas which were part of ancient Beringia. *Andromeda polifolia* has an incomplete circumboreal range, being replaced by *Andromeda glaucophylla* in most of eastern North America and West Greenland, which indicates the probable existence of other refugial centres along the Atlantic coast.

The results of the phytogeographical analysis, chiefly the prevalence of 'rigid' Beringian species, and the fact that vast areas in northern Yukon have never been extensively affected by glaciation (Douglas 1970), lead to the assumption that this *Salix pulchra* community represents a rare example of a preglacial vegetation which could have survived glaciation almost 'in situ'.

The different degrees of expansion from the Beringial refugial centre which are shown by the plants of groups 2–5 could be the object of much speculation, if explanations are sought in historical facts. The concept of 'conservative' or 'rigid' species (Fernald 1925; Hultén 1937) itself is based on historical speculation, i.e. on the assumption that the intense selection of the harsh periglacial environments may gradually reduce the genetic diversity of the survivors to the extent that significant post-glacial expansion is impossible. Murray (1981) rejects these assumptions on the basis of recent knowledge in the field of micro-evolution. Fortunately, as shown by Young (1971) the limits of distribution of several arctic species radiating from Beringia can be explained on the basis of actualistic principles: he demonstrated that some major floristic discontinuities in the Bering Sea region can be attributed to ecological (chiefly climatical) factors. It could be possible to test whether this applies for the species treated in this paper by studying their micro-climatical behaviour in other plant communities of the study area.

Conclusions

The example presented in this paper is limited to a single plant community: it demonstrates that the quantitative analysis of plant ranges can be a useful tool for the phytogeographical analysis of vegetation. When the analysis is extended to more communities with known ecological requirements, the results allow us to relate ecological and phytogeographical facts (Nimis 1981). I think that an extension of this approach to the field of classical phytosociology could prove very useful. Syntaxonomy, in particular, is a field in which chorological and ecological variables should be always considered jointly; there are many reasons why it is unwise to consider species with very different distribution patterns as characteristic of the same syntaxon; on the one hand this precludes the recognition of the syntaxon over large areas, on the other hand, species with different distributions are likely to differ also in their ecological requirements (Nimis 1984; Lausi & Nimis 1985). Unfortunately, there is a high number of syntaxa which have been created solely on the basis of regional studies, without considering the whole distributional range of the species. A more precise phytogeographical analysis of plant communities, along the lines suggested in this paper, could help to clarify many 'metaphysical' problems in syntaxonomy.

Acknowledgements

This study is part of a broader cooperative research

project co-directed by D. Lausi (Trieste) and L. Orlóci (London, Ont.). Partial funding was received on the grants to the directors respectively by the Italian C.N.R. and the Canadian N.S.E.R.C.

References

Crovello, J. J. 1981. Quantitative biogeography: an overview. Taxon 30: 563–575.

Douglas, R. J. W. (ed.) 1970. Geology and economic minerals of Canada. Dept. of Energy, Mines and Resources. Ottawa.

Fernald, M. L. 1925. Persistence of plants in unglaciated areas of boreal America. Mem. Amer. Acad. Arts 15: 239–342.

Gjaerevoll, O. 1980. A comparison between the alpine plant communities of Alaska and Scandinavia. Acta Phytogeogr. Suec. 68: 83–88.

Hopkins, D. M. (ed.) 1967. The Bering Land Bridge. Stanford Univ. Press. Stanford, Ca.

Hopkins, D. M. 1972. The palaeography and climatic history of Beringia during late Cenozoic time. Inter Nord 12: 121–150.

Hultén, E. 1937. Outline of the history of arctic and boreal biota during the Quaternary period. Stockholm.

Hultén, E. 1968. Flora of Alaska and neighbouring territories. Stanford Univ. Press, Stanford, Ca.

Hultén, E. & Fries, M. 1986. Atlas of North European vascular plants north of the Tropic of Cancer. Koeltz, Koenigstein.

Lausi, D. & Nimis, P. L. 1985. Quantitative phytogeography of the Yukon Territory (NW Canada) on a chorological–phytosociological basis. Vegetatio 59: 9–20.

Meusel, H., Jaeger, E. & Weinert, E. 1965. Vergleichende Chorologie der Zentraleuropäische Flora. Fischer, Jena.

Murray, D. F. 1981. The role of Arctic refugia in the evolution of the Arctic vascular flora. A Beringian perspective. In: Scudder, G. E. & Reveal, J. L. (eds.), Evolution today, pp. 11–20.

Murray, D. F., Murray, B. M. Yurtsev, B. A. & Howenstein, R. 1983. Biogeographic significance of steppe vegetation in subarctic Alaska. Proc. IV Int. Conf. Permafrost, Nat. Acad. Press, Washington, D.C., pp. 883–888.

Nimis, P. L. 1982. Phytogeography of periglacial steppes in the Yukon Territory. Coll. Phytosociol. 11: 2–13.

Nimis, P. L. 1984. Contributions to quantitative phytogeography of Sicily. III: correlations between phytogeographical categories, elevation and environment-types. Arch. Bot. Biogeogr. Ital. 60, 3–4: 11–40.

Nimis, P. L. 1985. Phytogeography and ecology of epiphytic lichens at the southern rim of the Clay Belt (N-Ontario, Canada). Bryologist 88: 315–324.

Oswald E. T. & Senyk, J. P. 1977. Ecoregions of the Yukon Territory. Can. Dept. Environ. Can. For. Serv., Victoria B.C.

Young, S. B. 1971. The vascular flora of St. Lawrence Island with special reference to floristic zonation in the Arctic regions. Contr. Gray Herb. 201: 11–115.

Vitt, D. H., Horton, D. & Pickard, J. 1987. An annotated list and the phytogeography of the Bryophytes of Keele Peak, Yukon, an isolated granitic mountain. Mem. New York Bot. Garden 45: 198–210.

Wildi, O. & Orlóci, L. 1980. Management and multivariate analysis of vegetation data. Swiss Fed. Inst. For. Res. Rep. No 215.

www.ingramcontent.com/pod-product-compliance
Lightning Source LLC
Chambersburg PA
CBHW081214180926
51613CB00003B/234
* 9 7 8 0 7 9 2 3 0 3 8 8 6 *